Springer Proceedings in Mathematics & Statistics

Volume 501

This book series features volumes composed of selected contributions from workshops and conferences in all areas of current research in mathematics and statistics, including data science, operations research and optimization. In addition to an overall evaluation of the interest, scientific quality, and timeliness of each proposal at the hands of the publisher, individual contributions are all refereed to the high quality standards of leading journals in the field. Thus, this series provides the research community with well-edited, authoritative reports on developments in the most exciting areas of mathematical and statistical research today.

Hemanta Kalita · Tomás Pérez Becerra ·
Kiran Kumar Behera
Editors

Nonlinear Analysis and Computational Techniques

ICNACT 2024, Bhopal, India, August 8–10

Springer

Editors
Hemanta Kalita
Mathematics Division
School of Advanced Sciences
and Languages (SASL)
VIT Bhopal University
Bhopal, Madhya Pradesh, India

Tomás Pérez Becerra
The Institute of Physics and Mathematics
Technological University of the Mixteca
Heroica Ciudad de Huajuapan de Leon
Oaxaca, Mexico

Kiran Kumar Behera
Mathematics Division
School of Advanced Sciences
and Languages (SASL)
VIT Bhopal University
Bhopal, Madhya Pradesh, India

ISSN 2194-1009 ISSN 2194-1017 (electronic)
Springer Proceedings in Mathematics & Statistics
ISBN 978-981-96-6037-7 ISBN 978-981-96-6038-4 (eBook)
https://doi.org/10.1007/978-981-96-6038-4

Mathematics Subject Classification: 40A05, 26A33, 47B02, 26D10, 47A08, 47A10, 40A35, 26A51

This Springer imprint is published by the registered company Springer Nature Singapore Pte Ltd.
The registered company address is: 152 Beach Road, #21-01/04 Gateway East, Singapore 189721, Singapore

Preface

Nonlinear analysis is a fascinating topic in and of itself, yet it also lays the groundwork for many branches of pure and applied mathematics. Various mathematical theories that are applicable to real-world issues are being actively analysed and developed by researchers worldwide. The International Conference on Nonlinear Analysis and Computational Techniques (ICNACT-2024) organized by the Mathematics Division, SASL, VIT Bhopal, Madhya Pradesh, India, in collaboration with the Institute of Physics and Mathematics, Technological University of the Mixteca Oaxaca, Mexico, aims to enhance knowledge in these domains and promote professional interchange. This conference will facilitate the exchange of latest developments in several areas of mathematical analysis. In this area, we strive to establish productive research avenues and partnerships.

The book has 13 selected chapters. In the first chapter "Generalization Orlicz Sequence Spaces with Rao and Ren's Norm", construction of generalized Orlicz sequence spaces by extending traditional Orlicz spaces through the use of Rao and Rens norm and a broader class of functions, known as s-Young functions, are discussed. Here, a different version of generalized Orlicz space has been constructed by replacing the Young function with an s-Young function.

In the second chapter "Strongly λ-Summable Functions Defined by Orlicz Functions", by taking a real-valued function that is measurable on (1; 1), concept of strongly-summable functions defined by Orlicz functions and also some considerable results are analyzed.

In the third chapter "Lacunary Strong Almost Summability of Weight ρ with Respect to Seminorm q", the concept of q-lacunary almost statistical convergence of weight, and creating some relations between q-lacunary almost statistical convergence of weight and q-lacunary strong almost summability of weight are discussed.

In the fourth chapter "New Integral Inequalities for Weighted Integrals and Some Fractional Derivations", a new integral inequalities in the framework of weighted integrals, which generalizes and extends Ostrowski-type inequalities, for differentiable $(h; m)$-convex functions of the second type are discussed. Some consequences for the Caputo and Atangana–Baleanu fractional integral operators are derived from the obtained results.

In the fifth chapter "Fractal Nonlinear Klein-Gordon Equation", the fractal nonlinear Klein–Gordon equation by using the homotopy perturbation method (HPM) is solved. The method effectively demonstrates how the solution evolves over time in relation to the fractal dimensions of space and time. To illustrate the approach, several examples are presented, showcasing the details of the solution process. These examples highlight the applicability of HPM for addressing complex fractal differential equations and provide insights into the behavior of solutions within fractal geometries.

In the sixth chapter "Survey on Different Near Linear Topological Spaces", some relationships among different types of near linear spaces, such as spaces with non-negative scalars, weak linear spaces, almost linear spaces, quasilinear spaces, semi-linear spaces and conlinear spaces are discussed. Some examples, properties and results on their topological structures are also established.

In the seventh chapter "Cayley Transform of Operators on the Hilbert Space", some algebraic properties such as unitariness, normality, additive inverse law, forward order law, backward order law and absorption law for Cayley transform of operators and operator matrices on the Hilbert space are discussed.

In the eighth chapter "Solvability of Hybrid Differential Equation in Hölder Space", the solvability of hybrid differential equations is covered. Schauder fixed point theorem fixed point theorem is used to establish the existence result. The investigations of the chapter are placed in the space of continuous and tempered functions on closed intervals $[0; T]$. A couple of examples are placed to verify the applicability of the above mentioned result.

In the ninth chapter "The Exponential Stability of Nonlinear Shallow Water Equations Within a Trapezoidal Channel", discusses the problem of exponential stability of the nonlinear Saint–Venant equation for a non-prismatic and non-rectangular channel. They consider the general case of systems containing not only both arbitrary friction and spatially varying slopes but also spatially varying channel dimensions (width and lateral slope), leading to non-uniform stationary states. An explicit quadratic Lyapunov function is constructed as a weighting function for steady-state small perturbations. Further local exponential stability of the nonlinear Saint–Venant equation for a trapezoidal channel can be guaranteed in the H2—a norm by an appropriate choice of boundary feedback control is shown. Finally, numerical simulations are discussed explicitly.

In the tenth chapter "On Some Fixed Point Results of Banach Type Contraction in Controlled *G*-Metric Spaces", a new type of metric spaces called controlled *G*-metric spaces has been introduced. Some convergence properties on this new space are discussed. In order to further illustrate the usefulness of this new structure, a generalised Banach contraction principle is introduced and new fixed point results are developed and proved. Some examples are presented to support the main result proved therein.

In the eleventh chapter "Some Common Coupled Fixed Point Theorems of Integral Type Conditions for Occasionally w-Compatible Mappings in S-Metric Spaces and an Application", certain common coupled fixed point theorems with integral type conditions for a pair of occasionally w-compatible mappings are covered. An example is provided to justify the result. Finding results generalize, expand, and enhance some previously published well-known conclusions from the current literature. Finally, an application of the Fredholm integral equation is provided.

In the twelfth chapter "Fixed Point Theorems that are Applicable to Compatible Mappings of Type (P) in Dynamic Programming", a few fixed point theorems for compatible mappings of type (P), which are extensions of compatible mapping are discussed.

In the thirteenth chapter "Impulsive Fractal Sturm–Liouville Equations", an impulsive fractal Sturm-Liouville problems are considered. Firstly, the symmetric operator corresponding to the problem is obtained. Later, an existence theorem is proved. Finally, an eigenfunction expansion is obtained.

Each chapter of the current book has been carefully peer-reviewed by the team of experts. The proof-reading has been done by all the editors.

Bhopal, India Hemanta Kalita
Heroica Ciudad de Huajuapan de Leon, Mexico Tomás Pérez Becerra
Bhopal, India Kiran Kumar Behera

Contents

Generalization Orlicz Sequence Spaces with Rao and Ren's Norm

Lianna Rizky Gunawan, Al Azhary Masta, Siti Fatimah, and Muh Nur

Abstract This paper introduces about construction of generalized Orlicz sequence spaces by extending traditional Orlicz spaces through the use of Rao and Ren's norm and a broader class of functions known as s-Young functions. By replacing the Young function with an s-Young function, we aim to construct a different version of generalized Orlicz sequence spaces. Utilizing the definitions and properties established in previous research on Orlicz sequence spaces, we develop a new generalized sequence space denoted as $l_{\Phi_s^*}$. The methodology used is the definition of Rao and Ren's generalized Orlicz space and the characteristics of S-Young functions. We then proceed to define a norm similar to the Rao and Ren's norm but using s-Young function. The resulting generalized Orlicz sequence space retains the desirable properties of Orlicz sequence spaces in previous research while offering greater flexibility and adaptability of the space. This research aims to expand the theoretical understanding and potential applications of Orlicz sequence spaces, paving the way for future research and practical implementations in various mathematical and applied field.

Keywords s-Convex function · s-Young function · Orlicz sequence space

L. R. Gunawan (✉) · A. A. Masta · S. Fatimah
Mathematics Study Program, Faculty of Mathematics and Natural Sciences Education, Indonesia University of Education, Bandung, Indonesia
e-mail: liannalrg66@gmail.com

A. A. Masta
e-mail: alazhari.masta@upi.edu

S. Fatimah
e-mail: sitifatimah@upi.edu

M. Nur
Department of Mathematics, Faculty of Mathematics and Natural Science, Hasanuddin University, Makassar, Indonesia
e-mail: muhammadnur@unhas.ac.id

H. Kalita et al. (eds.), *Nonlinear Analysis and Computational Techniques*,
Springer Proceedings in Mathematics & Statistics 501,
https://doi.org/10.1007/978-981-96-6038-4_1

1 Introduction and Preliminaries

Lebesgue spaces are the most important of measurable functions, as Castillo and Rafeiro say that in some ways, Lebesgue spaces are the prototype of all function spaces [1]. Lebesgue space introduced by Henri Lebesgue is a function space equipped with a norm derived from the absolute integrand of the function. Many mathematicians have studied Lebesgue space until now. One of the results of the study of Lebesgue space is Orlicz Space. Orlicz space L_Φ was first introduced in 1931 by Z. W. Birnbaum and W. Orlicz, which is an extension of Lebesgue space L_p with $1 \leq p < \infty$ [2]. There are two different versions of Orlicz spaces, namely continuous Orlicz spaces and Orlicz sequence spaces [3, 4]. Researchers who studied continuous Orlicz spaces include L. Maligranda, A. A. Masta, and other previous researchers [5–11]. Meanwhile, the space of Orlicz lines was first introduced in 1971 by J. Lindenstrauss and L. Tzafriri by summarizing the sequence space of summable-p for $1 \leq p < \infty$. Afterward, L. Maligranda, and M. Mastylo, Awad A. Bakery and Rafaf, R., P. S. Prayoga and S. Fatimah, etc. studied this Orlicz sequence space [12–16, 20]. Other studies on Orlicz spaces can also be seen in [21, 22]. In this research, the author will focus on discussing the Orlicz sequence space by referring to the results that have been obtained by researchers.

The definition of the Orlicz function used by [14] in defining the Orlicz sequence space is similar to the definition of the Young function on the Orlicz space, namely a function $\Phi : \mathbb{R}_0^+ \to \mathbb{R}_0^+$ is said to be Young's function if Φ is convex, $\Phi(0) = 0$, $\lim_{t\to\infty} \Phi(t) = \infty$, and Φ is continuous [7]. Furthermore, Prayoga used this definition to define the Orlicz sequence space in his research on the inclusion properties and the generality of Hölder's inequality on the Orlicz sequence space [14]. The properties of Young's functions that have been obtained by previous researchers are used to prove the inclusion properties of Orlicz spaces. The inclusion property is an interesting thing to study because it examines the relationship between spaces.

Meanwhile, the study of the generality of Orlicz spaces is also interesting to explore, as Rao and Ren in 1991 have studied it in the last chapter of their book entitled "Theory of Orlicz spaces" [8]. The method used by Rao and Ren was to modify Luxemburg's norm. Furthermore, Dermawan also studied the generality of Orlicz spaces by modifying the Young functions with extended Young functions known as s-Young function defined by Dermawan [11]. Then, in 2023, Dasep et al. studied the generality of Orlicz sequence spaces using a similar method to that of Dermawan [20]. The following is the definition of the generalized Orlicz space defined by Rao and Ren.

Let X be sequence space. The generalized Orlicz space denoted by $L_\Phi(X)$ is the set of all measurable functions $f{:}X \to R$ satisfying the property $\int_X \phi(\alpha|f|)\,dx < \infty$ for $\alpha > 0$, where Φ is increasing function, continuous, and $\Phi(0) = 0$, equipped with the norm

$$\|f\|_\phi = \inf\left\{k > 0 : \int_X \rho_\phi\left(\frac{|f|}{k}\right)\,dx \leq k\right\}$$

.

Based on the explanation above and the studies that have been done by previous researchers regarding the generality of Orlicz spaces, the author will construct a generalized Orlicz sequence space by modifying the Luxemburg norm almost similar to the Rao and Ren's norm on the generalized Orlicz space [8]. Then modify it again by changing the Young function using s-Young function.

2 Research Methods

To prove the results, we pay attention to the Young and s-Young functions properties, Rao and Ren's norm on the definition generalization Orlicz space, and definition and properties of generalized Orlicz sequence space. Noted, in this study we used Φ to represent the Young function and Φ_s for the s-Young function. The s-Young function is used to construct the generalization of Orlicz space.

2.1 s-Convex and s-Young Functions

Definition 1 Let $0 < s \leq 1$. A function $\Phi : \mathbb{R}_0^+ \to \mathbb{R}_0^+$ is said to be s-convex if

$$\Phi(\alpha x + \beta y) \leq \alpha^s \Phi(x) + \beta^s \Phi(y)$$

for any $x, y \in \mathbb{R}_0^+$ and $\alpha, \beta \geq 0$ with $\alpha^s + \beta^s = 1$.

Lemma 1 *If $\Phi : \mathbb{R}_0^+ \to \mathbb{R}_0^+$ is convex, then Φ is s-convex for $0 < s \leq 1$.*

Definition 2 The function $\Phi : \mathbb{R}_0^+ \to \mathbb{R}_0^+$ is said to be a Young function if it satisfies the following conditions:

1. Φ is convex,
2. $\Phi(0) = 0$,
3. $\lim_{t\to\infty} \Phi(t) = \infty$,
4. Φ is continuous.

Definition 3 The function $\Phi_s : \mathbb{R}_0^+ \to \mathbb{R}_0^+$ is said to be an s-Young function if it satisfies the following conditions:

1. Φ_s is s-convex,
2. $\Phi_s(0) = 0$,
3. $\lim_{t\to\infty} \Phi_s(t) = \infty$,
4. Φ_s is continuous.

Lemma 2 *Let $\Phi_s : \mathbb{R}_0^+ \to \mathbb{R}_0^+$ be an s-Young function. Then:*

1. *$\Phi_s(\alpha t) \leq \alpha^s \Phi_s(t)$ for every $t \geq 0$, $\alpha \in [0, 1]$, and $s \in (0, 1]$,*
2. *$\omega(t) = \frac{\Phi_s(t)}{t^s}$ is increasing for any $t > 0$ with $s \in (0, 1]$,*
3. *$\Phi_s(t)$ is increasing for any $t \geq 0$.*

2.2 Orlicz Sequence Space and Generalization Orlicz Sequence Space

Definition 4 Let $\Phi : \mathbb{R}_0^+ \to \mathbb{R}_0^+$ be an s-Young function. The Orlicz sequence space $l_\Phi^*(\mathbb{R})$ is defined as the set of all sequences $X = (x_n)$ of real numbers such that

$$\sum_{n=1}^{\infty} \Phi\left(\frac{|x_n|}{a}\right) < \infty$$

for some positive number a, and it is equipped with the norm

$$\|X\|_{l_\Phi^*(\mathbb{R})} = \inf\left\{k > 0 \mid \sum_{n=1}^{\infty} \Phi\left(\frac{|x_n|}{k}\right) \leq k\right\}.$$

Definition 5 Let Φ_s be an s-Young function. The generalized Orlicz sequence space $l_{\Phi_s}(\mathbb{R})$ consists of all sequences $X = (x_n)$ of real numbers such that

$$\sum_{n=1}^{\infty} \Phi_s\left(\frac{|x_n|}{a}\right) < \infty$$

for some positive number a. It is equipped with the norm

$$\|X\|_{l_{\Phi_s}(\mathbb{R})} = \inf\left\{k > 0 \mid \sum_{n=1}^{\infty} \Phi_s\left(\frac{|x_n|}{k}\right) \leq 1\right\}.$$

Lemma 3 *If $\|X\|_{l_{\Phi_s}(\mathbb{R})} \neq 0$ for all $X = (x_n) \in l_{\Phi_s}(\mathbb{R})$, then*

$$\sum_{n=1}^{\infty} \Phi_s\left(\frac{|x_n|}{\|X\|_{l_{\Phi_s}(\mathbb{R})}}\right) \leq 1.$$

Lemma 4 *Let $\Phi_s : \mathbb{R}_0^+ \to \mathbb{R}_0^+$ be an s-Young function and $X = (x_n) \in l_{\Phi_s}(\mathbb{R})$. Then*

$$\sum_{n=1}^{\infty} \Phi_s(|x_n|) \leq 1$$

if and only if

$$\|X\|_{l_{\Phi_s}(\mathbb{R})} \leq 1.$$

Lemma 5 *Let $\Phi_s : \mathbb{R}_0^+ \to \mathbb{R}_0^+$ be an s-Young function and $X = (x_n) \in l_{\Phi_s}(\mathbb{R})$. For any $\epsilon > 0$, the following holds:*

$$\sum_{n=1}^{\infty} \Phi_s\left(\frac{|x_n|}{\epsilon}\right) \leq 1$$

if and only if

$$\|X\|_{l_{\Phi_s}(\mathbb{R})} = 0.$$

Lemma 6 *Let $\Phi_s : \mathbb{R}_0^+ \to \mathbb{R}_0^+$ be an s-Young function and $X = (x_n) \in l_{\Phi_s}(\mathbb{R})$. Then*

$$\sum_{n=1}^{\infty} \Phi_s \left(\frac{|x_n|}{\alpha} \right) \le \frac{1}{\alpha^s}$$

for every $\alpha \ge 1$ with $0 < s \le 1$ if and only if

$$\|X\|_{l_{\Phi_s}(\mathbb{R})} \le 1.$$

Lemma 7 *Let $\Phi_s : \mathbb{R}_0^+ \to \mathbb{R}_0^+$ be an s-Young function and $X = (x_n) \in l_{\Phi_s}(\mathbb{R})$. Then $X = (x_n) = 0$ if and only if*

$$\|X\|_{l_{\Phi_s}(\mathbb{R})} = 0.$$

2.3 *Fréchet Norm*

Definition 6 Let X be a sequence space. The norm $\|\cdot\|$ in X is a Fréchet norm (F-norm) if it satisfies the following conditions for any $x, y \in X$:

1. $\|x\| \ge 0$ and $\|x\| = 0$ if and only if $x = 0$,
2. If $\alpha_n \to 0$ as $n \to \infty$, then $\|\alpha_n x\| \to 0$ as $n \to \infty$,
3. If $\|x_n\| \to 0$ as $n \to \infty$, then for any $\alpha \in \mathbb{R}$, $\|\alpha x_n\| \to 0$ as $n \to \infty$,
4. $\|x + y\| \le \|x\| + \|y\|$.

3 Results and Discussion

In this section, we will discuss about the generalized Orlicz sequence space with a different version from what previous researchers. We use Rao and Ren's norm and s-Young function to define the norm. In the following discussion, we will study the definition along with the properties of the generalized Orlicz sequence space.

Definition 7 Let $\Phi_s : \mathbb{R}_0^+ \to \mathbb{R}_0^+$ be an s-Young function. The generalized Orlicz sequence space $l^*_{\Phi_s}(\mathbb{R})$ is defined as the set of all sequences $X = (x_n)$ of real numbers such that

$$\sum_{n=1}^{\infty} \Phi_s \left(\frac{|x_n|}{a} \right) < \infty$$

for some positive number a. The space $l^*_{\Phi_s}(\mathbb{R})$ is equipped with the norm

$$\|X\|_{l^*_{\Phi_s}(\mathbb{R})} = \inf\left\{k > 0 \mid \sum_{n=1}^{\infty} \Phi_s\left(\frac{|x_n|}{k}\right) \leq k\right\}.$$

To prove that the function $\|\cdot\|_{l^*_{\Phi_s}(\mathbb{R})}$ is well-defined on $l^*_{\Phi_s}(\mathbb{R})$, it is sufficient to show that the set

$$\{k > 0 \mid \sum_{n=1}^{\infty} \Phi_s\left(\frac{|x_n|}{k}\right) \leq k\}$$

is nonempty. Furthermore, it must be shown that

$$\inf\left\{k > 0 \mid \sum_{n=1}^{\infty} \Phi_s\left(\frac{|x_n|}{k}\right) \leq k\right\}$$

exists.

Next, we will discuss the properties of the generalized Orlicz line space.

Lemma 8 *If $\|X\|_{l^*_{\Phi_s}(\mathbb{R})} \neq 0$ for all $X = (x_n) \in l^*_{\Phi_s}(\mathbb{R})$, then*

$$\sum_{n=1}^{\infty} \Phi_s\left(\frac{|x_n|}{\|X\|_{l^*_{\Phi_s}(\mathbb{R})}}\right) \leq b$$

where $b \in \left\{k > 0 \mid \sum_{n=1}^{\infty} \Phi_s\left(\frac{|x_n|}{k}\right) \leq k\right\}$.

Proof Let $\|X\|_{l^*_{\Phi_s}(\mathbb{R})} \neq 0$ for any $X = (x_n) \in l^*_{\Phi_s}(\mathbb{R})$ and let A be defined as

$$A := \left\{k > 0 \mid \sum_{n=1}^{\infty} \Phi_s\left(\frac{|x_n|}{k}\right) \leq k\right\}.$$

Take any $\epsilon > 0$ and $X = (x_n) \in l^*_{\Phi_s}(\mathbb{R})$ such that $\|X\|_{l^*_{\Phi_s}(\mathbb{R})} \neq 0$. Since $\|X\|_{l^*_{\Phi_s}(\mathbb{R})} = \inf A$, it follows that $\|X\|_{l^*_{\Phi_s}(\mathbb{R})} + \epsilon$ is not a lower bound of the set A. This means that there exists $b \in A$ such that

$$\|X\|_{l^*_{\Phi_s}(\mathbb{R})} + \epsilon \geq b \iff \frac{1}{\|X\|_{l^*_{\Phi_s}(\mathbb{R})} + \epsilon} \leq \frac{1}{b}.$$

Since $|x_n| \geq 0$ for all $n \in \mathbb{N}$, it follows that

$$\frac{|x_n|}{\|X\|_{l^*_{\Phi_s}(\mathbb{R})} + \epsilon} \leq \frac{|x_n|}{b}.$$

Next, since Φ_s is a monotonically increasing function, we have

$$\Phi_s\left(\frac{|x_n|}{\|X\|_{l^*_{\Phi_s}(\mathbb{R})}+\epsilon}\right)\leq\Phi_s\left(\frac{|x_n|}{b}\right).$$

Since this holds for every $n\in\mathbb{N}$ and $b\in A$, it can be written as

$$\sum_{n=1}^{\infty}\Phi_s\left(\frac{|x_n|}{\|X\|_{l^*_{\Phi_s}(\mathbb{R})}+\epsilon}\right)\leq\sum_{n=1}^{\infty}\Phi_s\left(\frac{|x_n|}{b}\right)\leq b.$$

Thus, since ϵ was arbitrary, it is proven that

$$\sum_{n=1}^{\infty}\Phi_s\left(\frac{|x_n|}{\|X\|_{l^*_{\Phi_s}(\mathbb{R})}}\right)\leq b.$$

Lemma 9 *Let $\Phi_s:\mathbb{R}_0^+\to\mathbb{R}_0^+$ be an s-Young function and $X=(x_n)\in l^*_{\Phi_s}(\mathbb{R})$. Then*

$$\sum_{n=1}^{\infty}\Phi_s\left(\frac{|x_n|}{\|X\|_{l^*_{\Phi_s}(\mathbb{R})}}\right)\leq\|X\|_{l^*_{\Phi_s}(\mathbb{R})}.$$

Proof Let $A:=\left\{k>0\mid\sum_{n=1}^{\infty}\Phi_s\left(\frac{|x_n|}{k}\right)\leq k\right\}$ and let $\|X\|_{l^*_{\Phi_s}(\mathbb{R})}\neq 0$ for all $X=(x_n)\in l^*_{\Phi_s}(\mathbb{R})$. From [Definition 3, 7], we have

$$\|X\|_{l^*_{\Phi_s}(\mathbb{R})}=\inf\left\{k>0\mid\sum_{n=1}^{\infty}\Phi_s\left(\frac{|x_n|}{k}\right)\leq k\right\}.$$

Since $\|X\|_{l^*_{\Phi_s}(\mathbb{R})}$ is the infimum of A, it is the largest lower bound of A. Thus, for any $\epsilon>0$, there exists $b=\|X\|_{l^*_{\Phi_s}(\mathbb{R})}+\epsilon\in A$ such that

$$\sum_{n=1}^{\infty}\Phi_s\left(\frac{|x_n|}{b}\right)\leq b$$

or equivalently

$$\sum_{n=1}^{\infty}\Phi_s\left(\frac{|x_n|}{\|X\|_{l^*_{\Phi_s}(\mathbb{R})}+\epsilon}\right)\leq\|X\|_{l^*_{\Phi_s}(\mathbb{R})}+\epsilon.$$

Since ϵ was arbitrary and $\|X\|_{l^*_{\Phi_s}(\mathbb{R})}\neq 0$, it follows that

$$\sum_{n=1}^{\infty}\Phi_s\left(\frac{|x_n|}{\|X\|_{l^*_{\Phi_s}(\mathbb{R})}}\right)\leq\|X\|_{l^*_{\Phi_s}(\mathbb{R})}.$$

Lemma 10 *Let $\Phi_s : \mathbb{R}_0^+ \to \mathbb{R}_0^+$ be an s-Young function and let $X = (x_n) \in l^*_{\Phi_s}(\mathbb{R})$. Then*

$$\sum_{n=1}^{\infty} \Phi_s\left(\frac{|x_n|}{k}\right) \leq k$$

if and only if

$$\|X\|_{l^*_{\Phi_s}(\mathbb{R})} \leq k.$$

Proof Let $\Phi_s : [0, \infty) \to [0, \infty)$ be an s-Young function, and let $X = (x_n) \in l^*_{\Phi_s}(\mathbb{R})$. Define

$$A := \left\{k > 0 \mid \sum_{n=1}^{\infty} \Phi_s\left(\frac{|x_n|}{k}\right) \leq k\right\}.$$

($\Rightarrow$) Suppose

$$\sum_{n=1}^{\infty} \Phi_s\left(\frac{|x_n|}{k}\right) \leq k,$$

where $k > 0$. We need to show that $\|X\|_{l^*_{\Phi_s}(\mathbb{R})} \leq k$.
Since

$$\sum_{n=1}^{\infty} \Phi_s\left(\frac{|x_n|}{k}\right) \leq k,$$

it follows that $k \in A$. Furthermore, by the definition of the norm in [Definition 3, 7],

$$\|X\|_{l^*_{\Phi_s}(\mathbb{R})} = \inf A,$$

so $\|X\|_{l^*_{\Phi_s}(\mathbb{R})} \leq k$.
($\Leftarrow$) Suppose $\|X\|_{l^*_{\Phi_s}(\mathbb{R})} \leq k$, where $k > 0$. We need to show that

$$\sum_{n=1}^{\infty} \Phi_s\left(\frac{|x_n|}{k}\right) \leq k.$$

Note that

$$\|X\|_{l^*_{\Phi_s}(\mathbb{R})} \leq k \implies \frac{|x_n|}{\|X\|_{l^*_{\Phi_s}(\mathbb{R})}} \leq \frac{|x_n|}{k} \implies \frac{|x_n|}{k} \leq \frac{|x_n|}{\|X\|_{l^*_{\Phi_s}(\mathbb{R})}}.$$

Since Φ_s is an increasing function, we have

$$\Phi_s\left(\frac{|x_n|}{k}\right) \leq \Phi_s\left(\frac{|x_n|}{\|X\|_{l^*_{\Phi_s}(\mathbb{R})}}\right). \tag{1}$$

Because this inequality (1) holds for every $n \in \mathbb{N}$, by [Lemma 3, 8], it follows that

$$\sum_{n=1}^{\infty} \Phi_s\left(\frac{|x_n|}{k}\right) \leq \sum_{n=1}^{\infty} \Phi_s\left(\frac{|x_n|}{\|X\|_{l^*_{\Phi_s}(\mathbb{R})}}\right) \leq k.$$

Thus, for $k > 0$, if $\|X\|_{l^*_{\Phi_s}(\mathbb{R})} \leq k$, then

$$\sum_{n=1}^{\infty} \Phi_s\left(\frac{|x_n|}{\|X\|_{l^*_{\Phi_s}(\mathbb{R})}}\right) \leq k.$$

Lemma 11 *Let $\Phi_s : \mathbb{R}_0^+ \to \mathbb{R}_0^+$ be an s-Young function and $X = (x_n) \in l^*_{\Phi_s}(\mathbb{R})$. For any $\epsilon > 0$, the following is true:*

$$\sum_{n=1}^{\infty} \Phi_s\left(\frac{|x_n|}{\epsilon}\right) \leq \epsilon$$

if and only if

$$\|X\|_{l^*_{\Phi_s}(\mathbb{R})} = 0.$$

Proof Let $\Phi_s : [0, \infty) \to [0, \infty)$ be an s-Young function, $X = (x_n) \in l^*_{\Phi_s}(\mathbb{R})$, and let A be defined as

$$A := \left\{k > 0 \mid \sum_{n=1}^{\infty} \Phi_s\left(\frac{|x_n|}{k}\right) \leq k\right\}.$$

($\Rightarrow$) Suppose that for any $\epsilon > 0$,

$$\sum_{n=1}^{\infty} \Phi_s\left(\frac{|x_n|}{\epsilon}\right) \leq \epsilon.$$

This implies $\epsilon \in A$. Based on the definition of the norm in [Definition 3, 7], we have

$$\|X\|_{l^*_{\Phi_s}(\mathbb{R})} = \inf A.$$

Since ϵ is an arbitrary positive number, and $\epsilon \in A$, it follows that

$$0 \leq \|X\|_{l^*_{\Phi_s}(\mathbb{R})} \leq \epsilon.$$

Taking $\epsilon \to 0$, we conclude that

$$\|X\|_{l^*_{\Phi_s}(\mathbb{R})} = 0.$$

($\Leftarrow$) Suppose $\|X\|_{l^*_{\Phi_s}(\mathbb{R})} = 0$. We want to show that for any $\epsilon > 0$,

$$\sum_{n=1}^{\infty} \Phi_s \left(\frac{|x_n|}{\epsilon} \right) \leq \epsilon.$$

Assume, for contradiction, that there exists $\epsilon_0 > 0$ such that

$$\sum_{n=1}^{\infty} \Phi_s \left(\frac{|x_n|}{\epsilon_0} \right) > \epsilon_0.$$

This means $\epsilon_0 \notin A$.

Take any $b \in A$. By definition of the infimum,

$$\|X\|_{l^*_{\Phi_s}(\mathbb{R})} \leq b.$$

Case (i): If $\epsilon_0 \geq b$ or $\frac{1}{\epsilon_0} \leq \frac{1}{b}$, then

$$\frac{|x_n|}{\epsilon_0} \leq \frac{|x_n|}{b} \quad \text{for all } n \in \mathbb{N}.$$

Since Φ_s is an increasing function, we obtain

$$\Phi_s \left(\frac{|x_n|}{\epsilon_0} \right) \leq \Phi_s \left(\frac{|x_n|}{b} \right).$$

Thus,

$$\sum_{n=1}^{\infty} \Phi_s \left(\frac{|x_n|}{\epsilon_0} \right) \leq \sum_{n=1}^{\infty} \Phi_s \left(\frac{|x_n|}{b} \right) \leq b \leq \epsilon_0,$$

which contradicts the assumption that

$$\sum_{n=1}^{\infty} \Phi_s \left(\frac{|x_n|}{\epsilon_0} \right) > \epsilon_0.$$

Case (ii): If $\epsilon_0 < b$, then ϵ_0 is a lower bound of A because k is an arbitrary member of A. Consequently,

$$\inf A = \|X\|_{l^*_{\Phi_s}(\mathbb{R})} \geq \epsilon_0 > 0.$$

This contradicts the fact that $\|X\|_{l^*_{\Phi_s}(\mathbb{R})} = 0$.

Therefore, the assumption that there exists $\epsilon_0 > 0$ such that

$$\sum_{n=1}^{\infty} \Phi_s \left(\frac{|x_n|}{\epsilon_0} \right) > \epsilon_0$$

is false. Thus, it must be that for any $\epsilon > 0$,

$$\sum_{n=1}^{\infty} \Phi_s\left(\frac{|x_n|}{\epsilon}\right) \leq \epsilon.$$

Lemma 12 *Let $\Phi_s : \mathbb{R}_0^+ \to \mathbb{R}_0^+$ be an s-Young function and let $X = (x_n) \in l_{\Phi_s}^*(\mathbb{R})$. Then*

$$\sum_{n=1}^{\infty} \Phi_s\left(\frac{|x_n|}{\alpha}\right) \leq \frac{k}{\alpha^s}$$

for every $\alpha \geq 1$ with $0 < s \leq 1$ if and only if $\|X\|_{l_{\Phi_s}(\mathbb{R})} \leq k$ for $k > 0$.

Proof Let $\Phi_s : [0, \infty) \to [0, \infty)$ be an s-Young function and let $X = (x_n) \in l_{\Phi_s}^*(\mathbb{R})$. ($\Rightarrow$) It is known that for every $\alpha \geq 1$ with $0 < s \leq 1$,

$$\sum_{n=1}^{\infty} \Phi_s\left(\frac{|x_n|}{\alpha k}\right) \leq \frac{k}{\alpha^s}.$$

We need to show $\|X\|_{l_{\Phi_s}^*(\mathbb{R})} \leq k$.
Note that $1 \leq \alpha^s \leq \alpha$, so

$$\frac{1}{\alpha} \leq \frac{1}{\alpha^s} \leq 1. \tag{2}$$

Since $k > 0$, (2) becomes

$$\frac{k}{\alpha} \leq \frac{k}{\alpha^s} \leq k.$$

Thus,

$$\sum_{n=1}^{\infty} \Phi_s\left(\frac{|x_n|}{\alpha k}\right) \leq \frac{k}{\alpha^s} \leq ak \Rightarrow \sum_{n=1}^{\infty} \Phi_s\left(\frac{|x_n|}{\alpha k}\right) \leq ak.$$

This means $\alpha k \in A$. Choose $\alpha = 1$, then

$$\sum_{n=1}^{\infty} \Phi_s\left(\frac{|x_n|}{k}\right) \leq k,$$

so $k \in A$. Since $\inf A = \|X\|_{l_{\Phi_s}^*(\mathbb{R})}$, it follows that $\|X\|_{l_{\Phi_s}^*(\mathbb{R})} \leq k$.
($\Leftarrow$) Given $\|X\|_{l_{\Phi_s}^*(\mathbb{R})} \leq k$, we will show that

$$\sum_{n=1}^{\infty} \Phi_s\left(\frac{|x_n|}{\alpha k}\right) \leq \frac{k}{\alpha^s}$$

for arbitrary $\alpha \geq 1$ and $0 < s \leq 1$.

Note that

$$\|X\|_{l^*_{\Phi_s}} \leq k \quad \text{or} \quad \frac{1}{k} \leq \frac{1}{\|X\|_{l^*_{\Phi_s}(\mathbb{R})}}.$$

Based on [Lemma 2.1, 2], for $\alpha \geq 1$ and $0 < s \leq 1$,

$$\sum_{n=1}^{\infty} \Phi_s\left(\frac{|x_n|}{\alpha k}\right) \leq \frac{1}{\alpha^s}\sum_{n=1}^{\infty} \Phi_s\left(\frac{|x_n|}{k}\right).$$

Since $\Phi_s\left(\frac{|x_n|}{k}\right) \leq \Phi_s\left(\frac{|x_n|}{\|X\|_{l^*_{\Phi_s}(\mathbb{R})}}\right)$ from inequality (1), then

$$\sum_{n=1}^{\infty} \Phi_s\left(\frac{|x_n|}{\alpha k}\right) \leq \frac{1}{\alpha^s}\sum_{n=1}^{\infty} \Phi_s\left(\frac{|x_n|}{k}\right) \leq \frac{1}{\alpha^s}\sum_{n=1}^{\infty} \Phi_s\left(\frac{|x_n|}{\|X\|_{l^*_{\Phi_s}(\mathbb{R})}}\right) \leq \frac{1}{\alpha^s}\cdot k = \frac{k}{\alpha^s}.$$

So, it is proved that for every $\alpha \geq 1$,

$$\sum_{n=1}^{\infty} \Phi_s\left(\frac{|x_n|}{\alpha}\right) \leq \frac{k}{\alpha^s}$$

with $0 < s \leq 1$.

Lemma 13 *Let $\Phi_s : \mathbb{R}_0^+ \to \mathbb{R}_0^+$ be an s-Young function and let $X = (x_n) \in l^*_{\Phi_s}(\mathbb{R})$. Then $X = (x_n) = 0$ if and only if $\|X\|_{l_{\Phi_s}(\mathbb{R})} = 0$.*

Proof Let $\Phi_s : [0, \infty) \to [0, \infty)$ be an s-Young function, $X = (x_n) \in l^*_{\Phi_s}(\mathbb{R})$, and define

$$A := \left\{k > 0 \,:\, \sum_{n=1}^{\infty} \Phi_s\left(\frac{|x_n|}{k}\right) \leq k\right\}.$$

($\Rightarrow$) Take any $\epsilon > 0$. If $x_n = 0$ for every $n \in \mathbb{N}$, then

$$\sum_{n=1}^{\infty} \Phi_s\left(\frac{|x_n|}{\epsilon}\right) = \sum_{n=1}^{\infty} \Phi_s\left(\frac{0}{\epsilon}\right) = \sum_{n=1}^{\infty} \Phi_s(0) = 0 < \epsilon.$$

Furthermore, according to [Lemma 3, 11], we have $\|X\|_{l^*_{\Phi_s}(\mathbb{R})} = 0$.
($\Leftarrow$) Since $\inf A = \|X\|_{l^*_{\Phi_s}(\mathbb{R})} = 0$, it follows that for any $\epsilon \in A$, we have $\|X\|_{l^*_{\Phi_s}(\mathbb{R})} < \epsilon$. Furthermore, based on the definition of A, note that

$$\sum_{n=1}^{\infty} \Phi_s\left(\frac{|x_n|}{\epsilon}\right) \leq \epsilon,$$

which implies that for every $n \in \mathbb{N}$, we have

$$0 < \Phi_s\left(\frac{|x_n|}{\epsilon}\right) \leq \epsilon.$$

Consequently, $\Phi_s\left(\frac{|x_n|}{\epsilon}\right) = 0$. Since $\epsilon \neq 0$, it follows that $|x_n| = 0$. Hence, $X = (x_n) = 0$.

Lemma 14 *Let $X = (x_n) \in l^*_{\Phi_s}(\mathbb{R})$. Then for all $\alpha \in \mathbb{R}$, it holds that*

$$\|\alpha X\|_{l^*_{\Phi_s}(\mathbb{R})} \leq \max\{1, |\alpha|\} \|X\|_{l^*_{\Phi_s}(\mathbb{R})}.$$

Proof Take any $X = (x_n) \in l^*_{\Phi_s}(\mathbb{R})$.
Case (i): For $\alpha = 0$, we have

$$\|\alpha X\|_{l^*_{\Phi_s}(\mathbb{R})} = \|0 \cdot X\|_{l^*_{\Phi_s}(\mathbb{R})} = \|0\|_{l^*_{\Phi_s}(\mathbb{R})} = 0 = 0 \cdot \|X\|_{l^*_{\Phi_s}(\mathbb{R})} = |\alpha| \|X\|_{l^*_{\Phi_s}(\mathbb{R})}.$$

Case (ii): For $\alpha \neq 0$ and $|\alpha| \leq 1$, from the definition of $\|\alpha X\|_{l^*_{\Phi_s}(\mathbb{R})}$, we obtain

$$\|\alpha X\|_{l^*_{\Phi_s}(\mathbb{R})} = \inf\left\{k > 0 \, : \, \sum_{n=1}^{\infty} \Phi_s\left(\frac{|\alpha x_n|}{k}\right) \leq k\right\}.$$

Since $|\alpha| \leq 1$,

$$\|\alpha X\|_{l^*_{\Phi_s}(\mathbb{R})} = \inf\left\{k > 0 \, : \, \sum_{n=1}^{\infty} \Phi_s\left(\frac{|\alpha||x_n|}{k}\right) \leq k\right\}.$$

Let $c = \frac{k}{|\alpha|}$, then

$$\|\alpha X\|_{l^*_{\Phi_s}(\mathbb{R})} = \inf\left\{c|\alpha| > 0 \, : \, \sum_{n=1}^{\infty} \Phi_s\left(\frac{|x_n|}{c}\right) \leq c|\alpha|\right\}.$$

Thus,

$$\|\alpha X\|_{l^*_{\Phi_s}(\mathbb{R})} = |\alpha| \inf\left\{c > 0 \, : \, \sum_{n=1}^{\infty} \Phi_s\left(\frac{|x_n|}{c}\right) \leq c\right\} = |\alpha| \|X\|_{l^*_{\Phi_s}(\mathbb{R})}.$$

Case (iii): For $|\alpha| > 1$, from the definition of $\|\alpha X\|_{l^*_{\Phi_s}(\mathbb{R})}$, we have

$$\|\alpha X\|_{l^*_{\Phi_s}(\mathbb{R})} = \inf\left\{k > 0 \, : \, \sum_{n=1}^{\infty} \Phi_s\left(\frac{|\alpha x_n|}{k}\right) \leq k\right\}.$$

Since $|\alpha| > 1$,

$$\|\alpha X\|_{l^*_{\Phi_s}(\mathbb{R})} = \inf\left\{k > 0 \,:\, \sum_{n=1}^{\infty} \Phi_s\left(\frac{|\alpha||x_n|}{k}\right) \le k\right\}.$$

Let $c = \frac{k}{|\alpha|}$, then

$$\|\alpha X\|_{l^*_{\Phi_s}(\mathbb{R})} = \inf\left\{c|\alpha| > 0 \,:\, \sum_{n=1}^{\infty} \Phi_s\left(\frac{|x_n|}{c}\right) \le c|\alpha|\right\}.$$

Let $C = \left\{c|\alpha| > 0 \,:\, \sum_{n=1}^{\infty} \Phi_s\left(\frac{|x_n|}{c}\right) \le c|\alpha|\right\}$. Define

$$C' = \left\{c|\alpha| > 0 \,:\, \sum_{n=1}^{\infty} \Phi_s\left(\frac{|x_n|}{c}\right) \le c\right\}.$$

For any $c|\alpha| \in C'$, we have

$$\sum_{n=1}^{\infty} \Phi_s\left(\frac{|x_n|}{c}\right) \le c \le c|\alpha|.$$

Thus, $c|\alpha| \in C$, so $C' \subseteq C$. Hence,

$$\inf C \le \inf C'.$$

Therefore,

$$\inf\left\{c|\alpha| > 0 \,:\, \sum_{n=1}^{\infty} \Phi_s\left(\frac{|x_n|}{c}\right) \le c|\alpha|\right\} \le \inf\left\{c|\alpha| > 0 \,:\, \sum_{n=1}^{\infty} \Phi_s\left(\frac{|x_n|}{c}\right) \le c\right\} \le$$

$$|\alpha| \inf\left\{c > 0 \,:\, \sum_{n=1}^{\infty} \Phi_s\left(\frac{|x_n|}{c}\right) \le c\right\}$$

Thus,

$$\|\alpha X\|_{l^*_{\Phi_s}(\mathbb{R})} \le |\alpha| \|X\|_{l^*_{\Phi_s}(\mathbb{R})}.$$

In summary, for $|\alpha| \le 1$, we have

$$\|\alpha X\|_{l^*_{\Phi_s}(\mathbb{R})} = |\alpha| \|X\|_{l^*_{\Phi_s}(\mathbb{R})},$$

and for $|\alpha| > 1$, we have

$$\|\alpha X\|_{l^*_{\Phi_s}(\mathbb{R})} \leq |\alpha| \|X\|_{l^*_{\Phi_s}(\mathbb{R})}.$$

In other words,

$$\|\alpha X\|_{l^*_{\Phi_s}(\mathbb{R})} \leq \max\{1, |\alpha|\} \|X\|_{l^*_{\Phi_s}(\mathbb{R})}.$$

Lemma 15 *The non-negative function* $\|\cdot\|_{l^*_{\Phi_s}(\mathbb{R})} : l^*_{\Phi_s}(\mathbb{R}) \to [0, \infty)$ *is the F-norm on* $l^*_{\Phi_s}(\mathbb{R})$.

Proof Take any $X = (x_k), Y = (y_k) \in l^*_{\Phi_s}(\mathbb{R})$ and scalar $\alpha \in \mathbb{R}$. By [Definition 3, 7], we know that

$$\|X\|_{l^*_{\Phi_s}(\mathbb{R})} = \inf\left\{k > 0 \,:\, \sum_{n=1}^{\infty} \Phi_s\left(\frac{|x_n|}{k}\right) \leq k\right\}.$$

Let $A = \left\{k > 0 \,:\, \sum_{n=1}^{\infty} \Phi_s\left(\frac{|x_n|}{k}\right) \leq k\right\}$. Since $\|X\|_{l^*_{\Phi_s}(\mathbb{R})} = \inf A$, it is obvious that $0 \leq \|X\|_{l^*_{\Phi_s}(\mathbb{R})} \leq k$ for all $k \in A$. Therefore, it is proved that $\|X\|_{l^*_{\Phi_s}(\mathbb{R})} \geq 0$.
Next, by [Lemma 3, 13], it is proved that $\|X\|_{l^*_{\Phi_s}(\mathbb{R})} = 0$ if and only if $X = 0$.
If $\alpha_n \to 0$ as $n \to \infty$, then $\|\alpha_n X\| \to 0$ as $n \to \infty$. Also, if $\|X_n\| \to 0$ as $n \to \infty$, then for every $\alpha \in \mathbb{R}$, $\|\alpha X_n\| \to 0$ as $n \to \infty$.

From $\lim_{n\to\infty} \alpha_n = 0$, it follows that for every $\epsilon > 0$, there exists $K_\epsilon \in \mathbb{N}$ such that for every $n \geq K_\epsilon$, $|\alpha_n| < \epsilon$. Since $\|X\|_{l^*_{\Phi_s}(\mathbb{R})} = \inf\left\{k > 0 \,:\, \sum_{n=1}^{\infty} \Phi_s\left(\frac{|x_n|}{k}\right) \leq k\right\}$, for every $\epsilon > 0$ it holds that $\|X\|_{l^*_{\Phi_s}(\mathbb{R})} \leq \epsilon$. Consequently,

$$\|\alpha_n X\|_{l^*_{\Phi_s}(\mathbb{R})} \leq \max\{1, |\alpha_n|\} \|X\|_{l^*_{\Phi_s}(\mathbb{R})} < \epsilon.$$

So, for every $\epsilon > 0$, there exists $K_\epsilon \in \mathbb{N}$ such that for every $n \geq K_\epsilon$, $\|\alpha_n X\|_{l^*_{\Phi_s}(\mathbb{R})} < \epsilon$. In other words,

$$\lim_{n\to\infty} \|\alpha_n X\|_{l^*_{\Phi_s}(\mathbb{R})} = 0.$$

There are two cases:

Case (i): For $\alpha = 0$, clearly

$$\lim_{n\to\infty} \|\alpha X_n\|_{l^*_{\Phi_s}(\mathbb{R})} = 0.$$

Case (ii): For $\alpha \neq 0$, from $\lim_{n\to\infty} \|X_n\|_{l^*_{\Phi_s}(\mathbb{R})} = 0$, meaning that for every $\epsilon > 0$, there exists $K_\epsilon \in \mathbb{N}$ such that for every $n \geq K_\epsilon$,

$$\|X_n\|_{l^*_{\Phi_s}(\mathbb{R})} < \frac{\epsilon}{|\alpha|}.$$

Note that

$$\|\alpha X_n\|_{l^*_{\Phi_s}(\mathbb{R})} \leq \max\{1, |\alpha|\}\|X_n\|_{l^*_{\Phi_s}(\mathbb{R})} < \max\{1, |\alpha|\} \cdot \frac{\epsilon}{|\alpha|} < \epsilon.$$

Thus, for every $\epsilon > 0$, there exists $K_\epsilon \in \mathbb{N}$ such that for every $n \geq K_\epsilon$, $\|\alpha X_n\|_{l^*_{\Phi_s}(\mathbb{R})} < \epsilon$. In other words,

$$\lim_{n\to\infty} \|\alpha X_n\|_{l^*_{\Phi_s}(\mathbb{R})} = 0.$$

Let $D = \left\{k > 0 : \sum_{n=1}^{\infty} \Phi_s\left(\frac{|x_n+y_n|}{k}\right) \leq k\right\}$. Assuming $\|X\|_{l^*_{\Phi_s}(\mathbb{R})}, \|Y\|_{l^*_{\Phi_s}(\mathbb{R})} > 0$, note that

$$\sum_{n=1}^{\infty} \Phi_s\left(\frac{|x_n + y_n|}{\|X\|_{l^*_{\Phi_s}(\mathbb{R})} + \|Y\|_{l^*_{\Phi_s}(\mathbb{R})}}\right) \leq \sum_{n=1}^{\infty} \Phi_s\left(\frac{|x_n| + |y_n|}{\|X\|_{l^*_{\Phi_s}(\mathbb{R})} + \|Y\|_{l^*_{\Phi_s}(\mathbb{R})}}\right)$$

$$= \sum_{n=1}^{\infty} \Phi_s\left(\frac{\|X\|_{l^*_{\Phi_s}(\mathbb{R})}}{\|X\|_{l^*_{\Phi_s}(\mathbb{R})} + \|Y\|_{l^*_{\Phi_s}(\mathbb{R})}} \cdot \frac{|x_n|}{\|X\|_{l^*_{\Phi_s}(\mathbb{R})}} + \frac{\|Y\|_{l^*_{\Phi_s}(\mathbb{R})}}{\|X\|_{l^*_{\Phi_s}(\mathbb{R})} + \|Y\|_{l^*_{\Phi_s}(\mathbb{R})}} \cdot \frac{|y_n|}{\|Y\|_{l^*_{\Phi_s}(\mathbb{R})}}\right)$$

$$\leq \left(\frac{\|X\|_{l^*_{\Phi_s}(\mathbb{R})}}{\|X\|_{l^*_{\Phi_s}(\mathbb{R})} + \|Y\|_{l^*_{\Phi_s}(\mathbb{R})}}\right)^s \sum_{n=1}^{\infty} \Phi_s\left(\frac{|x_n|}{\|X\|_{l^*_{\Phi_s}(\mathbb{R})}}\right) +$$

$$\left(\frac{\|Y\|_{l^*_{\Phi_s}(\mathbb{R})}}{\|X\|_{l^*_{\Phi_s}(\mathbb{R})} + \|Y\|_{l^*_{\Phi_s}(\mathbb{R})}}\right)^s \sum_{n=1}^{\infty} \Phi_s\left(\frac{|y_n|}{\|Y\|_{l^*_{\Phi_s}(\mathbb{R})}}\right)$$

$$\leq \left(\frac{\|X\|_{l^*_{\Phi_s}(\mathbb{R})}}{\|X\|_{l^*_{\Phi_s}(\mathbb{R})} + \|Y\|_{l^*_{\Phi_s}(\mathbb{R})}}\right)^s \cdot \|X\|_{l^*_{\Phi_s}(\mathbb{R})} + \left(\frac{\|Y\|_{l^*_{\Phi_s}(\mathbb{R})}}{\|X\|_{l^*_{\Phi_s}(\mathbb{R})} + \|Y\|_{l^*_{\Phi_s}(\mathbb{R})}}\right)^s \cdot \|Y\|_{l^*_{\Phi_s}(\mathbb{R})}$$

$$= \frac{(\|X\|_{l^*_{\Phi_s}(\mathbb{R})})^{s+1} + (\|Y\|_{l^*_{\Phi_s}(\mathbb{R})})^{s+1}}{(\|X\|_{l^*_{\Phi_s}(\mathbb{R})} + \|Y\|_{l^*_{\Phi_s}(\mathbb{R})})^s} \leq \|X\|_{l^*_{\Phi_s}(\mathbb{R})} + \|Y\|_{l^*_{\Phi_s}(\mathbb{R})}.$$

Since $\|X\|_{l^*_{\Phi_s}(\mathbb{R})}, \|Y\|_{l^*_{\Phi_s}(\mathbb{R})} > 0$ and

$$\sum_{n=1}^{\infty} \Phi_s\left(\frac{|x_n + y_n|}{\|X\|_{l^*_{\Phi_s}(\mathbb{R})} + \|Y\|_{l^*_{\Phi_s}(\mathbb{R})}}\right) \leq \|X\|_{l^*_{\Phi_s}(\mathbb{R})} + \|Y\|_{l^*_{\Phi_s}(\mathbb{R})},$$

then by the definition of set D, $\|X\|_{l^*_{\Phi_s}(\mathbb{R})} + \|Y\|_{l^*_{\Phi_s}(\mathbb{R})} \in D$. Consequently,

$$\inf D = \|X + Y\|_{l^*_{\Phi_s}(\mathbb{R})} \leq \|X\|_{l^*_{\Phi_s}(\mathbb{R})} + \|Y\|_{l^*_{\Phi_s}(\mathbb{R})}.$$

So, the triangle inequality holds.

Recall [Lemma 2.2, 5], which defines the space $l_{\Phi_s}(\mathbb{R})$. The space $l^*_{\Phi_s}(\mathbb{R})$ can be considered an extension of the space $l_{\Phi_s}(\mathbb{R})$ when $k > 1$. Specifically, let $k \in \{b > 0 : \sum_{n=1}^{\infty} \Phi_s\left(\frac{|x_n|}{b}\right) \leq b\}$. This property is illustrated by the following lemma.

Lemma 16 *If $k > 1$, then the space $l_{\Phi_s}(\mathbb{R})$ is contained in the space $l^*_{\Phi_s}(\mathbb{R})$, i.e.,*

$$l_{\Phi_s}(\mathbb{R}) \subseteq l^*_{\Phi_s}(\mathbb{R}).$$

Proof Take any sequence $X = (x_n) \in l_{\Phi_s}(\mathbb{R})$. By [Lemma 2.2, 5], for the sequence $X = (x_n) \in \mathbb{R}$, there exists $\alpha > 0$ such that

$$\sum_{n=1}^{\infty} \Phi_s\left(\frac{|x_n|}{\alpha}\right) < \infty,$$

with

$$\|X\|_{l_{\Phi_s}(\mathbb{R})} = \inf\left\{k > 0 : \sum_{n=1}^{\infty} \Phi_s\left(\frac{|x_n|}{k}\right) \leq 1\right\}.$$

Define

$$A := \left\{k > 0 : \sum_{n=1}^{\infty} \Phi_s\left(\frac{|x_n|}{k}\right) \leq 1\right\}$$

and

$$B := \left\{k > 0 : \sum_{n=1}^{\infty} \Phi_s\left(\frac{|x_n|}{k}\right) \leq k\right\}.$$

Take any $a_0 \in A$. This implies

$$\sum_{n=1}^{\infty} \Phi_s\left(\frac{|x_n|}{a_0}\right) \leq 1.$$

Since $a_0 \in A$ and $a_0 < b$, we have

$$\sum_{n=1}^{\infty} \Phi_s\left(\frac{|x_n|}{b}\right) \leq 1 < b,$$

which implies $a_0 \in B$. Therefore, $A \subseteq B$ and

$$\inf A \leq \inf B.$$

Note that

$$\inf A = \|X\|_{l_{\Phi_s}(\mathbb{R})}$$

and

$$\inf B = \|X\|_{l^*_{\Phi_s}(\mathbb{R})}.$$

Thus,

$$\|X\|_{l^*_{\Phi_s}(\mathbb{R})} \leq \|X\|_{l_{\Phi_s}(\mathbb{R})} < \infty.$$

Therefore,

$$\|X\|_{l^*_{\Phi_s}(\mathbb{R})} < \infty.$$

In other words, for any $X = (x_n) \in l_{\Phi_s}(\mathbb{R})$, it follows that $X = (x_n) \in l^*_{\Phi_s}(\mathbb{R})$. Hence,

$$l_{\Phi_s}(\mathbb{R}) \subseteq l^*_{\Phi_s}(\mathbb{R}).$$

Furthermore, if Φ is a convex function and $\Phi(0) = 0$, then Φ is an s-convex function. Consequently, $l^*_{\Phi}(\mathbb{R}) = l^*_{\Phi_s}(\mathbb{R})$, which means the space $l^*_{\Phi_s}(\mathbb{R})$ is a generalization of the Orlicz sequence space when $s = 1$.

4 Conclusions

The conclusion from this discussion is that the generalized Orlicz sequence space constructed by modifying the norm as stated in [Definition 3, 7] is wider than the space of generalized Orlicz sequences in previous research (see [Lemma 2.2, 5]) when $k > 1$. Additionally, the properties of the generalized Orlicz sequence space also apply to the generalized Orlicz sequence space with Rao and Ren's norm, although with some adjustments. The validity of the norm on this constructed space is the Fréchet norm, which is more flexible and adaptive. In previous research, it was shown that the generalized Orlicz sequence space does not satisfy the triangle inequality, but in the generalized Orlicz sequence space with Rao and Ren's norm, the triangle inequality holds.

Acknowledgements This research was funded by Hibah Kompetensi and UPI Research Grant.

References

1. Castillo, R.E., Rafeiro, H.: An Introductory Course in Lebesgue Spaces. Springer, Bogotá (2016)
2. Orlicz, W.: Linear Functional Analysis (Series in Real Analysis Volume 4). World Scientific, Singapore (1992)
3. Kamthan, P.K., Gupta, M.: Sequence Spaces and Series. Marcel Dekker Inc., New York (1981)
4. Lindenstrauss, J., Tzafriri, L.: On Orlicz sequence spaces I. Israel J. Math. **10**, 379–390 (1971)
5. Maligranda, L.: Orlicz Spaces and Interpolation, Thesis, Departamento de Matemática, Universidade Estadual de Campinas (1989)
6. Masta, A.A.: Sifat Inklusi pada Ruang Orlicz-Morrey. Institut Teknologi Bandung, Indonesia (2018)
7. Masta, A.A., Gunawan, H., Budhi, W.S.: Inclusion property of Orlicz and weak Orlicz spaces. J. Math. Fundam. Sci. **48**(3), 193–203 (2016)
8. Rao, M.M., Ren, Z.D.: Theory of Orlicz Spaces. Marcel Dekker Inc., New York (1991)
9. Harjulehto, P., Hästö, P.: Generalized Orlicz Spaces. pp. 47–78. Springer (2019)
10. Ifronika, Masta, A.A., Nur, M., Gunawan, H.: Generalized Hölder's inequality in Orlicz spaces. Proc. Jangjeon Math. Soc. **22**(1), 25–34 (2018)
11. Dermawan, R.: Sifat Inklusi dan Ketaksamaan Hölder pada Ruang Orlicz Diperumum. Universitas Pendidikan Indonesia, Indonesia (2022)
12. Maligranda, L., Mastylo, M.: Inclusion mappings between Orlicz sequence spaces. J. Funct. Anal. **176**, 264–279 (2000)
13. Bakery, A.A., Afaf, R.: Some properties of pre-quasi norm on Orlicz sequence space. J. Inequal. Appl. **55** (2020)
14. Prayoga, P.S.: Sifat Inklusi dan Perumuman Ketaksamaan Hölder pada Ruang Barisan Orlicz. Universitas Pendidikan Indonesia, Indonesia (2020)
15. Prayoga, P.S., Masta, A.A., Fatimah, S.: Sifat Inklusi dan Perumuman Ketaksamaan Hölder Pada Ruang Barisan Orlicz. Jurnal EurekaMatika **8**(2) (2020)
16. Dasep: Sifat Inklusi dan Ketaksamaan Hölder pada Ruang Barisan Orlicz Diperumum. Universitas Pendidikan Indonesia, Indonesia (2023)
17. Hudzik, H., Maligranda, L.: Some remarks on s-convex functions. Aequationes Mathematicae **48**, 100–111 (1994)
18. Sumiaty, E., Kurniadi, Y.: Ruang Barisan Musielak-Orlicz. Seminar Nasional Matematika, Indonesia (2008)
19. Gunawan, L.R.: Perumuman Ruang Barisan Orlicz Menggunakan Norma versi Rao dan Ren Beserta Sifat Inklusinya. Universitas Pendidikan Indonesia, Indonesia (2024)
20. Kustiawan, C., Dasep, Sumiaty, E., Fatimah, S., Hazmy, S.A.: GENERALIZED ORLICZ SEQUENCE SPACES. BAREKENG: Jurnal Ilmu Matematika dan Terapan **17**(1), 0427–0438 (2023)
21. Hazarika, B., Kalita, H.: Henstock-Orlicz space and its dense space. Asian Eur. J. Math. **14**(2), 1–17 (2020)
22. Kalita, H., Hazarika, B.: Countable additivity of Henstock-Dunford Integrable functions and Orlicz Space. Anal. Math. Phys. **11**(2), 1–13 (2021)

Strongly λ-Summable Functions Defined by Orlicz Functions

Rabia Savaş

Abstract The opinion of statistical convergence is substantial in the summability theory. Later Borwein (J London Math Soc 40:628–634, 1965, [1]) introduced and studied strongly summable functions. In this paper, by taking a real-valued function that is measurable on $(1, \infty)$ we hold forth the concept of strongly λ-Summable functions defined by Orlicz functions and also some considerable results will be analyzed.

Keywords Orlicz functions · Strongly summability · Statistical convergence

1 Introduction

Fast put forward the conception of statistical convergence in [2] and Borwein [1] extended this concept to strongly summable functions. Following Borwein's result Nuray [4] introduced λ-statistically convergent functions by considering nonnegative real-valued Lebesgue measurable function on $(1, \infty)$ instead of sequences. Let $\lambda = (\lambda_\delta)$ be a non-decreasing sequence of positive numbers going to ∞ and $\lambda_{\delta+1} \leq \lambda_\delta + 1$, $\lambda_1 = 1$ and Δ state the class of all such sequences. For a sequence $\xi = (\xi_\upsilon)$ the generalized de la Vallee-Poussin mean is shown by

$$t_\delta(\xi) = \frac{1}{\lambda_\delta} \sum_{\upsilon \in V_\delta} \xi_\upsilon,$$

where $V_\delta = [\delta - \lambda_\delta + 1, \delta]$.

Nuray [4] presented the following definitions.

R. Savaş (✉)
Department of Mathematics and Science Education, Istanbul Medeniyet University, Istanbul, Turkey
e-mail: rabiasavass@hotmail.com; abc@uni-heidelberg.de; lncs@uni-heidelberg.de
URL: http://www.springer.com/gp/computer-science/lncs

ABC Institute, Rupert-Karls-University Heidelberg, Heidelberg, Germany

H. Kalita et al. (eds.), *Nonlinear Analysis and Computational Techniques*,
Springer Proceedings in Mathematics & Statistics 501,
https://doi.org/10.1007/978-981-96-6038-4_2

Definition 1 Let $\lambda \in \Delta$ and $\varsigma(p)$ be a real valued function, if

$$\lim_{\delta\to\infty}\frac{1}{\lambda_\delta}\int_{\delta-\lambda_\delta+1}^{\delta}|\varsigma(p)-\varpi|\,dp=0.$$

Whenever this occurs $\varsigma(p)$ is λ-strongly summable to ϖ. It will be indicated by $[\widetilde{W},\lambda]-\lim\varsigma(p)=\varpi$.

Definition 2 Let $\lambda \in \Delta$ and $\varsigma(p)$ be a real valued function, if for every $\varepsilon > 0$

$$\lim_{\delta\to\infty}\frac{1}{\lambda_\delta}\left|\{p\in V_\delta : |\varsigma(p)-\varpi|\geq\varepsilon\}\right|=0$$

Whenever this happens $\varsigma(p)$ is λ-statistically convergent to ϖ. It will be indicated by $S_\lambda-\lim\varsigma(p)=\varpi$.

In another direction, Lindenstrauss and Tzafriri [3] introduced the idea of Orlicz function to identify the sequence space

$$\ell_M=\left\{\xi\in\omega:\sum_{\upsilon=1}^{\infty}Z\left(\frac{|\xi_\upsilon|}{\eta}\right)<\infty \text{ for } \eta>0\right\}.$$

The space ℓ_M with the norm

$$||\xi||=\inf\left\{\eta>0:\sum_{\upsilon=1}^{\infty}Z\left(\frac{|\xi_\upsilon|}{\eta}\right)\leq 1\right\}$$

occurs a Banach space which is called an Orlicz sequence space. The space ℓ_M is nearly connected with the space ℓ_p which is an Orlicz sequence with $Z(\xi)=\xi^r$ for $1\leq r<\infty$. An Orlicz is a function $Z:[0,\infty)\to[0,\infty)$, which is a continuous, non-decreasing, and convex function described for $\xi\geq 0$ such that $Z(0)=0$ and $Z(\xi)>0$ for $\xi>0$. Orlicz function Z is a convex function and $Z(0)=0$, then $Z(t\xi)\leq tM(\xi)$ for all t with $0<t<1$.

Definition 3 [5] An Orlicz function Z is said to satisfy Δ_2-condition for all values of g, if there exists constant $\varphi>0$, such that $Z(2g)\leq\varphi Z(g)\ \ (g\geq 0)$.

The Δ_2-condition is counterbalance the inequality $Z(\zeta g)\leq\varphi\cdot\zeta Z(g)$ for all values of g and for $\varphi>2$ and $\zeta>1$. The following inequality will be used and we consider $\varsigma(p)$ as a real valued function which is measurable in the Lebesgue sense on $(1,\infty)$ during this paper. Let $r=(r_t)$ be a sequence of positive real numbers with $0<m=\inf r_t\leq r_t\leq\sup r_t=M$, and let $D=\max(1,2^{M-1})$. Then for $\phi_t,\varrho_t\in\mathbb{C}$ and for all $t\in\mathbb{N}$, we have the following Δ_2-condition and let $0<\mu<1$.

$$|\phi_t+\varrho_t|^{r_t}\leq M\left\{|\phi_t|^{r_t}+|\varrho_t|^{r_t}\right\}.$$

2 Main Results

Let us define the main ideals that were used in the main results of the paper.

Definition 4 Let Z be an Orlicz function, Ψ be locally convex Hausdorff topological linear space and $r = r_t$ be a sequence of positive real numbers, $\lambda \in \Delta$, $\varsigma(p)$ be a real valued function, $w(\Psi)$ indicates the space of all measurable functions $\varsigma(p)$, where $\varsigma(p) \in \Psi$. Then, the following sequence spaces are defined

$$[\widetilde{W}, \lambda, Z, r]_1 = \left\{ \varsigma(p) : \frac{1}{\lambda_\delta} \int_{p \in V_\delta} \left[Z\left(\frac{|\varsigma(p) - \varpi|}{\eta} \right) \right]^{r_t} \to 0 \text{ for } \eta > 0 \text{ and } \varpi > 0 \right\},$$

$$[\widetilde{W}, \lambda, Z, r]_0 = \left\{ \varsigma(p) : \frac{1}{\lambda_\delta} \int_{p \in V_\delta} \left[Z\left(\frac{|\varsigma(p)|}{\eta} \right) \right]^{r_t} \to 0 \text{ for some } \eta > 0 \right\},$$

$$[\widetilde{W}, \lambda, Z, r]_\infty = \left\{ \varsigma(p) : \sup_\delta \frac{1}{\lambda_\delta} \int_{p \in V_\delta} \left[Z\left(\frac{|\varsigma(p)|}{\eta} \right) \right]^{r_t} < \infty \text{ for some } \eta > 0 \right\}.$$

We indicate $[\widetilde{W}, \lambda, Z, r]_1$, $[\widetilde{W}, \lambda, Z, r]_0$ and $[\widetilde{W}, \lambda, Z, r]_\infty$ as $[\widetilde{W}, \lambda, Z]_1$, $[\widetilde{W}, \lambda, Z]_0$ and $[\widetilde{W}, \lambda, Z]_\infty$ when $r_t = 1$ for all t. If $Z(\xi) = \xi$, $r_t = 1$ for all t, then $[\widetilde{W}, \lambda, Z, r]_1$, $[\widetilde{W}, \lambda, Z, r]_0$,and $[\widetilde{W}, \lambda, Z, r]_\infty$ will reduce to $[\widetilde{W}, \lambda]_1$, $[\widetilde{W}, \lambda]_0$ and $[\widetilde{W}, \lambda]_\infty$, respectively. Throughout the paper, R will be denote one of the notation $0, 1$ or ∞.

Theorem 1 *Let $\lambda \in \Delta$, and $\varsigma(p)$ be a real-valued function. For any Orlicz function Z, let the sequence (r_t) be bounded, then $[\widetilde{W}, \lambda, Z, r]_R$ are linear spaces over the class of $\mathbb{C}$.*

Proof We shall demonstrate only for $[\widetilde{W}, \lambda, Z, r]_0$. The other conclusions can be shown in similar manner. Let $\varsigma(p), \sigma(p) \in [\widetilde{W}, \lambda, Z, r]_0$ and $\alpha, \beta \in \mathbb{C}$. For some $\eta_1, \eta_2 \in \mathbb{Z}^+$, we state

$$\lim_{\delta \to \infty} \frac{1}{\lambda_\delta} \int_{p \in V_\delta} \left[Z\left(\frac{|\varsigma(p)|}{\eta_1} \right) \right]^{r_t} = 0$$

and

$$\lim_{\delta\to\infty}\frac{1}{\lambda_\delta}\int_{p\in V_\delta}\left[Z\left(\frac{|\sigma(p)|}{\eta_2}\right)\right]^{r_t}=0.$$

Let $\eta_3=\max(2|\alpha|\eta_1, 2|\beta|\eta_2)$. We obtain the following

$$\begin{aligned}
&\frac{1}{\lambda_\delta}\int_{p\in V_\delta}\left[Z\left(\frac{|\alpha\varsigma(p)+\beta\sigma(p)|}{\eta_3}\right)\right]^{r_t}\\
&\leq\frac{1}{\lambda_\delta}\int_{p\in V_\delta}\left[Z\left(\frac{|\alpha\varsigma(p)|}{\eta_3}+\frac{|\beta\sigma(p)|}{\eta_3}\right)\right]^{r_t}\\
&\leq\frac{1}{\lambda_\delta}\int_{p\in V_\delta}\frac{1}{2^{r_t}}\left[Z\left(\frac{|\varsigma(p)|}{\eta_1}\right)+Z\left(\frac{|\sigma(p)|}{\eta_2}\right)\right]^{r_t}\\
&\leq\frac{1}{\lambda_\delta}\int_{p\in V_\delta}\left[Z\left(\frac{|\varsigma(p)|}{\eta_1}\right)+Z\left(\frac{|\sigma(p)|}{\eta_2}\right)\right]^{r_t}\\
&\leq D\frac{1}{\lambda_\delta}\int_{p\in V_\delta}\left[Z\left(\frac{|\varsigma(p)|}{\eta_1}\right)\right]^{r_t}+D\frac{1}{\lambda_\delta}\int_{p\in I_\delta}\left[Z\left(\frac{|\sigma(p)|}{\eta_2}\right)\right]^{r_t}\to 0
\end{aligned}$$

as $\delta\to\infty$. This shows that $\left[\widetilde{W},\lambda,Z,r\right]_0$ is linear.

Theorem 2 *Let Z be an Orlicz function, $\lambda\in\Delta$, and $\varsigma(p)$ be a real valued function. We are granted $\left[\widetilde{W},\lambda,r\right]_R\subseteq\left[\widetilde{W},\lambda,Z,r\right]_R$.*

Proof (i) The first conclusion will be demonstrated for $R=0$ and other circumstances will pursue on carrying out similar methods. Let $\varsigma(p)\in\left[\widetilde{W},\lambda,r\right]_R$ so that

$$T_\delta=\frac{1}{\lambda_\delta}\int_{p\in V_\delta}|\varsigma(p)|^{r_t}\to 0$$

as $\delta\to\infty$ and let $\varepsilon>0$ and pick δ with $0<\delta<1$ such that $Z(p)<\varepsilon$ for $0\leq p\leq\mu$. Consider $\sigma(p)=Z\left(\frac{\varsigma(p)}{\eta}\right)$ and

$$\frac{1}{\lambda_\delta}\int_{p\in V_\delta}[Z|\sigma(p)|]^{r_t}=\int_1[Z|\sigma(p)|]^{r_t}+\int_2[Z|\sigma(p)|]^{r_t},$$

where the first summation is over $\sigma(p)\leq\mu$ and the second summation is over $\sigma(p)>\mu$. Since Z is continuous, we obtain

$$\int_1[Z\sigma(p)]^{r_t}<\varepsilon^M\lambda_\delta$$

For $\sigma(p) > \mu$, we use the fact that

$$|\sigma(p)| < \frac{\sigma(p)}{\mu} \leq 1 + \left(\frac{\sigma(p)}{\mu}\right).$$

Since Z is non-decreasing and convex, it pursues that

$$Z|\sigma(p)| < Z\left(1 + \frac{\sigma(p)}{\mu}\right) < \frac{1}{2}Z(2) + \frac{1}{2}Z\left(2\frac{\sigma(p)}{\mu}\right),$$

then there is a constant $\varphi > 2$ such that

$$Z\sigma(p) < \frac{1}{2}\varphi\frac{\sigma(p)}{\mu}Z(2) + \frac{1}{2}\varphi\frac{\sigma(p)}{\mu}Z(2) = \varphi\sigma(p)\frac{Z(2)}{\mu}.$$

Hence,

$$\int_{2} [Z(\sigma(p))]^{r_t} \leq \max\left(1, \left(\varphi\mu^{-1}Z(2)\right)^{M}\right)\lambda_\delta T_\delta$$

By (2.1) and (2.2) we obtain $\sigma(p) \in \left[\widetilde{W}, \lambda, Z, r\right]_0$.

The other inclusions can be shown similarly. Consequently, we obtain $\left[\widetilde{W}, \lambda, r\right]_\infty \subseteq \left[\widetilde{W}, \lambda, Z, r\right]_\infty$.

Theorem 3 *Let Z be an Orlicz function, $\lambda \in \Delta$, and $\varsigma(p)$ be a real-valued function. Then $\left[\widetilde{W}, \lambda, Z, r\right]_0 \subset \left[\widetilde{W}, \lambda, Z, r\right]_1 \subset \left[\widetilde{W}, \lambda, Z, r\right]_\infty$.*

Proof The relation $\left[\widetilde{W}, \lambda, Z, r\right]_0 \subset \left[\widetilde{W}, \lambda, Z, r\right]_1$ is obvious. Now let $\varsigma(p) \in \left[\widetilde{W}, \lambda, Z, r\right]_1$. Then we obtain

$$\begin{aligned}
&\frac{1}{\lambda_\delta}\int_{p\in V_\delta}\left[Z\left(\frac{|\varsigma(p) - \varpi|}{\eta}\right)\right]^{r_t} \\
&\leq \frac{D}{\lambda_\delta}\int_{p\in V_\delta}\left[Z\left(\frac{|\varsigma(p) - \varpi|}{\eta}\right)\right]^{r_t} + \frac{D}{\lambda_\delta}\int_{p\in V_\delta}\left[Z\left(\frac{|\varpi|}{\eta}\right)\right]^{r_t} \\
&\leq \frac{D}{\lambda_\delta}\int_{p\in V_\delta}\left[Z\left(\frac{|\varsigma(p) - \varpi|}{\eta}\right)\right]^{r_t} + D\left[\varphi\mu^{-1}\frac{|\varpi|}{\eta}Z(2)\right]^{M}.
\end{aligned}$$

Hence, $\varsigma(p) \in \left[\widetilde{W}, \lambda, Z, r\right]_\infty$.

Corollary 1 $\left[\widetilde{W}, \lambda, Z, r\right]_0$ *and* $\left[\widetilde{W}, \lambda, Z, r\right]_1$ *are nowhere dense subclass of* $\left[\widetilde{W}, \lambda, Z, r\right]_\infty$.

Theorem 4 *Let Z be an Orlicz function, $\lambda \in \Delta$ and $\varsigma(p)$ be a real-valued function. Then $\left[\widetilde{W}, \lambda, Z, r\right]_1 \subset S_\lambda$.*

Proof Let $\varsigma(p) \in \left[\widetilde{W}, \lambda, Z, r\right]_1$. Then there exists $\eta > 0$ such that

$$\frac{1}{\lambda_\delta} \int\limits_{p \in V_\delta} \left[Z\left(\frac{|\varsigma(p) - \varpi|}{\eta}\right)\right]^{r_t} = 0.$$

Then

$$\begin{aligned}
&\frac{1}{\lambda_\delta} \int\limits_{p \in V_\delta} \left[Z\left(\frac{|\varsigma(p) - \varpi|}{\eta}\right)\right]^{r_t} \\
=&\frac{1}{\lambda_\delta} \int\limits_{p \in V_\delta, (|\varsigma(p) - \varpi| \geq \varepsilon} \left[Z\left(\frac{|\varsigma(p) - \varpi|}{\eta}\right)\right]^{r_t} \\
&+ \frac{1}{\lambda_\delta} \int\limits_{p \in V_\delta, |\varsigma(p) - \varpi|) < \varepsilon} \left[Z\left(\frac{|\varsigma(p) - \varpi|}{\eta}\right)\right]^{r_t} \\
\geq&\frac{1}{\lambda_\delta} \int\limits_{p \in V_\delta, |\varsigma(p) - \varpi|) < \varepsilon} \left[Z\left(\frac{|\varsigma(p) - \varpi|}{\eta}\right)\right]^{r_t} \\
\geq&\frac{1}{\lambda_\delta} \int\limits_{p \in V_\delta} [Z(\varepsilon)]^{r_t} \\
\geq&\frac{1}{\lambda_\delta} \int\limits_{p \in V_\delta, |\varsigma(p) - \varpi|) < \varepsilon} \min\left\{[Z(\varepsilon)]^m, [Z(\varepsilon)]^M\right\} \\
\geq&\frac{1}{\lambda_\delta} \{p \in V_\delta : |\varsigma(p) - \varpi| \geq \varepsilon\} \min\left\{[Z(\varepsilon)]^m, [Z(\varepsilon)]^M\right\}
\end{aligned}$$

Hence $\varsigma(p) \in S_\lambda$.

Theorem 5 *Let* Z *be an Orlicz function and* $\varsigma(p) \in \ell_\infty$ *, where* $\ell_\infty = \{\varsigma : \sup \varsigma(p) < \infty\}$. *Then* $S_\lambda \subset \left[\widetilde{W}, \lambda, Z, r\right]_R$.

Proof Assume that $\varsigma(p) \in \ell_\infty$ and $\varsigma(p) \longrightarrow \varpi$. Since $\varsigma(p) \in \ell_\infty$, there is a constant $\varphi > 0$ such that $|\varsigma(p) - \varpi| \leq \varphi$ for p. Given $\varepsilon > 0$ we obtain the following

$$\begin{aligned}
&\frac{1}{\lambda_\delta} \int\limits_{p \in V_\delta} \left[Z\left(\frac{|\varsigma(p) - \varpi|}{\eta}\right)\right]^{r_t} \\
=&\frac{1}{\lambda_\delta} \int\limits_{p \in V_\delta, (|\varsigma(p) - \varpi| \geq \varepsilon} \left[Z\left(\frac{|\varsigma(p) - \varpi|}{\eta}\right)\right]^{r_t} \\
&+ \frac{1}{\lambda_\delta} \int\limits_{p \in V_\delta, (|\varsigma(p) - \varpi| < \varepsilon} \left[Z\left(\frac{|\varsigma(p) - \varpi|}{\eta}\right)\right]^{r_t}
\end{aligned}$$

$$\leq \frac{1}{\lambda_\delta} \int_{p\in V_\delta,(|\varsigma(p)-\varpi|\geq\varepsilon} \max\left\{\left[Z\left(\frac{\varphi}{\eta}\right)\right]^m, \left[Z\left(\frac{\varphi}{\eta}\right)\right]^M\right\}$$

$$+\frac{1}{\lambda_\delta} \int_{p\in V_\delta,(|\varsigma(p)-\varpi|<\varepsilon} \left[Z\left(\frac{\varepsilon}{\eta}\right)\right]^{r_t}$$

$$\leq \max\left\{[Z(T)]^m, [Z(T)]^M\right\} \frac{1}{\lambda_\delta} |\{p \in V_\delta : |\varsigma(p) - \varpi| \geq \varepsilon\}|$$

$$+ \max\left\{\left[Z\left(\frac{\varepsilon}{\eta}\right)\right]^m, \left[Z\left(\frac{\varepsilon}{\eta}\right)\right]^M\right\}$$

where $\frac{\varphi}{\eta} = T$. Consequently, $\varsigma(p) \in \left[\widetilde{W}, \lambda, Z, r\right]_R$.

Theorem 6 $S_\lambda \cap \ell_\infty = \left[\widetilde{W}, \lambda, Z\right]_1 \cap \ell_\infty$

Proof By previous theorem, we need only demonstrate that $S_\lambda \cap \ell_\infty \subset \left[\widetilde{W}, \lambda, M,\right]_Z \cap \ell_\infty$. Let $\sigma(p) = |\varsigma(p) - \varpi| \to 0\,(S_\lambda)$. Since $\varsigma(p) \in \ell_\infty$, so there exists $\varphi > 0$ such that

$$Z\left(\frac{|\sigma(p)|}{\eta}\right) \leq \varphi.$$

Then for $\varepsilon > 0$ and for each $\delta \in \mathbb{N}$, we obtain

$$\frac{1}{\lambda_\delta} \int_{p\in V_\delta} \left[Z\left(\frac{|\sigma(p)|}{\eta}\right)\right] = \frac{1}{\lambda_\delta} \int_{p\in V_\delta,(|\varsigma(p)-\varpi|\geq\varepsilon} \left[Z\left(\frac{|\sigma(p)|}{\eta}\right)\right]$$

$$+\frac{1}{\lambda_\delta} \int_{p\in V_\delta,(|\varsigma(p)-\varpi|<\varepsilon} \left[Z\left(\frac{|\sigma(p)|}{\eta}\right)\right]$$

$$\leq \frac{K}{\lambda_\delta} |\{p \in V_\delta : |\sigma(p)| \geq \varepsilon\}| + Z\left(\frac{\varepsilon}{\eta}\right).$$

As a result, $\varsigma(p) \in \left[\widetilde{W}, \lambda, M\right]_1 \cap \ell_\infty$

References

1. Borwein, D.: Linear functionals with strong Cesaro summability. J. London Math. Soc. **40**, 628–634 (1965)
2. Fast, H.: Sur la convergence statistiqu. Colloquium Mathematicum **2**, 241–244 (1951)
3. Lindenstrauss, J., Tzafriri, L.: On Orlicz sequence spaces. Israel J. Math. **10**(3), 379–390 (1971)
4. Nuray, F.: λ-strongly summable and λ-statistically convergent functions. Iranian J. Sci. Technol. Trans. A **34**(A4) (2010)
5. Parashar, S.D., Choudhary, B.: Sequence spaces defined by Orlicz functions. Indian J. Pure Appl. Math. **25**(4), 419–428 (1994)

Lacunary Strong Almost Summability of Weight ρ with Respect to Seminorm q

Ekrem Savaş

Abstract The present work aims to consider the concept of q-lacunary almost statistical convergence of weight ρ and create some relations between q-lacunary almost statistical convergence of weight ρ and q-lacunary strong almost summability of weight ρ. We believe that provided results of this article can be useful in further developments of sequence spaces.

Keywords Weight function ρ · Statistical convergence · Strong almost convergence · Modulus function

1 Introduction

Researchers are engaged in improving existing techniques and technology and creating new ones. Convergence is a concept used to acquire information from existing data. There are several procedures for convergence, both analytical and non-analytical. Alternative procedures will be tried when classical methods fail to provide solutions. Statistical convergence is applied when the usual convergence concept is unable to give a solution by Fast [5]. Later on generalizations of statistical convergence are emerging in many articles by various authors (see, [2–4, 7, 9, 12–15]).

The natural density of the set U, where $U \subset \mathbf{N}$ has been given by

$$\delta(U) = lim_{n\to\infty}\frac{1}{n}|\{k < n : k \in U\}|;$$

where $|.|$ used for the order of an enclosed set. Moreover, sequence $s = (s_k)$ statistically convergent to number s_0, provided

$$U(\varepsilon) = \{k \in \mathbf{N} : |s_k - s_0| > \varepsilon\}$$

E. Savaş (✉)
Department of Mathematics, Usak University, Usak, Turkey
e-mail: ekremsavas@yahoo.com

H. Kalita et al. (eds.), *Nonlinear Analysis and Computational Techniques*,
Springer Proceedings in Mathematics & Statistics 501,
https://doi.org/10.1007/978-981-96-6038-4_3

has zero natural density [7]. Write that if $U \subset \mathbf{N}$ is finite set , then $\delta(U) = 0$, and for any set $U \subset \mathbf{N}$, $\delta(U^c) = 1 - \delta(U)$. It is represented symbolically by S, the set of all statistically convergent sequences.

In this work, the idea of q-lacunary f-statistical convergence of weight ρ is explored by fully characterizing the connection between q-lacunary f-statistical convergence of weight ρ and q-lacunary strong f-summability of weight ρ. Actually it is fundamentally extending the scope of lacunary statistical convergence under the effect of a parameter ρ. This acts as an intermediate moment between typical and lacunary statistical convergence of the sequences.

2 Preliminaries

In summability theory, the concept of almost convergence was first introduced by Banach [1] as following: A linear functional σ on m (the set of the bounded sequences) is called to be a Banach limit if it has the following properties:

1. $\sigma(s) \geq 0$ if $n \geq 0$ (i.e. $s_n \geq 0$ for all n),
2. $\sigma(e) = 1$, where $e = (1, 1, \ldots)$,
3. $\sigma(\varphi s) = \sigma(s)$,

where the shift operator φ is defined by $\varphi(s_n) = \{s_{n+1}\}$. A sequence $s \in m$ is said to be almost convergent if all Banach limits of s coincide. Let f denote the space of almost convergent sequences. Lorentz [10] has shown that

$$f = \left\{ s \in m : \lim_m t_{m,n}(s) \text{ exists uniformly in } n \right\}$$

where

$$t_{m,n}(s) = \frac{s_n + s_{n+1} + s_{n+2} + \cdots + s_{n+m}}{m+1}.$$

Maddox [11] presented the space $[f]$ of strongly almost convergent sequences as follows:

$$[f] = \left\{ s \in m : \lim_m t_{m,n}(|s - s_0|) = 0, \text{ uniformly in } n, \text{ for some } s_0 \right\}.$$

Using lacunary sequences in the setting of the classic concepts of statistical convergence and also the strong Cesàro convergence, it is also possible to open or close the degree of precision in filtering data sets. Freeman and Sember [6], in their 1978 visionary paper studied the concept of strong lacunary convergence. These studies were refined by Fridy and Orhan [8] by relating them to the new concept introduced by them: lacunary statistical convergence.

A sequence of natural numbers $\lambda = (\gamma_r)$; $r = 0, 1, 2, \ldots$ such that $\gamma_0 = 0$ and $\upsilon_r = \gamma_r - \gamma_{r-1}$ we will be called a lacunary sequence and let us denote the intervals $J_r = (\gamma_{r-1}, \gamma_r]$. For example $\lambda = (r)$ is not a lacunary sequence because $\gamma_0 = 0$ but $\upsilon_r = \gamma_r - \gamma_{r-1} = 1$. for all $r = 0, 1, \ldots$.

The sequence (s_k) is lacunary statistical convergent to s_0 if for any given number $\varepsilon > 0$,

$$\frac{1}{\upsilon_r}|\{k \in J_r : |s_k - s_0| \geq \varepsilon\}| \to 0$$

as $r \to \infty$. In [16] Das and Savaş applied these new convergence methods in the study of I-convergence of order α. Analogously we extend those result to q-lacunary f-statistically convergent of weight ρ, where functions $\rho : [0, \infty) \to [0, \infty)$ such that $\rho(s_n) \to \infty$ if $s_n \to \infty$. Δ denotes the set of all such functions throughout this article.

3 Main Results

Accordingly, we are ready to give the following definition.

Definition 1 Let E be a seminormed space with the seminorm q and $w(s)$ be the space of all sequences $s = (s_k)$, where $s_k \in E$. A sequence $s = (s_k) \in w(s)$ is called to be q-lacunary f-statistically convergent of weight ρ if there is a complex number s_0 such that

$$\lim_{r\to\infty} \frac{1}{\rho(\upsilon_r)} |\{k \in I_r : q(s_{k+m} - s_0) \geq \varepsilon\}| = 0$$

uniformly in m, where $\lambda = (\gamma_r)$ is lacunary sequence. It is represented symbolically by $\hat{S}^{\rho}(\lambda, q) - \lim s_k = s_0$.

The q-lacunary f-statistically convergent sequences of weight ρ is shown by $\hat{S}^{\rho}(\lambda, q)$.

Remark 1 If we consider $\rho(r) = r^{\varrho}$, $0 < \varrho \leq 1$ we have the notion of q-lacunary f-statistical convergence of order ϱ. If $\lambda = 2^r$ and $\rho(r) = r^{\varrho}$ then we have the notion of q-almost convergence of order ϱ.

Theorem 1 *Let $s = (s_k)$, $r = (r_k)$ be sequences, then we have*

(i) If $\hat{S}^{\rho}(\lambda, q) - \lim s_k = s_0$ and $c \in \mathbb{C}$, then $\hat{S}^{\rho}(\lambda, q) - \lim cs_k = cs_0$.
(ii) If $\hat{S}^{\rho}(\lambda, q) - \lim s_k = s_0$ and $\hat{S}^{\rho}(\lambda, q) - \lim r_k = r_0$, then $\hat{S}^{\rho}(\lambda, q) - \lim(s_k + r_k) = s_0 + r_0$.

Proof (i) The proof is clear for $c = 0$. In the case of $c \neq 0$, we have

$$\frac{1}{\rho(\upsilon_r)} |\{k \in J_r : q(cs_{k+m} - cs_0) \geq \varepsilon\}| = \frac{1}{\rho(\upsilon_r)} \left|\left\{k \in J_r : q(s_{k+m} - s_0) \geq \frac{\varepsilon}{|c|}\right\}\right|.$$

(ii) It follows from the fact that

$$\frac{1}{\rho(\upsilon_r)}\left|\{k \in J_r : q(s_{k+m} + r_{k+m} - (s_0 + r_0)) \geq \varepsilon\}\right|$$
$$\leq \frac{1}{\rho(\upsilon_r)}\left|\{k \in J_r : q(s_{k+m} - s_0) \geq \frac{\varepsilon}{2}\}\right| + \frac{1}{\rho(\upsilon_r)}\left|\{k \in J_r : q(r_{k+m} - r_0) \geq \frac{\varepsilon}{2}\}\right|.$$

Definition 2 Suppose E is a seminormed space with the seminorm q and $w(s)$ is the space of all sequences $s = (s_k)$, where $s_k \in E$. If there is a complex number s_0 such that

$$\lim_{r\to\infty} \frac{1}{\rho(\upsilon_r)} \sum_{k\in J_r} q(s_{k+m} - s_0)^p = 0$$

uniformly in m, we call that the sequence $s = (s_k)$ is strongly q-lacunary f-summable of weight ρ, where λ is a lacunary sequence, p is a positive real number.

By $\left[\hat{N}_p^{\rho}(\lambda, q)\right]$, we denote the set of the strongly q-lacunary almost summable of weight ρ.

Remark 2 In this case, $\rho(r) = r^{\varrho}$, $0 < \varrho \leq 1$ this notion coincides with the notion of strongly q-lacunary f-summability of order ϱ. For $\rho(r) = r$, $p = 1$, $\varrho = 1$, we obtain the notion of strongly q-lacunary f-summable.

Theorem 2 *Suppose $\rho_1, \rho_2 \in \Delta$. Then for any given number $\tau > 0$ and $n \in \mathbb{N}$ such that $\rho_1(\upsilon_r)/\rho_2(\upsilon_r) \leq \tau$ for all $r \geq n$, we have $\hat{S}^{\rho_1}(\lambda, q) \subseteq \hat{S}^{\rho_2}(\lambda, q)$.*

Proof Write that,

$$\frac{1}{\rho_2(\upsilon_r)}\left|\{k \in J_r : q(s_{k+m} - s_0) \geq \varepsilon\}\right| = \frac{\rho_1(\upsilon_r)}{\rho_2(\upsilon_r)} \cdot \frac{1}{\rho_1(\upsilon_r)}\left|\{k \in J_r : q(s_{k+m} - s_0) \geq \varepsilon\}\right|$$
$$\leq \tau \cdot \frac{1}{\rho_1(\upsilon_r)}\left|\{k \in J_r : q(s_{k+m} - s_0) \geq \varepsilon\}\right|$$

for all $r \geq n$. If $s = (s_k) \in \hat{S}^{\rho_1}(\lambda, q)$, then we have

$$\frac{1}{\rho_2(\upsilon_r)}\left|\{k \in J_r : q(s_{k+m} - s_0) \geq \varepsilon\}\right| = 0$$

uniformly in m. Finally we have $\hat{S}^{\rho_1}(\lambda, q) \subseteq \hat{S}^{\rho_2}(\lambda, q)$ as we desired.

Hence the theorem.

Corollary 1 *As a special case, if $\rho \in \Delta$ and for any given number $\tau > 0$ and a $n \in \mathbb{N}$ such that $r/\rho(\upsilon_r) \leq \tau$ for all $r \geq n$, we have $\hat{S}^{\rho}(\lambda, q) \subseteq \hat{S}(\lambda, q)$.*

Theorem 3 *Suppose $\rho_1, \rho_2 \in \Delta$. Then for any given number $\tau > 0$ and $n \in \mathbb{N}$ such that $\rho_1(\upsilon_r)/\rho_2(\upsilon_r) \leq \tau$ for all $r \geq n$ we have $\left[\hat{N}_p^{\rho_1}(\lambda, q)\right] \subseteq \left[\hat{N}_p^{\rho_2}(\lambda, q)\right]$.*

Proof The proof is straightforward, hence omitted.

Theorem 4 *For any lacunary sequence λ, q-almost statistical convergence of weight ρ implies strongly q-almost lacunary statistical convergence of weight ρ if*

$$\liminf_{r} \frac{\rho\left(\upsilon_r\right)}{\rho\left(k_r\right)} > 1.$$

Proof Since $\liminf_{r} \frac{\rho(\upsilon_r)}{\rho(k_r)} > 1$, so we can find a $\tau > 1$ such that for sufficiently large r we have

$$\frac{\rho\left(\upsilon_r\right)}{\rho\left(k_r\right)} \geq \tau.$$

Since $s_k \rightarrow s_0\left(\hat{S}_q^{\rho}\right)$, hence for every $\varepsilon > 0$ and sufficiently large r we get

$$\begin{aligned}\frac{1}{\rho\left(k_r\right)}\left|\left\{k \leq k_r : q(s_{k+m} - s_0) \geq \varepsilon\right\}\right| &\geq \frac{1}{\rho\left(k_r\right)}\left|\left\{k \in J_r : q(s_{k+m} - s_0) \geq \varepsilon\right\}\right| \\ &\geq \tau \cdot \frac{1}{\rho\left(\upsilon_r\right)}\left|\left\{k \in J_r : q(s_{k+m} - s_0) \geq \varepsilon\right\}\right|.\end{aligned}$$

we have $x_k \rightarrow s_0\left(V^{\rho}(\lambda, q)\right)$.

Theorem 5 *If $0 < p < s < \infty$ and $g \in \Delta$ then $\left[\hat{N}_s^{\rho}(\lambda, q)\right] \subset \left[\hat{N}_p^{\rho}(\lambda, q)\right]$.*

By using Holder's inequality we can get proof easily.

Theorem 6 *Let $\rho_1, \rho_2 \in \Delta$ be such that there exist $\tau > 0$ and $n \in \mathbb{N}$ such that $\rho_1(\upsilon_r)/\rho_2(\upsilon_r) \leq \tau$ for all $r \geq n$ and let $0 < p < \infty$. If the sequence $s = (s_k)$ is strongly q-lacunary almost summable of weight ρ_1 to s_0 then it is q-lacunary almost statistically convergent of weight ρ_2 to s_0 i.e $\left[\hat{N}_p^{\rho_1}(\lambda, q)\right] \subset \hat{S}^{\rho_2}(\lambda, q)$.*

Proof Let $s = (s_k) \in \left[\hat{N}_p^{\rho_1}(\lambda, q)\right]$ and let $\varepsilon > 0$ be given. Observe that

$$\begin{aligned}\sum_{k \in J_r} q(s_{k+m} - s_0)^p &= \sum_{\substack{k \in J_r \\ q(s_{k+m} - s_0) \geq \varepsilon}} q(s_{k+m} - s_0)^p + \sum_{\substack{k \in J_r \\ q(s_{k+m} - s_0) < \varepsilon}} q(s_{k+m} - s_0)^p \\ &\geq \sum_{\substack{k \in J_r \\ q(s_{k+m} - s_0) \geq \varepsilon}} q(x_{k+m} - s_0)^p \\ &\geq \left|\left\{k \in J_r : q(s_{k+m} - s_0) \geq \varepsilon\right|.\varepsilon^p.\end{aligned}$$

Now it follows that

$$\frac{1}{\rho_1(\upsilon_r)} \sum_{k \in J_r} q(s_{k+m} - s_0)^p \geq \frac{1}{\rho_1(\upsilon_r)}\left|\left\{k \in J_r : q(s_{k+m} - s_0) \geq \varepsilon\right|.\varepsilon^p$$

$$= \frac{\rho_2(\upsilon_r)}{\rho_1(\upsilon_r)} \cdot \frac{1}{\rho_2(\upsilon_r)} \left|\{k \in J_r : q(x_{k+m} - s_0) \geq \varepsilon\right| .\varepsilon^p$$
$$\geq \frac{1}{\tau} \cdot \frac{1}{\rho_2(\upsilon_r)} \left|\{k \in J_r : q(s_{k+m} - s_0) \geq \varepsilon\right| .\varepsilon^p$$

for all $r \geq n$. If $s \to s_0(\left[\hat{N}_p^{\rho_1}(\lambda, q)\right])$ then it is easy to see that $s \to s_0(\hat{S}^{\rho_2}(\lambda, q))$ as we desired.

References

1. Banach, S.: Théorie des Opérations lieaires. Chelsea Publ. Co., New York (1955)
2. Çakalli, H.: Statistical quasi-Cauchy sequences. Math. Comput. Model. **54**(5–6), 1620–1624 (2011)
3. Colak, R.: Statistical convergence of order α. Modern methods in Analysis and its Applications, New Delhi, India, Anamaya Pub., (2010), pp. 121–129
4. Colak, R., Tripathy, B.C., Et, M.: Lacunary strongly summable sequences and q-lacunary almost statistical convergence. Vietnam J. Math. **34**(2), 129–138 (2006)
5. Fast, H.: Sur la convergence statistique. Colloq. Math. **2**, 41–44 (1951)
6. Freedman, A.R., Sember, J.J., Rapnael, M.: Some Cesaro type summability spaces. Proc. London. Math. Soc. **37**, 508–520 (1978)
7. Fridy, J.A.: On statistical convergence. Analysis **5**, 301–313 (1985)
8. Fridy, J.A., Orhan, C.: Lacunary statistical convergence. Pacific J. Math. **160**, 43–51 (1993)
9. Gurdal, M., Pehlivan, S.: The statistical convergence in 2-Banach spaces. Thai J. Math. **2**(1), 107–113 (2004)
10. Lorentz, G.G.: A contribution to the theory of divergent sequences. Acta Math. **80**, 167–190 (1948)
11. Maddox, I.J.: Spaces of strongly summable sequences. Quart. J. Math. **18**, 345–355 (1967)
12. Mursaleen, M., Debnath, S., Rakshit, D.: I-statistical limit superior and I-statistical limit inferior. Filomat **31**(7), 2103–2108 (2017)
13. Šalát, T.: On statistically convergent sequences of real numbers. Math. Slovaca **30**, 139–150 (1980)
14. Savaş, R.: Lambda strongly bivariate summable functions of weight g. Suleyman Demirel Univ. Fen Edebiyat Fak. Dergisi **15**(1), 80–89 (2020)
15. Savaş, R.: On some multiple statistical convergence of order θ defined by almost Lacunary sequences. J. Anal. Number Theory **8**(1), 7–10 (2020)
16. Das, P., Savaş, E.: On I-statistical and I-lacunary statistical convergence of order α. Bull. Iranian Math. Soc. **40**(2), 459–472 (2014)

New Integral Inequalities for Weighted Integrals and Some Fractional Derivations

Lucas Gómez and Juan Eduardo Nápoles Valdés

Abstract Fractional integral operators have emerged as a modern foundation for establishing novel integral inequalities, particularly due to their ability to extend various classical results. This work presents new integral inequalities within the context of weighted integrals, broadening and refining Ostrowski-type inequalities for differentiable (h, m)-convex functions modified of the second type. Additionally, as a consequence of our findings, we derive specific results for the Caputo and Atangana-Baleanu fractional integral operators.

Keywords Integral inequalities · (h, m)-convex functions · Weighted integrals · Atangana Baleanu integral fractionals

1 Introduction

The Inequality Theory is an area of great importance in Mathematical Sciences for several key reasons:

(i) Foundation in Mathematical Analysis: Inequalities are essential tools in mathematical analysis. They allow for establishing limits and comparisons between different expressions, which is fundamental for proving theorems and results in various areas.
(ii) Practical Applications: Inequalities have applications in many disciplines, such as statistics, probability theory, optimization, economics, and physics. For example, they are used to establish properties of estimators in statistics or to solve optimization problems.

L. Gómez—Student of the Bachelor's Degree in Mathematics.

L. Gómez · J. E. Nápoles Valdés (✉)
UNNE FaCENA, Corrientes, Argentina
e-mail: jnapoles@exa.unne.edu.ar

J. E. Nápoles Valdés
UTN-FRRE, Resistencia, Chaco, Argentina

H. Kalita et al. (eds.), *Nonlinear Analysis and Computational Techniques*,
Springer Proceedings in Mathematics & Statistics 501,
https://doi.org/10.1007/978-981-96-6038-4_4

(ii) Development of New Theories: Research in inequalities has led to the development of new theories and results, such as convexity theory, which has profound implications in optimization and functional analysis.
(iv) Interdisciplinarity: Inequality Theory is applied in multiple fields, which encourages collaboration between mathematicians and scientists from other disciplines. This enriches both mathematical theory and its practical application.
(v) Complex Problem Solving: Inequalities allow us to tackle complex problems by providing bounds and estimates that simplify analysis. This is especially useful in situations where exact solutions are difficult to obtain.
(vi) Research Advances: Inequality Theory remains an active area of research, with many open problems and new inequalities constantly being discovered. This keeps this area interesting and relevant in academia.
(vii) Mathematics Education: Inequalities are a central topic in mathematics education, as they help students develop critical reasoning skills and understand fundamental concepts in mathematics.

In summary, Inequality Theory is crucial in Mathematical Sciences due to its wide applicability, its role in theoretical development, and its ability to solve complex problems.

In mathematics, the concept of convex functions plays a significant role due to its numerous applications and theoretical connections with various scientific fields, such that: they have optimal properties (something fundamental in optimization problems), economic applications, in Game Theory, in Mathematical Analysis (they facilitate the study of continuity and derivability, for example), in Numerical Methods, etc. This has led to the classical notion of convexity being generalized and extended in several directions (see [34] for more details).

Within this area, Integral Inequalities have earned a rightful place and are today one of the most dynamic areas of the field. One of the most fundamental integral inequalities for convex functions is the well-known Hermite-Hadamard inequality (see [19, 20]):

$$\varphi\left(\frac{\alpha+\beta}{2}\right) \leq \frac{1}{\beta-\alpha}\int_{\alpha}^{\beta} \varphi(t)\,dt \leq \frac{\varphi(\alpha)+\varphi(\beta)}{2}. \tag{1}$$

This inequality holds for any convex function ψ on the interval $[\nu_1, \nu_2]$ and provides an estimate of the mean value of a convex function, Another way of interpreting this double inequality is to say that the mean value of a convex function interpolates the value of the midpoint of the interval and the average of the images of the ends of said interval. The Hermite-Hadamard inequality is of great importance and has attracted considerable attention in recent years, as it serves as a fundamental tool in various topics, in particular:

- Theoretical Foundation: It is a fundamental result in mathematical analysis that connects the convexity of functions with properties of integrals. This makes it a solid foundation for the study of many areas of mathematics.

- Wide Applications: The inequality has applications in various disciplines, such as statistics, optimization, probability theory, and economics. This makes it relevant to researchers in different fields looking for tools to tackle complex problems.
- Development of New Inequalities: The Hermite-Hadamard Inequality has inspired the development of many other inequalities and results in mathematics. Researchers often seek to extend or generalize these types of results, which can lead to new discoveries.
- Interest in Convexity: Convexity is a central topic in mathematics and has implications in game theory, economics, and optimization. The Hermite-Hadamard Inequality provides a framework for studying properties of convex functions, which is of great interest to many mathematicians.
- Open Problems: Despite its age, there are still open problems and areas of research related to the Hermite-Hadamard Inequality. This keeps it interesting for researchers looking for new perspectives and applications.
- Interdisciplinarity: The inequality is applied in interdisciplinary contexts, allowing researchers to collaborate and apply mathematical concepts in areas such as physics, biology, and engineering.

In short, the Hermite-Hadamard Inequality is a rich and versatile topic that remains relevant in contemporary mathematical research, which explains why many researchers focus on it. Readers interested in this variety of applications and pure developments can consult [2, 3, 7, 8, 16, 18, 27, 32, 33, 35, 44] and the many recommendations given.

As a graphical detail, we illustrate below the Hermite-Hadamard Inequality for the convex function $\varphi = t^2$ on the interval $[-1, 2]$ (Fig. 1). We have used the equivalent expression of (1):

$$(\beta - \alpha)\varphi\left(\frac{\alpha+\beta}{2}\right) \le \int_{\alpha}^{\beta} \varphi(t)\,dt \le (\beta - \alpha)\left(\frac{\varphi(\alpha)+\varphi(\beta)}{2}\right)$$

So we have $(\beta - \alpha)\varphi\left(\frac{\alpha+\beta}{2}\right) = \frac{3}{4}$, $\int_{\alpha}^{\beta} \varphi(t)\,dt = 3$, $(\beta - \alpha)\left(\frac{\varphi(\alpha)+\varphi(\beta)}{2}\right) = \frac{15}{2}$.

In [7], the following definitions were introduced.

Definition 1 Let $h : [0, 1] \to \mathbb{R}$ be a nonnegative function such that $h \neq 0$ and let $\varphi : I = [0, +\infty) \to [0, +\infty)$. The function φ is called (h, m)-convex modified of the first type on I if it satisfies

$$\varphi\left(\mu\lambda_1 + m(1-\mu)\lambda_2\right) \le h^s(\mu)\varphi(\lambda_1) + m(1 - h^s(\mu))\varphi(\lambda_2), \tag{2}$$

for all $\lambda_1, \lambda_2 \in I$ and $\mu \in [0, 1]$, where $m \in [0, 1]$ and $s \in [-1, 1]$.

Definition 2 Let $h : [0, 1] \to \mathbb{R}$ be a nonnegative function such that $h \neq 0$ and let $\varphi : I = [0, +\infty) \to [0, +\infty)$. The function φ is called (h, m)-convex modified of the second type on I if it satisfies

$$\varphi\left(\mu\lambda_1 + m(1-\mu)\lambda_2\right) \le h^s(\mu)\varphi(\lambda_1) + m(1 - h(\mu))^s\varphi(\lambda_2), \tag{3}$$

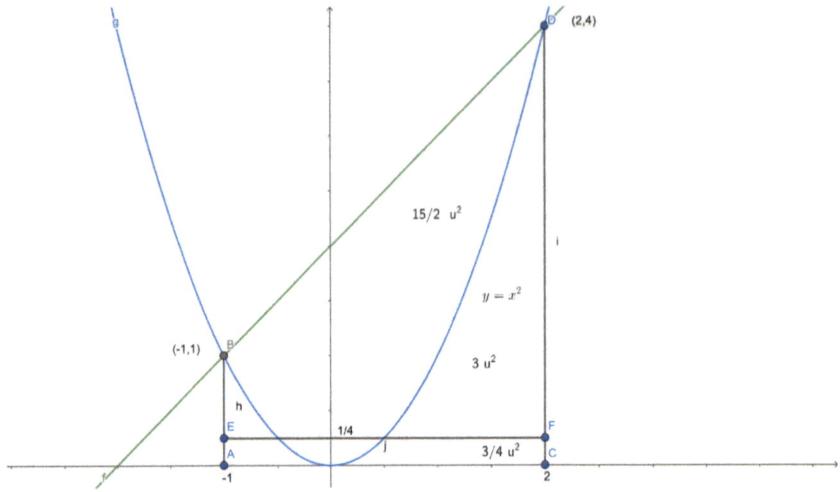

Fig. 1 Hermite-Hadamard Inequality for $y = x^2$ on $[-1, 2]$

for all $\lambda_1, \lambda_2 \in I$ and $\mu \in [0, 1]$, where $m \in [0, 1]$ and $s \in [-1, 1]$.

Remark 1 From Definitions 1 and 2, we define $N^s_{h,m}[\alpha, \beta]$, where $\alpha, \beta \in [0, +\infty)$, as the set of (h, m)-convex modified functions satisfying $\varphi(\alpha) \geq 0$, characterized by the triple $(h(\mu), m, s)$. Notable cases include:

1. $(h(\mu), 0, 0)$ corresponds to increasing functions (see [11]).
2. $(\mu, 0, s)$ corresponds to s-starshaped functions (see [11]).
3. $(\mu, 0, 1)$ corresponds to starshaped functions (see [11]).
4. $(\mu, 1, 1)$ corresponds to convex functions on $[0, +\infty)$ (see [11]).
5. $(1, 1, s)$ corresponds to P-convex functions on $[0, +\infty)$ (see [13]).
6. $(\mu, m, 1)$ corresponds to m-convex functions on $[0, +\infty)$ (see [45]).
7. $(\mu, 1, s)$ for $s \in (0, 1]$ corresponds to s-convex functions on $[0, +\infty)$ (see [10, 21]).
8. $(\mu, 1, s)$ for $s \in [-1, 1]$ corresponds to s-convex extended functions on $[0, +\infty)$ (see [49]).
9. (μ, m, s) for $s \in (0, 1]$ corresponds to (s, m)-convex extended functions on $[0, +\infty)$ (see [38]).
10. $(\mu^a, 1, s)$ with $a \in (0, 1]$ corresponds to (a, s)-convex functions on $[0, +\infty)$ (see [9]).
11. $(\mu^a, m, 1)$ with $a \in (0, 1]$ corresponds to (a, m)-convex functions on $[0, +\infty)$ (see [29]).
12. (μ^a, m, s) with $a \in (0, 1]$ corresponds to s-(a, m)-convex functions on $[0, +\infty)$ (see [50]).
13. $(h(\mu), m, 1)$ represents a variant of the (h, m)-convex function on $[0, +\infty)$ (see [37]).

A novel approach to defining an integral operator and taking an initial step towards generalizing a known result involves incorporating a specific weight function within the integral operator's definition, as follows:

Definition 3 (see [7]) Let $\phi \in L_1[a_1, a_2]$ and let $w : I \to \mathbb{R}$ be a continuous, positive function whose first derivative is integrable in the interior of I. The weighted fractional integral operators are introduced as follows (right and left respectively):

$$J^w_{a_1+}\phi(\tau) = \int_{a_1}^{\tau} w'\left(\frac{a_2-\xi}{a_2-a_1}\right)\phi(\xi)\,d\xi, \quad \tau > a_1, \tag{4}$$

$$J^w_{a_2-}\phi(\tau) = \int_{\tau}^{a_2} w'\left(\frac{\xi-a_1}{a_2-a_1}\right)\phi(\xi)\,d\xi, \quad \tau < a_2. \tag{5}$$

Remark 2 The inclusion of the first derivative of the weight function w arises from the inherent nature of the problem. Alternatively, the second derivative may also be considered.

Remark 3 To better illustrate the scope of Definition 3, consider specific instances of the weight function w':

(a) Setting $w'(t) \equiv 1$ recovers the classical Riemann integral.
(b) Choosing $w'(t) = \frac{t^{\eta-1}}{\Gamma(\eta)}$ leads to the Riemann-Liouville fractional integral.
(c) By selecting appropriate weight functions w', various fractional integral operators can be derived, such as the k-Riemann-Liouville integrals [31], right-sided fractional integrals of a function φ relative to another function h on $[\alpha, \beta]$ [1], and integral operators introduced in [22, 23, 26, 39].
(d) Additional well-known integral operators, fractional or otherwise, can be retrieved as particular cases of the above formulation. Interested readers may consult [28, 46].

The Atangana-Baleanu (AB) integral operators are formulated as follows:

Definition 4 (see [6]) The left-sided fractional integral associated with the novel fractional derivative featuring a non-local kernel, for a function $\varphi \in H^1(\alpha, \beta)$, is defined as:

$$^{AB}\mathrm{I}^{\eta}_{\alpha+}\varphi(t) = \frac{1-\eta}{B(\eta)}\varphi(t) + \frac{\eta}{B(\eta)\Gamma(\eta)}\int_{\alpha}^{t}\frac{\varphi(u)}{(u-\alpha)^{1-\eta}}\,du, \tag{6}$$

where $\beta > \alpha$, $\eta \in [0, 1]$, $B(\eta)$ is a normalization function and $\Gamma(\eta)$ denotes the well-known Gamma function.

Definition 5 (see [5]) The right-sided fractional integral corresponding to the fractional derivative with a non-local kernel, for a function $\varphi \in H^1(\alpha, \beta)$, is given by:

$$
{}^{AB}\mathrm{I}^{\eta}_{\beta-}\varphi(t) = \frac{1-\eta}{B(\eta)}\varphi(t) + \frac{\eta}{B(\eta)\Gamma(\eta)}\int_t^{\beta} \frac{\varphi(u)}{(\beta-u)^{1-\eta}}\,du, \tag{7}
$$

where $\beta > \alpha$, $\eta \in [0, 1]$, $B(\eta)$ is a normalization function and $\Gamma(\eta)$ represents the Gamma function.

This work establishes novel integral inequalities within the framework of second-type modified (h, m)-convex functions by employing weighted fractional integrals. Various implications for fractional integrals of the Caputo and Atangana-Baleanu types are explored throughout the study.

2 First Result

In the following, we analyze the Hermite-Hadamard inequality through the application of weighted integral operators.

Theorem 1 *Let $\alpha, \beta \in \mathbb{R}$ with $\alpha < \beta$ and let $\varphi : [\alpha, \beta] \to \mathbb{R}$ be a function that satisfies the modified (h, m)-convexity condition of the second type. Consider a weight function $w : [0, 1] \to \mathbb{R}$. Suppose that $w'(t) \geq 0$ for all $t \in (0, 1)$. Under this assumptions we have:*

$$
\begin{aligned}
&\varphi\left(\frac{\alpha+\beta}{2}\right)(w(1)-w(0)) \leq \frac{(r+1)}{(\beta-\alpha)}\left[h^s\left(\frac{1}{2}\right) J^{w}_{\left(\frac{\alpha+r\beta}{r+1}\right)-}\varphi(\alpha) + \left(1-h\left(\frac{1}{2}\right)\right)^s J^{w}_{\left(\frac{\alpha+r\beta}{r+1}\right)+}\varphi(\beta)\right] \\
&\leq \int_0^1 w'(t)h^s\left(\frac{t}{r+1}\right)dt\left[h^s\left(\frac{1}{2}\right)\varphi(\beta) + \left(1-h\left(\frac{1}{2}\right)\right)^s\varphi(\alpha)\right] \\
&+ m\int_0^1 w'(t)\left(1-h\left(\frac{t}{r+1}\right)\right)^s dt\left[h^s\left(\frac{1}{2}\right)\varphi\left(\frac{\alpha}{m}\right) + \left(1-h\left(\frac{1}{2}\right)\right)^s\varphi\left(\frac{\beta}{m}\right)\right],
\end{aligned} \tag{8}
$$

for all $r \in \mathbb{N}_0$.

Proof For $\alpha < \beta$, $t = \frac{1}{2}$ and $m = 1$, we get

$$
\varphi\left(\frac{\alpha+\beta}{2}\right) \leq h^s\left(\frac{1}{2}\right)\varphi(\alpha) + \left(1-h\left(\frac{1}{2}\right)\right)^s\varphi(\beta)
$$

Making $x = \frac{t}{r+1}\beta + \frac{r+1-t}{r+1}\alpha$, $y = \frac{t}{r+1}\alpha + \frac{r+1-t}{r+1}\beta$, with $t \in [0, 1]$ we have

$$
\begin{aligned}
\varphi\left(\frac{\alpha+\beta}{2}\right) &= \varphi\left(\frac{x+y}{2}\right) \\
&\leq h^s\left(\frac{1}{2}\right)\varphi\left(\frac{t}{r+1}\beta + \frac{r+1-t}{r+1}\alpha\right) \\
&+ \left(1-h\left(\frac{1}{2}\right)\right)^s\varphi\left(\frac{t}{r+1}\alpha + \frac{r+1-t}{r+1}\beta\right)
\end{aligned}
$$

Multiplying both members of the previous inequality by $w'(t)$, integrating over the interval [0, 1] with respect to t, we obtain

$$\varphi\left(\frac{\alpha+\beta}{2}\right)\int_0^1 w'(t)\,dt \le h^s\left(\frac{1}{2}\right)\int_0^1 w'(t)\varphi\left(\frac{t}{r+1}\beta+\frac{r+1-t}{r+1}\alpha\right)dt$$
$$+\left(1-h\left(\frac{1}{2}\right)\right)^s\int_0^1 w'(t)\varphi\left(\frac{t}{r+1}\alpha+\frac{r+1-t}{r+1}\beta\right)dt$$

Applying the respective changes of variables $u=\frac{t}{r+1}\beta+\frac{r+1-t}{r+1}\alpha$ and $v=\frac{t}{r+1}\alpha+\frac{r+1-t}{r+1}\beta$, we find

$$\varphi\left(\frac{\alpha+\beta}{2}\right)(w(1)-w(0)) \le h^s\left(\frac{1}{2}\right)\frac{(r+1)}{(\beta-\alpha)}\int_\alpha^{\frac{r\alpha+\beta}{r+1}} w'\left[\frac{(\xi-\alpha)}{\frac{(\beta-\alpha)}{(r+1)}}\right]\varphi(u)\,du$$
$$+\left(1-h\left(\frac{1}{2}\right)\right)^s\frac{(r+1)}{(\beta-\alpha)}\int_{\frac{\alpha+r\beta}{r+1}}^{\beta} w'\left[\frac{(\beta-v)}{\frac{(\beta-\alpha)}{(r+1)}}\right]\varphi(v)\,dv$$
$$=\frac{(r+1)}{(\beta-\alpha)}\left[h^s\left(\frac{1}{2}\right)J^w_{\left(\frac{\alpha+r\beta}{r+1}\right)-}\varphi(\alpha)+\left(1-h\left(\frac{1}{2}\right)\right)^s J^w_{\left(\frac{\alpha+r\beta}{r+1}\right)+}\varphi(\beta)\right]$$

In this way we have the first part of the desired inequality.

Now utilizing the (h,m)-convexity of φ, we have

$$\varphi\left(\frac{t}{r+1}\beta+\frac{r+1-t}{r+1}\alpha\right)=\varphi\left(\frac{t}{r+1}\beta+m\left(\frac{r+1-t}{r+1}\right)\frac{\alpha}{m}\right)$$
$$\le h^s\left(\frac{t}{r+1}\right)\varphi(\beta)+m\left(1-h\left(\frac{t}{r+1}\right)\right)^s\varphi\left(\frac{\alpha}{m}\right) \tag{9}$$

In an analogous way,

$$\varphi\left(\frac{t}{r+1}\alpha+\frac{r+1-t}{r+1}\beta\right)=\varphi\left(\frac{t}{r+1}\alpha+m\left(\frac{r+1-t}{r+1}\right)\frac{\beta}{m}\right)$$
$$\le h^s\left(\frac{t}{r+1}\right)\varphi(\alpha)+m\left(1-h\left(\frac{t}{r+1}\right)\right)^s\varphi\left(\frac{\beta}{m}\right) \tag{10}$$

Multiplying (9) by $h^s\left(\frac{1}{2}\right)$ and (10) by $\left(1-h\left(\frac{1}{2}\right)\right)^s$ and adding both results leads us to

$$h^s\left(\frac{1}{2}\right)\varphi\left(\frac{t}{r+1}\beta+\frac{r+1-t}{r+1}\alpha\right)+\left(1-h\left(\frac{1}{2}\right)\right)^s\varphi\left(\frac{t}{r+1}\alpha+\frac{r+1-t}{r+1}\beta\right)$$

$$\leq h^s\left(\frac{1}{2}\right)\left[h^s\left(\frac{t}{r+1}\right)\varphi(\beta)+m\left(1-h\left(\frac{t}{r+1}\right)\right)^s\varphi\left(\frac{\alpha}{m}\right)\right]$$
$$+\left(1-h\left(\frac{1}{2}\right)\right)^s\left[h^s\left(\frac{t}{r+1}\right)\varphi(\alpha)+m\left(1-h\left(\frac{t}{r+1}\right)\right)^s\varphi\left(\frac{\beta}{m}\right)\right]$$

Multiplying by $w'(t)$, integrating over the interval [0, 1] with respect to t, and suitably rearranging the terms, we reach the right-hand side of (8). This concludes the proof. □

Remark 4 By setting $w'(t) = 1$, $r = 0$, $m = s = 1$ and $h(t) = t$ in Eq. (8), the well-known Hermite-Hadamard Inequality (1) is recovered.

Remark 5 Substituting $w(t) = \frac{t^\eta}{\eta}$, $r = 0$, $m = s = 1$ and $h(t) = t$ in the previous result leads to the following inequality for the Riemann-Liouville fractional integral (refer to Theorem 2 in [40]):

$$\varphi\left(\frac{\alpha+\beta}{2}\right) \leq \frac{\Gamma(\eta+1)}{2(\beta-\alpha)^\eta}\left[I^\eta_{\beta-}\varphi(\alpha) + I^\eta_{\alpha+}\varphi(\beta)\right] \leq \frac{\varphi(\alpha)+\varphi(\beta)}{2}$$

Remark 6 With $w(t) = \frac{t^\eta}{\eta}$, $r = m = s = 1$ and $h(t) = t$, the previous result simplifies to Theorem 4 in [41].

For s-convex functions, using $w(t) = \frac{t^\eta}{\eta}$ and $r = 0$, we recover Theorem 2.1 from [47]. Additionally, Theorem 3 in [43], for $w(t) = t^\eta$, provides further results. In this work, Theorem 2 for m-convex functions is also established under similar conditions and can be derived with ease.

Remark 7 Theorem 5 in [48] (also see Theorem 1 in [36]), which is based on k-Riemann-Liouville fractional integrals, can be obtained from Theorem 1 by setting $w(t) = t^{\frac{\eta}{k}}$, $r = 0$, $m = s = 1$ and $h(t) = t$.

The above results form the foundation for deriving other inequalities using different types of integral operators, as demonstrated in the following Remark.

Remark 8 By assigning $r = 0$, $m = s = 1$ and $h(t) = t$ in (8), which corresponds to working with convex functions and choosing $w'(t) = \frac{\eta t^{\frac{\eta}{k}-1}}{kB(\eta)\Gamma_k(\eta)}$, the left-hand side yields:

$$\varphi\left(\frac{\alpha+\beta}{2}\right)\frac{1}{B(\eta)\Gamma_k(\eta)} \leq \frac{1}{2(\beta-\alpha)^{\frac{\eta}{k}}}\left[\frac{\eta}{kB(\eta)\Gamma_k(\eta)}\int_\alpha^\beta \frac{\varphi(u)}{(u-\alpha)^{1-\frac{\eta}{k}}}du\right.$$
$$\left.+\frac{\eta}{kB(\eta)\Gamma_k(\eta)}\int_\alpha^\beta \frac{\varphi(u)}{(\beta-u)^{1-\frac{\eta}{k}}}du\right]$$

$$\varphi\left(\frac{\alpha+\beta}{2}\right)\frac{2(\beta-\alpha)^{\frac{\eta}{k}}}{B(\eta)\Gamma_k(\eta)} \leq \frac{\eta}{kB(\eta)\Gamma_k(\eta)}\int_\alpha^\beta \frac{\varphi(u)}{(u-\alpha)^{1-\frac{\eta}{k}}}du \tag{11}$$

$$+\frac{\eta}{kB(\eta)\Gamma_k(\eta)}\int_{\alpha}^{\beta}\frac{\varphi(u)}{(\beta-u)^{1-\frac{\eta}{k}}}\,du \tag{12}$$

Adding the term $\frac{1-\eta}{B(\eta)}(\varphi(\alpha)+\varphi(\beta))$ on both sides of (11) and considering that $\varphi\left(\frac{\alpha+\beta}{2}\right)\leq\frac{\varphi(\alpha)+\varphi(\beta)}{2}$, we obtain:

$$\varphi\left(\frac{\alpha+\beta}{2}\right)\left[\frac{2(\beta-\alpha)^{\frac{\eta}{k}}}{B(\eta)\Gamma_k(\eta)}+\frac{(1-\eta)}{B(\eta)}\right]\leq^{AB}\mathrm{I}_{\alpha+}^{\eta}\varphi(\beta)+^{AB}\mathrm{I}_{\beta-}^{\eta}\varphi(\alpha) \tag{13}$$

A similar approach applied to the right-hand side of (8) gives:

$$\frac{1}{2(\beta-\alpha)^{\frac{\eta}{k}}}\left[\frac{\eta}{kB(\eta)\Gamma_k(\eta)}\int_{\alpha}^{\beta}\frac{\varphi(u)}{(u-\alpha)^{1-\frac{\eta}{k}}}\,du+\frac{\eta}{kB(\eta)\Gamma_k(\eta)}\int_{\alpha}^{\beta}\frac{\varphi(u)}{(\beta-u)^{1-\frac{\eta}{k}}}\,du\right]$$
$$\leq\frac{1}{B(\eta)\Gamma_k(\eta)}\left(\frac{\varphi(\alpha)+\varphi(\beta)}{2}\right)$$

Multiplying both sides by $2(\beta-\alpha)^{\frac{\eta}{k}}$, adding $\frac{(1-\eta)}{B(\eta)}(\varphi(\alpha)+\varphi(\beta))$ and rearranging terms, we arrive at:

$$^{AB}\mathrm{I}_{\alpha+}^{\eta}\varphi(\beta)+^{AB}\mathrm{I}_{\beta-}^{\eta}\varphi(\alpha)\leq\left[\frac{2(\beta-\alpha)^{\frac{\eta}{k}}}{B(\eta)\Gamma_k(\eta)}+\frac{(1-\eta)}{B(\eta)}\right]\left(\frac{\varphi(\alpha)+\varphi(\beta)}{2}\right) \tag{14}$$

By combining (13) and (14), we obtain a relation that closely resembles Theorem 6 in [25]. Moreover, setting $k=1$ in this expression yields a result comparable to Proposition 2.1 in [17].

3 Main Results

The subsequent identity will serve as a fundamental tool moving forward.

Lemma 1 *Let be a function f defined over a closed interval $I\subset\mathbb{R}$, which it is differentiable in the interior I^o. Consider a weight function $w:[0,1]\to\mathbb{R}$. Suppose that $f'\in L[\alpha,\beta]$, then we have the following equality:*

$$\left(\frac{\beta-\xi}{r+1}\right)^{\alpha}\left[w(1)\varphi\left(\frac{\xi+r\beta}{r+1}\right)-w(0)\varphi(\beta)\right]+\left(\frac{\xi-\alpha}{r+1}\right)^{\alpha}\left[w(1)\varphi\left(\frac{\xi+r\alpha}{r+1}\right)-w(0)\varphi(\alpha)\right]$$
$$-\left(\frac{\beta-\xi}{r+1}\right)^{\alpha+1}J^{w}_{\left(\frac{\xi+r\beta}{r+1}\right)^{+}}\varphi(\beta)-\left(\frac{\xi-\alpha}{r+1}\right)^{\alpha+1}J^{w}_{\left(\frac{\xi+r\alpha}{r+1}\right)^{-}}\varphi(\alpha)$$
$$=\left(\frac{\xi-\alpha}{r+1}\right)^{\alpha+1}\int_0^1 w(t)\varphi'\left(\frac{t}{r+1}\xi+\frac{r+1-t}{r+1}\alpha\right)dt$$

$$-\left(\frac{\beta-\xi}{r+1}\right)^{\alpha+1}\int_0^1 w(t)\varphi'\left(\frac{t}{r+1}\xi+\frac{r+1-t}{r+1}\beta\right)dt, \tag{15}$$

with $\xi \in [\alpha, \beta] \subset I^o$ *and* $r \in \mathbb{N}_0$.

Proof Taking into account the Definition 3, we find

$$\left(\frac{\xi-\alpha}{r+1}\right)^{\eta+1} J^w_{\left(\frac{\xi+r\alpha}{r+1}\right)^-}\varphi(\alpha)+\left(\frac{\beta-\xi}{r+1}\right)^{\eta+1} J^w_{\left(\frac{\xi+r\beta}{r+1}\right)^+}\varphi(\beta)$$
$$=\left(\frac{\xi-\alpha}{r+1}\right)^{\eta+1}\int_\alpha^{\left(\frac{\xi+r\alpha}{r+1}\right)} w'\left(\frac{u-\alpha}{\frac{\xi+r\alpha}{r+1}-\alpha}\right)\varphi(u)\,du+\left(\frac{\beta-\xi}{r+1}\right)^{\eta+1}\int_{\left(\frac{\xi+r\beta}{r+1}\right)}^{\beta} w'\left(\frac{\beta-v}{\beta-\frac{\xi+r\beta}{r+1}}\right)\varphi(v)\,dv \tag{16}$$

By making a change of variables $u=\frac{\xi}{r+1}t+\frac{r+1-t}{r+1}\alpha$ and $v=\frac{\xi}{r+1}t+\frac{r+1-t}{r+1}\beta$ where $t\in[0,1]$ in (16), we have

$$\left(\frac{\xi-\alpha}{r+1}\right)^{\eta+1} J^w_{\left(\frac{\xi+r\alpha}{r+1}\right)^-}\varphi(\alpha)+\left(\frac{\beta-\xi}{r+1}\right)^{\eta+1} J^w_{\left(\frac{\xi+r\beta}{r+1}\right)^+}\varphi(\beta)$$
$$=\left(\frac{\xi-\alpha}{r+1}\right)^{\eta}\int_0^1 w'(t)\varphi\left(\frac{\xi}{r+1}t+\frac{r+1-t}{r+1}\alpha\right)dt+\left(\frac{\beta-\xi}{r+1}\right)^{\eta}\int_0^1 w'(t)\varphi\left(\frac{\xi}{r+1}t+\frac{r+1-t}{r+1}\beta\right)dt$$
$$=\left(\frac{\xi-\alpha}{r+1}\right)^{\eta}N_1+\left(\frac{\beta-\xi}{r+1}\right)^{\eta}N_2 \tag{17}$$

Applying integration by parts, we obtain for N_1

$$N_1=w(1)\varphi\left(\frac{\xi+r\alpha}{r+1}\right)-w(0)\varphi(\alpha)-\left(\frac{\xi-\alpha}{r+1}\right)\int_0^1 w(t)\varphi'\left(\frac{\xi}{r+1}t+\frac{r+1-t}{r+1}\alpha\right)dt \tag{18}$$

In an analogous way,

$$N_2=w(1)\varphi\left(\frac{\xi+r\beta}{r+1}\right)-w(0)\varphi(\beta)+\left(\frac{\beta-\xi}{r+1}\right)\int_0^1 w(t)\varphi'\left(\frac{\xi}{r+1}t+\frac{r+1-t}{r+1}\beta\right)dt \tag{19}$$

By combining (17)–(19) and conveniently grouping the terms, it follows (15). □

It might be thought that this equality is not very significant result. Below we present a couple of Remarks that will help understand its generality.

Remark 9 Let's put in $\left(\frac{\xi-\alpha}{r+1}\right)^{\eta+1}\int_0^1 w'(t)\varphi\left(\frac{t}{r+1}\xi+\frac{r+1-t}{r+1}\alpha\right)dt$, $\xi=\beta$, $w(t)=t$, $\alpha=1$ and $r=0$, so we have

$$\left(\frac{\xi-\alpha}{r+1}\right)^{\eta+1}\int_0^1 w'(t)\varphi\left(\frac{t}{r+1}\xi+\frac{r+1-t}{r+1}\alpha\right)dt=(\beta-\alpha)^2\int_0^1\varphi(t\beta+(1-t)\alpha)\,dt$$
$$=(\beta-\alpha)\varphi(\beta)-\int_\alpha^\beta\varphi(z)\,dz=(\beta-\alpha)^2\int_0^1 t\varphi'(t\beta+(1-t)\alpha)\,dt \tag{20}$$

On the other hand

$$\int_0^1 t\varphi'(t\beta + (1-t)\alpha)\,dt = \int_0^1 (1-t)\varphi'(t\alpha + (1-t)\beta)\,dt \tag{21}$$

Using (21) in (20), we obtain

$$(\beta-\alpha)\int_0^1 \varphi(t\beta + (1-t)\alpha)\,dt = (\beta-\alpha)\varphi(\beta) - \int_\alpha^\beta \varphi(z)\,dz$$
$$= (\beta-\alpha)^2 \int_0^1 (1-t)\varphi'(t\alpha + (1-t)\beta)\,dt \tag{22}$$

Similarly, putting in $\left(\frac{\beta-\xi}{r+1}\right)\int_0^1 w'(t)\varphi\left(\frac{\xi}{r+1}t + \frac{r+1-t}{r+1}\beta\right)dt, \xi = \alpha, w(t) = t, \alpha =$ 1 and $r = 0$ we get

$$\left(\frac{\beta-\xi}{r+1}\right)\int_0^1 w'(t)\varphi\left(\frac{\xi}{r+1}t + \frac{r+1-t}{r+1}\beta\right)dt = (\beta-\alpha)\int_0^1 \varphi(t\alpha + (1-t)\beta)\,dt$$
$$= -(\beta-\alpha)\varphi(\alpha) + \int_\alpha^\beta \varphi(z)\,dz = (\beta-\alpha)^2 \int_0^1 t\varphi'(t\alpha + (1-t)\beta)\,dt \tag{23}$$

If we subtract (23) from (22), we derive

$$(\beta-\alpha)\varphi(\beta) - \int_\alpha^\beta \varphi(z)\,dz + (\beta-\alpha)\varphi(\alpha) - \int_\alpha^\beta \varphi(z)\,dz$$
$$= (\beta-\alpha)^2 \int_0^1 (1-t)\varphi'(t\alpha + (1-t)\beta)\,dt - (\beta-\alpha)^2 \int_0^1 t\varphi'(t\alpha + (1-t)\beta)\,dt$$

Dividing by $2(\beta-\alpha)$ both members, we finally obtain

$$\frac{\varphi(\beta)+\varphi(\alpha)}{2} - \frac{1}{\beta-\alpha}\int_\alpha^\beta \varphi(z)\,dz$$
$$= \frac{\beta-\alpha}{2}\int_0^1 (1-2t)\varphi'(t\alpha + (1-t)\beta)\,dt$$

As presented in Lemma 2.1 of [12], one of the most important results in integral inequalities.

Remark 10 A related result is established in [42], which states the following:
Lemma 2.1 Let $\varphi : [\alpha, \beta] \to \mathbb{R}$ be a differentiable function on (α, β) with $\alpha < \beta$. The following identity holds for Atangana-Baleanu fractional integral operators:

$$
\begin{aligned}
&{}^{AB}\mathrm{I}^{\eta}_{\alpha+}\varphi(\xi) + {}^{AB}\,\mathrm{I}^{\eta}_{\beta-}\varphi(\xi) - \frac{(\xi-\alpha)^{\eta}\varphi(\alpha) + (\beta-\xi)^{\eta}\varphi(\beta)}{B(\eta)\Gamma(\eta)} - \frac{2(1-\eta)\varphi(\xi)}{B(\eta)}\\
&= \frac{(\xi-\alpha)^{\eta+1}}{B(\eta)\Gamma(\eta)}\int_0^1 (1-t)^{\eta}\varphi'(t\xi + (1-t)\alpha)\,dt\\
&\quad -\frac{(\beta-\xi)^{\eta+1}}{B(\eta)\Gamma(\eta)}\int_0^1 t^{\eta}\varphi'(t\beta + (1-t)\xi)\,dt,
\end{aligned}
$$

where $\eta \in (0, 1]$, $\xi \in [\alpha, \beta]$, $t \in [0, 1]$, and $\Gamma(.)$ represents the Gamma function.

By considering Remark 9, it becomes evident that this result follows as a special case of Lemma 1 by choosing $w(t) = t^{\eta}$.

Remark 11 If we consider $w(t) = \frac{t^{\eta}}{B(\eta)\Gamma(\eta)}$, $r = 0$ after some tedious algebraic work, we can obtain the Lemma 2.1 of [4]:

$$
\begin{aligned}
&\frac{\varphi(\xi)}{B(\eta)\Gamma(\eta)}\left[(\xi-\alpha)^{\eta} + (\beta-\xi)^{\eta}\right] - \left({}^{AB}\mathrm{I}^{\eta}_{\alpha+}\varphi(\beta) + {}^{AB}\,\mathrm{I}^{\eta}_{\beta-}\varphi(\alpha)\right) + \frac{2(1-\eta)}{B(\eta)}\varphi(\xi)\\
&= \frac{(\xi-\alpha)^{\eta+1}}{B(\eta)\Gamma(\eta)}\int_0^1 t^{\eta}\varphi'(t\xi + (1-t)\alpha)\,dt - \frac{(\beta-\xi)^{\eta+1}}{B(\eta)\Gamma(\eta)}\int_0^1 t^{\eta}\varphi'(t\xi + (1-t)\beta)\,dt
\end{aligned}
$$

We must point out that there is an error in said work since instead of $\frac{2(1-\eta)}{B(\eta)}\varphi(\xi)$ appears $\frac{(1-\eta)}{B(\eta)}(\varphi(\alpha) + \varphi(\beta))$.

Remark 12 Taking into account the Remark 9, if instead of using $w(t) = t$ in (22) and (23) we put $w(t) = t^{\frac{\eta}{k}}$ and $r = 0$, we obtain the Lemma 1 of [25].

Let denote us:

$\mathbb{S}(\alpha, \beta, \varphi, w)$ denotes the absolute value of the left-hand side of Eq. (15),

$$
\mathbb{T} = \int_0^1 w(t)\left(\frac{t}{r+1}\right)^{s} dt,
$$

$$
\mathbb{F} = \int_0^1 w(t)\left(\frac{r+1-t}{r+1}\right)^{s} dt,
$$

$$
\mathbb{G} = \int_0^1 w(t)\,h^{s}\left(\frac{t}{r+1}\right) dt,
$$

$$
\mathbb{H} = \int_0^1 w(t)\left(1 - h\left(\frac{r+1-t}{r+1}\right)\right)^{s} dt,
$$

$$
\mathbb{I} = \int_0^1 h^{s}\left(\frac{t}{r+1}\right) dt,
$$

$$\mathbb{J} = \int_0^1 \left(1 - h\left(\frac{r+1-t}{r+1}\right)\right)^s dt,$$

$$\mathbb{W} = \int_0^1 \left(\frac{t}{r+1}\right)^s dt,$$

$$\mathbb{L} = \int_0^1 \left(\frac{r+1-t}{r+1}\right)^s dt,$$

$$\mathbb{U}_p = \left(\int_0^1 w^p(t)\, dt\right)^{\frac{1}{p}},$$

$$\mathbb{V}_q = \left(\int_0^1 w(t)\, dt\right)^{1-\frac{1}{q}}$$

with $r \in \mathbb{N}_0$.

Applying Lemma 1, we derive the following conclusions.

Theorem 2 *Let be a function φ defined over a closed interval $I \subset \mathbb{R}$, which it is differentiable in the interior I^o. Consider a weight function $w : [0, 1] \to \mathbb{R}$. Suppose that $|\varphi'|$ satisfies s-convexity condition over $[\alpha, \beta] \subset I^o$ and that $\varphi' \in L[\alpha, \beta]$, then it holds that*

$$\mathbb{S}(\alpha, \beta, \varphi, w) \leq \left[\left(\frac{\xi - \alpha}{r+1}\right)^{\alpha+1} + \left(\frac{\beta - \xi}{r+1}\right)^{\alpha+1}\right] |\varphi'(\xi)|\, \mathbb{T}$$
$$+\mathbb{F}\left[\left(\frac{\xi - \alpha}{r+1}\right)^{\alpha+1} |\varphi'(\alpha)| + \left(\frac{\beta - \xi}{r+1}\right)^{\alpha+1} |\varphi'(\beta)|\right] \tag{24}$$

where $\xi \in [\alpha, \beta]$ and $r \in \mathbb{N}_0$.

Proof We proceed by applying Lemma 1 and using the triangle inequality, we get

$$\mathbb{S}(\alpha, \beta, \varphi, w) \leq \left(\frac{\xi - \alpha}{r+1}\right)^{\eta+1} \int_0^1 w(t) \left|\varphi'\left(\frac{t}{r+1}\xi + \frac{r+1-t}{r+1}\alpha\right)\right| dt$$
$$+ \left(\frac{\beta - \xi}{r+1}\right)^{\eta+1} \int_0^1 w(t) \left|\varphi'\left(\frac{t}{r+1}\xi + \frac{r+1-t}{r+1}\beta\right)\right| dt$$

Utilizing the s-convexity of $|\varphi'|$, we find

$$\left(\frac{\xi - \alpha}{r+1}\right)^{\eta+1} \int_0^1 w(t) \left|\varphi'\left(\frac{t}{r+1}\xi + \frac{r+1-t}{r+1}\alpha\right)\right| dt$$
$$\leq \left(\frac{\xi - \alpha}{r+1}\right)^{\eta+1} \int_0^1 w(t) \left[\left(\frac{t}{r+1}\right)^s |\varphi'(\xi)| + \left(\frac{r+1-t}{r+1}\right)^s |\varphi'(\alpha)|\right] dt$$

$$= \left(\frac{\xi-\alpha}{r+1}\right)^{\eta+1}\left[\left|\varphi'(\xi)\right|\mathbb{T}+\left|\varphi'(\alpha)\right|\mathbb{F}\right] \tag{25}$$

Analogously, we have

$$\left(\frac{\beta-\xi}{r+1}\right)^{\eta+1}\int_0^1 w(t)\left|\varphi'\left(\frac{t}{r+1}\xi+\frac{r+1-t}{r+1}\beta\right)\right|dt$$
$$\leq \left(\frac{\beta-\xi}{r+1}\right)^{\eta+1}\left[\left|\varphi'(\xi)\right|\mathbb{T}+\left|\varphi'(\beta)\right|\mathbb{F}\right] \tag{26}$$

By combining (25)–(26) and conveniently grouping the terms, it follows (24). □

Theorem 3 *Let be a function φ defined over a closed interval $I \subset \mathbb{R}$, which it is differentiable in the interior I^o. Consider a weight function $w : [0, 1] \to \mathbb{R}$. Suppose that $|\varphi'|$ satisfies the modified (h, m)-convexity condition of the second type over $[\alpha, \frac{\beta}{m}] \subset I^o$ and that $\varphi' \in L[\alpha, \beta]$, then it holds that:*

$$\mathbb{S}(\alpha, \beta, \varphi, w) \leq \left[\left(\frac{\xi-\alpha}{r+1}\right)^{\alpha+1}+\left(\frac{\beta-\xi}{r+1}\right)^{\alpha+1}\right]|\varphi'(\xi)|\,\mathbb{G}$$
$$+m\,\mathbb{H}\left[\left(\frac{\xi-\alpha}{r+1}\right)^{\alpha+1}\left|\varphi'\left(\frac{\alpha}{m}\right)\right|+\left(\frac{\beta-\xi}{r+1}\right)^{\alpha+1}\left|\varphi'\left(\frac{\beta}{m}\right)\right|\right] \tag{27}$$

where $r \in \mathbb{N}_0$ and $\xi \in [\alpha, \beta]$.

Proof By applying Lemma 1 and using the triangle inequality, we get

$$\mathbb{S}(\alpha, \beta, \varphi, w) \leq \left(\frac{\xi-\alpha}{r+1}\right)^{\eta+1}\int_0^1 w(t)\left|\varphi'\left(\frac{t}{r+1}\xi+\frac{r+1-t}{r+1}\alpha\right)\right|dt$$
$$+\left(\frac{\beta-\xi}{r+1}\right)^{\eta+1}\int_0^1 w(t)\left|\varphi'\left(\frac{t}{r+1}\xi+\frac{r+1-t}{r+1}\beta\right)\right|dt$$

Utilizing (h, m)-convexity of $|\varphi'|$, we find

$$\left(\frac{\xi-\alpha}{r+1}\right)^{\eta+1}\int_0^1 w(t)\left|\varphi'\left(\frac{t}{r+1}\xi+\frac{r+1-t}{r+1}\alpha\right)\right|dt$$
$$\leq \left(\frac{\xi-\alpha}{r+1}\right)^{\eta+1}\int_0^1 w(t)\left[h^s\left(\frac{t}{r+1}\right)\left|\varphi'(\xi)\right|+m\left(1-h\left(\frac{r+1-t}{r+1}\right)\right)^s\left|\varphi'(\alpha)\right|\right]dt$$
$$= \left(\frac{\xi-\alpha}{r+1}\right)^{\eta+1}\left[\left|\varphi'(\xi)\right|\int_0^1 w(t)h^s\left(\frac{t}{r+1}\right)dt\right.$$
$$\left.+\,m\left|\varphi'\left(\frac{\alpha}{m}\right)\right|\int_0^1 w(t)\left(1-h\left(\frac{r+1-t}{r+1}\right)\right)^s dt\right]$$

$$= \left(\frac{\xi-\alpha}{r+1}\right)^{\eta+1}\left[\left|\varphi'(\xi)\right|\mathbb{G}+m\left|\varphi'\left(\frac{\alpha}{m}\right)\right|\mathbb{H}\right] \tag{28}$$

Analogously, we have

$$\left(\frac{\beta-\xi}{r+1}\right)^{\eta+1}\int_0^1 w(t)\left|\varphi'\left(\frac{t}{r+1}\beta+\frac{r+1-t}{r+1}\alpha\right)\right|dt \le \left(\frac{\beta-\xi}{r+1}\right)^{\eta+1}\left[\left|\varphi'(\xi)\right|\mathbb{G}+m\left|\varphi'\left(\frac{\beta}{m}\right)\right|\mathbb{H}\right] \tag{29}$$

By combining (28)–(29) and conveniently grouping the terms, it follows (27). □

Putting $w(t)=t^{n-\frac{\eta}{k}}$ we have the following remark.

Remark 13 Setting $r=0$, one can verify that the initial part of Theorem 2.7 in [30] applies to (h,m)-convex functions with $n=1$. For $r=1$, Theorem 7 from [14] (with $q=1$) for convex functions, along with Theorem 5 from [15] and Theorem 3.2 from [24] (both considering the case $q=1$ for convex functions), are all special instances of the preceding theorem.

A refinement of this result is possible by imposing further constraints on $|f'|^q$.

Theorem 4 *Let be a function φ defined over a closed interval $I\subset\mathbb{R}$, which it is differentiable in the interior I^o. Consider a weight function $w:[0,1]\to\mathbb{R}$. Suppose that $|\varphi'|^q$ satisfies the s-convexity condition over $[\alpha,\beta]\subset I^o$ and that $\varphi'\in L[\alpha,\beta]$, then it holds that:*

$$\begin{aligned}\mathbb{S}(\alpha,\beta,\varphi,w) &\le \mathbb{U}_p\Bigg[\left(\frac{\xi-\alpha}{r+1}\right)^{\eta+1}\left(|\varphi'(\xi)|^q\,\mathbb{W}+\left|\varphi'(\alpha)\right|^p\mathbb{L}\right)^{\frac{1}{q}}\\ &+\left(\frac{\beta-\xi}{r+1}\right)^{\eta+1}\left(|\varphi'(\xi)|^q\,\mathbb{W}+\left|\varphi'(\beta)\right|^p\mathbb{L}\right)^{\frac{1}{q}}\Bigg]\end{aligned} \tag{30}$$

where $p,q>1$, $\frac{1}{p}+\frac{1}{q}=1$, $\xi\in[\alpha,\beta]$ and $r\in\mathbb{N}_0$.

Proof Following an approach similar to that used in a previous theorem, we obtain

$$\begin{aligned}\mathbb{S}(\alpha,\beta,\varphi,w) &\le \left(\frac{\xi-\alpha}{r+1}\right)^{\eta+1}\int_0^1 w(t)\left|\varphi'\left(\frac{t}{r+1}\xi+\frac{r+1-t}{r+1}\alpha\right)\right|dt\\ &+\left(\frac{\beta-\xi}{r+1}\right)^{\eta+1}\int_0^1 w(t)\left|\varphi'\left(\frac{t}{r+1}\xi+\frac{r+1-t}{r+1}\beta\right)\right|dt\end{aligned} \tag{31}$$

Applying Hölder's inequality to (31), we derive

$$\mathbb{S}(\alpha,\beta,\varphi,w) \le \mathbb{U}_p\Bigg[\left(\frac{\xi-\alpha}{r+1}\right)^{\eta+1}\left(\int_0^1\left|\varphi'\left(\frac{t}{r+1}\xi+\frac{r+1-t}{r+1}\alpha\right)\right|^q dt\right)^{\frac{1}{q}}$$

$$+\left(\frac{\beta-\xi}{r+1}\right)^{\eta+1}\left(\int_0^1\left|\varphi'\left(\frac{t}{r+1}\xi+\frac{r+1-t}{r+1}\beta\right)\right|^q dt\right)^{\frac{1}{q}}\Bigg] \tag{32}$$

where $\frac{1}{p}+\frac{1}{q}=1$ and $p, q>1$.

Utilizing the s-convexity of $|\varphi'|^q$, we refine (32) as follows

$$\mathbb{S}(\alpha,\beta,\varphi,w)\le \mathbb{U}_p\Bigg[\left(\frac{\xi-\alpha}{r+1}\right)^{\eta+1}\left(|\varphi'(\xi)|^q\,\mathbb{W}+|\varphi'(\alpha)|^p\,\mathbb{L}\right)^{\frac{1}{q}}$$
$$+\left(\frac{\beta-\xi}{r+1}\right)^{\eta+1}\left(|\varphi'(\xi)|^q\,\mathbb{W}+|\varphi'(\beta)|^p\,\mathbb{L}\right)^{\frac{1}{q}}\Bigg] \tag{33}$$

Thus, the proof is completed. □

Theorem 5 *Let be a function φ defined over a closed interval $I\subset\mathbb{R}$, which it is differentiable in the interior I^o. Consider a weight function $w:[0,1]\to\mathbb{R}$. Suppose that $|\varphi'|^q$ satisfies the modified (h,m)-convexity condition of the second type over $[\alpha,\frac{\beta}{m}]\subset I^o$ and that $\varphi'\in L[\alpha,\beta]$, then it holds that:*

$$\mathbb{S}(\alpha,\beta,\varphi,w)\le \mathbb{U}_p\Bigg[\left(\frac{\xi-\alpha}{r+1}\right)^{\eta+1}\left(|\varphi'(\xi)|^q\,\mathbb{G}+m\left|\varphi'\left(\frac{\alpha}{m}\right)\right|^q\mathbb{H}\right)^{\frac{1}{q}}$$
$$+\left(\frac{\beta-\xi}{r+1}\right)^{\eta+1}\left(|\varphi'(\xi)|^q\,\mathbb{G}+m\left|\varphi'\left(\frac{\beta}{m}\right)\right|^q\mathbb{H}\right)^{\frac{1}{q}}\Bigg] \tag{34}$$

where $p,q>1$, $\frac{1}{p}+\frac{1}{q}=1$ and $\xi\in[\alpha,\beta]$.

Proof Following an approach similar to that used in a previous theorem, we have

$$\mathbb{S}(\alpha,\beta,\varphi,w)\le\left(\frac{\xi-\alpha}{r+1}\right)^{\eta+1}\int_0^1 w(t)\left|\varphi'\left(\frac{t}{r+1}\xi+\frac{r+1-t}{r+1}\alpha\right)\right|dt$$
$$+\left(\frac{\beta-\xi}{r+1}\right)^{\eta+1}\int_0^1 w(t)\left|\varphi'\left(\frac{t}{r+1}\xi+\frac{r+1-t}{r+1}\beta\right)\right|dt \tag{35}$$

Applying Hölder's inequality in (35), we get

$$\mathbb{S}(\alpha,\beta,\varphi,w)\le \mathbb{U}_p\Bigg[\left(\frac{\xi-\alpha}{r+1}\right)^{\eta+1}\left(\int_0^1\left|\varphi'\left(\frac{t}{r+1}\xi+\frac{r+1-t}{r+1}\alpha\right)\right|^q dt\right)^{\frac{1}{q}}$$
$$+\left(\frac{\beta-\xi}{r+1}\right)^{\eta+1}\left(\int_0^1\left|\varphi'\left(\frac{t}{r+1}\xi+\frac{r+1-t}{r+1}\beta\right)\right|^q dt\right)^{\frac{1}{q}}\Bigg] \tag{36}$$

for $\frac{1}{p}+\frac{1}{q}=1$ and $p, q>1$.

Utilizing the (h, m)-convexity of $\left|\varphi'\right|^q$, from (36) we obtain

$$\mathbb{S}(\alpha, \beta, \varphi, w) \leq \mathbb{U}_p\left[\left(\frac{\xi-\alpha}{r+1}\right)^{\eta+1}\left(|\varphi'(\xi)|^q\, \mathbb{G}+m\left|\varphi'\left(\frac{\alpha}{m}\right)\right|^q \mathbb{H}\right)^{\frac{1}{q}}\right.$$
$$\left.+\left(\frac{\beta-\xi}{r+1}\right)^{\eta+1}\left(|\varphi'(\xi)|^q\, \mathbb{G}+m\left|\varphi'\left(\frac{\beta}{m}\right)\right|^q \mathbb{H}\right)^{\frac{1}{q}}\right]$$

Which was to be proved. □

As before, setting $w(t)=t^{n-\frac{\eta}{k}}$ yields:

Remark 14 Theorem 8 ($q>1$) from [14], Theorem 6 in [15], the latter part of Theorem 2.7 in [30], and Theorem 3.2 ($q>1$) of [24] follow as particular instances of the above result, under different values of r and varying convexity notions.

Theorem 6 *Let be a function φ defined over a closed interval $I \subset \mathbb{R}$, which it is differentiable in the interior I^o. Consider a weight function $w:[0,1] \rightarrow \mathbb{R}$. Suppose that $|\varphi'|^q$ satisfies the s-convexity condition over $[\alpha, \beta] \subset I^o$ and that $\varphi' \in L[\alpha, \beta]$, then it holds that:*

$$\mathbb{S}(\alpha, \beta, \varphi, w) \leq \mathbb{V}_q\left[\left(\frac{\xi-\alpha}{r+1}\right)^{\eta+1}\left(|\varphi'(\xi)|^q\, \mathbb{G}+\left|\varphi'(\alpha)\right|^q \mathbb{H}\right)^{\frac{1}{q}}\right. \tag{37}$$
$$\left.+\left(\frac{\beta-\xi}{r+1}\right)^{\eta+1}\left(|\varphi'(\xi)|^q\, \mathbb{G}+\left|\varphi'(\beta)\right|^q \mathbb{H}\right)^{\frac{1}{q}}\right]$$

where $q \geq 1$ and $\xi \in[\alpha, \beta]$.

Proof We know that

$$\mathbb{S}(\alpha, \beta, \varphi, w) \leq\left(\frac{\xi-\alpha}{r+1}\right)^{\eta+1} \int_0^1 w(t)\left|\varphi'\left(\frac{t}{r+1} \xi+\frac{r+1-t}{r+1} \alpha\right)\right| d t$$
$$+\left(\frac{\beta-\xi}{r+1}\right)^{\eta+1} \int_0^1 w(t)\left|\varphi'\left(\frac{t}{r+1} \xi+\frac{r+1-t}{r+1} \beta\right)\right| d t \tag{38}$$

Applying the mean power inequality in (38), we find

$$\mathbb{S}(\alpha, \beta, \varphi, w) \leq \mathbb{V}_q\left[\left(\frac{\xi-\alpha}{r+1}\right)^{\eta+1}\left(\int_0^1 w(t)\left|\varphi'\left(\frac{t}{r+1} \xi+\frac{r+1-t}{r+1} \alpha\right)\right|^q d t\right)^{\frac{1}{q}}\right.$$
$$\left.+\left(\frac{\beta-\xi}{r+1}\right)^{\eta+1}\left(\int_0^1 w(t)\left|\varphi'\left(\frac{t}{r+1} \xi+\frac{r+1-t}{r+1} \beta\right)\right|^q d t\right)^{\frac{1}{q}}\right] \tag{39}$$

with $q \geq 1$ and $\xi \in [\alpha, \beta]$. Using the s-convexity of $\left|\varphi'\right|^q$, from (39) we obtain

$$\mathbb{S}(\alpha, \beta, \varphi, w) \leq \mathbb{V}_q \Bigg[\left(\frac{\xi - \alpha}{r+1}\right)^{\eta+1} \left(|\varphi'(\xi)|^q \,\mathbb{G} + \left|\varphi'(\alpha)\right|^q \mathbb{H}\right)^{\frac{1}{q}} + \left(\frac{\beta - \xi}{r+1}\right)^{\eta+1} \left(|\varphi'(\xi)|^q \,\mathbb{G} + \left|\varphi'(\beta)\right|^q \mathbb{H}\right)^{\frac{1}{q}} \Bigg]$$

Which was to be proved. □

Theorem 7 *Let be a function φ defined over a closed interval $I \subset \mathbb{R}$, which it is differentiable in the interior I^o. Consider a weight function $w : [0, 1] \to \mathbb{R}$. Suppose that $|\varphi'|^q$ satisfies the modified (h, m)-convexity condition of the second type over $[\alpha, \frac{\beta}{m}] \subset I^o$ and that $\varphi' \in L[\alpha, \beta]$, then it holds that:*

$$\mathbb{S}(\alpha, \beta, \varphi, w) \leq \mathbb{V}_q \left[\left(\frac{\xi - \alpha}{r+1}\right)^{\eta+1} \left(|\varphi'(\xi)|^q \,\mathbb{G} + m \left|\varphi'\left(\frac{\alpha}{m}\right)\right|^q \mathbb{H}\right)^{\frac{1}{q}} + \left(\frac{\beta - \xi}{r+1}\right)^{\eta+1} \left(|\varphi'(\xi)|^q \,\mathbb{G} + m \left|\varphi'\left(\frac{\beta}{m}\right)\right|^q \mathbb{H}\right)^{\frac{1}{q}} \right]$$

where $q \geq 1$, $m \in [0, 1]$ and $\xi \in [\alpha, \beta]$.

Proof We know that

$$\mathbb{S}(\alpha, \beta, \varphi, w) \leq \left(\frac{\xi - \alpha}{r+1}\right)^{\eta+1} \int_0^1 w(t) \left|\varphi'\left(\frac{t}{r+1}\xi + \frac{r+1-t}{r+1}\alpha\right)\right| dt + \left(\frac{\beta - \xi}{r+1}\right)^{\eta+1} \int_0^1 w(t) \left|\varphi'\left(\frac{t}{r+1}\xi + \frac{r+1-t}{r+1}\beta\right)\right| dt \tag{40}$$

Applying the mean power inequality in (40), we get

$$\mathbb{S}(\alpha, \beta, \varphi, w) \leq \mathbb{V}_q \Bigg[\left(\frac{\xi - \alpha}{r+1}\right)^{\eta+1} \left(\int_0^1 \left|\varphi'\left(\frac{t}{r+1}\xi + \frac{r+1-t}{r+1}\alpha\right)\right|^q dt\right)^{\frac{1}{q}} + \left(\frac{\beta - \xi}{r+1}\right)^{\eta+1} \left(\int_0^1 \left|\varphi'\left(\frac{t}{r+1}\xi + \frac{r+1-t}{r+1}\beta\right)\right|^q dt\right)^{\frac{1}{q}} \Bigg] \tag{41}$$

with $q \geq 1$. Utilizing the (h, m)-convexity of $\left|\varphi'\right|^q$ and from (41), we obtain

$$\mathbb{S}(\alpha, \beta, \varphi, w) \leq \mathbb{V}_q \Bigg[\left(\frac{\xi - \alpha}{r+1}\right)^{\eta+1} \left(|\varphi'(\xi)|^q \,\mathbb{G} + m \left|\varphi'\left(\frac{\alpha}{m}\right)\right|^q \mathbb{H}\right)^{\frac{1}{q}}$$

$$+ \left(\frac{\beta - \xi}{r+1}\right)^{\eta+1} \left(|\varphi'(\xi)|^q \,\mathbb{G} + m \left|\varphi'\left(\frac{\beta}{m}\right)\right|^q \mathbb{H}\right)^{\frac{1}{q}} \Bigg]$$

Which was to be proved. □

Remark 15 The previous result, under the assumption on $w(t)$, generalizes Theorem 7 (with $q > 1$) from [14], Theorem 5 from [15] (with $q > 1$), and Theorem 3.3 from [24], considering different values of r and others notions of convexity.

Remark 16 All previous results contain many of those reported in the literature, particularizing the function w as well as for different notions of convexity.

4 Conclusions

This study presents new formulations of the classic Hermite-Hadamard Inequality in the context of weighted fractional integrals of Caputo and Atangana-Baleanu. Throughout the work we have shown that several known results from the literature are particular cases of ours, which demonstrates their scope and strength. It is clear that the same methodology can be used with other known differential operators, in particular the Caputo Derivative and Caputo-Fabrizio.

We would like to highlight that the ideas presented here can be used in other Comprehensive Inequalities, which would increase the growth of research on this very current topic.

References

1. Akkurt, A., Yildirim, M.E., Yildirim, H.: On some integral inequalities for (k, h)-RiemannLiouville fractional integral. NTMSCI **4**(1), 138–146 (2016). https://doi.org/10.20852/ntmsci.2016217824
2. Ali, M.A., Valdés, J.E.N., Kashuri, A., Zhang, Z.: Fractional non conformable Hermite-Hadamard inequalities for generalized-convex functions. Fasciculi Mathematici, Nr 64, 5–16 (2020). https://doi.org/10.21008/j.0044-4413.2020.0007
3. Amlashi, M.E., Hassani, M.: More on the Hermite-Hadamard inequality. Int. J. Nonlinear Anal. Appl. **12**(2), 2153–2159 (2021). https://doi.org/10.22075/ijnaa.2021.21753.2293
4. Ardci, M.A., Akdemir, A.O., Onalan, H.K.: Integral inequalities for differentiable s-convex functions in the second sense via Atangana-Baleanu fractional integral operators. Filomat **37**(18), 6229–6244 (2023). https://doi.org/10.2298/FIL2318229A
5. Abdeljawad, T., Baleanu, D.: Integration by parts and its applications of a new nonlocal fractional derivative with Mittag-Leffler non-singular kernel. J. Nonlinear Sci. Appl. **10**, 1098–1107 (2017). https://doi.org/10.22436/jnsa.010.03.20

6. Atangana, A., Baleanu, D.: New fractional derivatives with non-local and non-singular kernel: theory and application to Heat transfer model. Therm. Sci. **20**, 763–769 (2016). https://doi.org/10.48550/arXiv.1602.03408
7. Bayraktar, B., Nápoles Valdés, J.E.: A note on Hermite-Hadamard integral inequality for (h, m)-convex modified functions in a generalized framework, submitted
8. Bermudo, S., Kórus, P., Nápoles Valdés, J.E.: On q-Hermite-Hadamard inequalities for general convex functions. Acta Math. Hungar. **162**, 364–374 (2020). https://doi.org/10.1007/s10474-020-01025-6
9. Bilal, M., Imtiaz, M., Khan, A.R., Khan, I.U., Zafran, M.: Generalized Hermite-Hadamard inequalities for s-convex functions in the mixed kind, submitted
10. Breckner, W.W.: Stetigkeitsaussagen fur eine Klasse verallgemeinerter konvexer funktionen in topologischen linearen Raumen. Publ. Inst. Math. **23**, 13–20 (1978)
11. Bruckner, A.M., Ostrow, E.: Some function classes related to the class of convex functions. Pacific J. Math. **12**, 1203–1215 (1962)
12. Dragomir, S.S., Agarwal, R.P.: Two inequalities for differentiable mappings and applications to special means of real numbers and to trapezoidal formula. Appl. Math. Lett. **11**(5), 91–95 (1998)
13. Dragomir, S.S., Pecaric, J., Persson, L.E.: Some inequalities of Hadamard type. Soochow J. Math. **21**, 335–241 (1995)
14. Farid, G., Javed, A., Rehman, A.U., Qureshi, M.I.: On Hadamard-type inequalities for differentiable functions via Caputo k-fractional derivatives. Cogent Math. **4**, 1355429 (2017). https://doi.org/10.1080/23311835.2017.1355429
15. Farid, G., Naqvi, S., Rehman, A.U.: A version of the Hadamard inequality for Caputo fractional derivatives and related results. RGMIA Research Report Collection, 11 pp, 20, Article 59 (2017)
16. Farissi, A.E.: Simple proof and refinement of Hermite-Hadamard inequality. J. Math. Inequal. **4**(3), 365–369 (2010)
17. Fernandez, A., Mohammed, P.: Hermite-Hadamard inequalities in fractional calculus defined using Mittag-Leffler kernels. Math. Meth. Appl. Sci. **44**, 8414–8431 (2021). https://doi.org/10.1002/mma.6188
18. Guzmán, P.M., Nápoles Valdés, J.E., Gasimov, Y.: Integral inequalities within the framework of generalized fractional integrals. Fract. Differ. Calc. **11**(1), 69–84 (2021). https://doi.org/10.7153/fdc-2021-11-05
19. Hadamard, J.: Étude sur les propriétés des fonctions entières et en particulier d'une fonction considérée par Riemann. J. Math. Pures App. **9**, 171–216 (1893)
20. Hermite, C.: Sur deux limites d'une intégrale définie. Mathesis **3**, 82 (1883)
21. Hudzik, H., Maligranda, L.: Some remarks on s-convex functions. Aequationes Math. **48**(1), 100–111 (1994)
22. Jarad, F., Abdeljawad, T., Shah, T.: On the weighted fractional operators of a function with respect to another function. Fractals **28**(8), 2040011 (12 pages) (2020). https://doi.org/10.1142/S0218348X20400113
23. Jarad, F., Ugurlu, E., Abdeljawad, T., Baleanu, D.: On a new class of fractional operators. Adv. Differ. Equ. **2017**, 247 (2017). https://doi.org/10.1186/s13662-017-1306-z
24. Kang, S.M., Farid, G., Nazeer, W., Naqvi, S.: A version of the Hadamard inequality for Caputo fractional derivatives and related results. J. Comput. Anal. Appl. **27**(6), 962–972 (2019). https://doi.org/10.30538/oms2021.0139
25. Kermausuor, S., Nwaeze, E.R.: New fractional integral inequalities via k-Atangana-Baleanu fractional integral operators. Fractal Fract. **7**(10), 740 (2023). https://doi.org/10.3390/fractalfract7100740
26. Khan, T.U., Khan, M.A.: Generalized conformable fractional integral operators. J. Comput. Appl. Math. **346**, 378–389 (2019). https://doi.org/10.1016/j.cam.2018.07.018
27. Marinescu, D.S., Monea, M.: A very short proof of the Hermite-Hadamard inequalities. Am. Math. Mon. **127**(9), 850–851 (2020). https://doi.org/10.1080/00029890.2020.1803648
28. Mehmood, S., Nápoles Valdés, J.E., Fatima, N., Shahid, B.: Some new inequalities using conformable fractional integral of order β. J. Math. Ext. **15**, SI-NTFCA, (2021) (33)1-22. https://doi.org/10.30495/JME.SI.2021.2184

29. Mihesan, V.G.: A generalization of the convexity, Seminar on Functional Equations, Approx. and Convex., Cluj-Napoca (Romania) (1993)
30. Mishra, L.N., Ain, Q.U., Farid, G., Rehman, A.U.: k-fractional integral inequalities for (h,m)-convex functions via Caputo k-fractional derivatives. Korean J. Math. **27**(2), 357–374 (2019). https://doi.org/10.11568/kjm.2019.27.2.357
31. Mubeen, S., Habibullah, G.M.: k-fractional integrals and applications. Int. J. Contemp. Math. Sci. **7**, 89–94 (2012)
32. Nápoles Valdés, J.E.: On the Hermite-Hadamard type inequalities involving generalized integrals. Contrib. Math. **5**, 45–51 (2022). https://doi.org/10.47443/cm.2022.020
33. Nápoles Valdés, J.E.: A review of Hermite-Hadamard inequality. Partners Univ. Int. Res. J. (PUIRJ) **01**(04), 98–101 (2022). https://doi.org/10.5281/zenodo.7492608
34. Nápoles Valdés, J.E., Rabossi, F., Samaniego, A.D.: Convex functions: Ariadne's thread or Charlotte's spiderweb? Adv. Math. Models Appl. **5**(2), 176–191 (2020)
35. Nápoles Valdés, J.E., Rodríguez, J.M., Sigarreta, J.M.: On Hermite-Hadamard type inequalities for non-conformable integral operators. Symmetry **11**, 1108 (2019)
36. Noor, M.A., Noor, K.I., Awan, M.U.: GENERALIZED FRACTIONAL HERMITE-HADAMARD INEQUALITIES, Miskolc Mathematical Notes **21**(2), 1001–1011 (2020). https://doi.org/10.18514/MMN.2020.1143
37. Özdemit, M.E., Akdemri, A.O., Set, E.: On (h, m)-convexity and Hadamard-type inequalities. Transylv. J. Math. Mech. **8**(1), 51–58 (2016)
38. Park, J.: Generalization of Ostrowski-type inequalities for differentiable real (s, m)-convex mappings. Far East J. Math. Sci. **49**, 157–171 (2011)
39. Sarikaya, M.Z., Ertugral, F.: On the generalized Hermite-Hadamard inequalities. Ann. Univ. Craiova, Math. Comput. Sci. Ser. **47**(1), 193–213 (2020)
40. Sarikaya, M.Z., Set, E., Yaldiz, H., Basak, N.: Hermite-Hadamard's inequalities for fractional integrals and related fractional inequalities. Math. Comput. Model. **57**, 2403–2407 (2013). https://doi.org/10.1016/j.mcm.2011.12.048
41. Sarikaya, M.Z., Yildirim, H.: ON HERMITE-HADAMARD TYPE INEQUALITIES FOR RIEMANN-LIOUVILLE FRACTIONAL INTEGRALS, Miskolc Mathematical Notes **17**(2), 1049–1059 (2017). https://doi.org/10.18514/MMN.2017.1197
42. Set, E., Akdemir, A.O., Özdemir, M.E., Karaoglan, A., Dokuyucu, M.A.: New integral inequalities for Atangana-Baleanu fractional integral operators and various comparisons via simulations. Filomat **37**(7), 2251–2267 (2023). https://doi.org/10.2298/FIL2307251S
43. Set, E., Sarikaya, M.Z., Özdemir, M.E., Yildirim, H.: The Hermite-Hadamard's inequality for some convex functions via fractional integrals and related results. JAMSI **10**(2), 69–83 (2014). https://doi.org/10.48550/arXiv.1112.6176
44. Simic, S.: Some refinements of Hermite-Hadamard inequality and an open problem. Kragujevac J. Math. **42**(3), 349–356 (2018)
45. Toader, G.: Some generalizations of the convexity. In: Proceedings of the Colloquium on Approximation and Optimization, University Cluj-Napoca (1985), pp. 329–338
46. Tomar, M., Set, E., Sarikaya, M.Z.: Hermite-Hadamard type Riemann-Liouville fractional integral inequalities for convex functions. AIP Conf. Proc. **1726**, 020035 (2016). https://doi.org/10.1063/1.4945861
47. Wang, J.R., Li, X., Zhou, Y.: Hermite-Hadamard inequalities involving Riemann-Liouville fractional integrals via s-convex functions and applications to special means. Filomat **30**(5), 1143–1150 (2016). https://doi.org/10.2298/FIL1605143W
48. Wu, S., Iqbal, S., Aamir, M., Samraiz, M., Younus, A.: On some Hermite-Hadamard inequalities involving k-fractional operators. J. Inequal. Appl. **2021**, 32 (2021). https://doi.org/10.1186/s13660-020-02527-1
49. Xi, B.Y., Qi, F.: Inequalities of Hermite-Hadamard type for extended s-convex functions and applications to means. J. Nonlinear Convex. Anal. **16**(5), 873–890 (2015). https://doi.org/10.48550/arXiv.1406.5409
50. Xi, B.Y., Gao, D.D., Qi, F.: Integral inequalities of Hermite-Hadamard type for (α, s)-convex and (α, s, m)-convex functions. Italian J. Pure Appl. Math. N.44-2020 (499-510)

Fractal Nonlinear Klein-Gordon Equation

Alireza Khalili Golmankhaneh and Lucero Damián

Abstract In this paper, we solve the fractal nonlinear Klein-Gordon equation using the Homotopy Perturbation Method (HPM). The method effectively demonstrates how the solution evolves over time in relation to the fractal dimensions of space and time. To illustrate the approach, several examples are presented, showcasing the details of the solution process. These examples highlight the applicability of HPM for addressing complex fractal differential equations and provide insights into the behavior of solutions within fractal geometries.

Keywords Fractal nonlinear Klein-Gordon equation · Homotopy perturbation method · Fractal dimensions · Fractal differential equations

1 Introduction

Fractal geometry has been extensively employed to study and describe recurring patterns found in nature, such as blood vessels, mountain ranges, clouds, and coastlines. This topic explores the unique features and measurements of these formations. Common traits of fractals include fractional dimensions, self-similarity, and fractal dimensions that exceed their topological dimensions [10, 11, 46, 47]. Various analytical approaches are used in fractal analysis, including harmonic analysis [12, 42, 45, 57, 60], fractional calculus, stochastic processes [3], and fractal space theory [55]. Several Riemann-type integrals, such as s-Riemann, s-HK, and s-first-return integrals, have been developed by modifying the Fundamental Theorem of Calculus. Scholars have investigated the connections between these integrals and the Lebesgue integral associated with the Hausdorff measure [4–6, 41]. Fractals, with

A. K. Golmankhaneh (✉)
Department of Physics, Ur. C., Islamic Azad University, Urmia, West Azerbaijan, Iran
e-mail: alireza.khalili@iau.ac.ir

L. Damián
Department, Universidad Politécnica de Yucatán, Carretera Mérida-Tetiz, Ucú, Mexico
e-mail: lucero.damian@upy.edu.mx

H. Kalita et al. (eds.), *Nonlinear Analysis and Computational Techniques*,
Springer Proceedings in Mathematics & Statistics 501,
https://doi.org/10.1007/978-981-96-6038-4_5

their intricate geometric patterns and self-similarity across different scales [43], often require non-standard mathematical methods for analysis. By extending traditional calculus, a new framework for fractal calculus has emerged, offering an algorithmic, geometric, and physically meaningful approach. This generalization allows for the analysis of functions with fractal properties, such as the Koch curve and Cantor set [21, 49, 50, 52]. Similar to how ordinary calculus is expanded through Riemann-Liouville and Caputo generalizations, non-local fractal calculus has been introduced to model incompressible viscous fluids in fractal media and systems with memory effects [1, 2, 18, 33].

Fractal local derivatives have been employed to preserve the central limit theorem and locality in research on sub-diffusion and super-diffusion [16, 24]. Fractal calculus has also been applied in physics to create new models of fractal space and time [9, 30, 48, 51, 54, 58, 59]. These models frequently exhibit power law behaviors and self-similarity. Fractal Laplace, Sumudu, and Fourier transforms have been used to investigate the dynamics of systems governed by fractal time [15, 20, 22, 23, 26, 34]. Additionally, fractal calculus has been employed to compute the derivatives and integrals of various functions, such as the Weierstrass function [35].

Nonstandard analysis, incorporating hyperreal numbers and hyperintegers, has provided tools for defining derivative structures and bounds on fractal curves, as well as fractal integral and differential forms [31]. Gauge function generalizations have further enabled the application of fractal calculus to unbounded functions [17]. The stability of fractal differential equations has been examined, establishing conditions for the existence of unique and stable solutions [28]. An analogue of Noether's theorem for fractal sets has been developed, introducing a corresponding conserved quantity [44]. This paper also lays out a framework for studying non-differentiable fractal manifolds, introducing mathematical tools such as the Metric Tensor, Curvature Tensors, Analogue Arc Length, and Inner Product, which facilitate the study of complex, self-similar patterns across different dimensions and scales [25, 29].

Solutions to fractal retarded, neutral, and renewal delay differential equations with constant coefficients have been obtained using the steps method and the Laplace transform [27]. A new approach has been developed to incorporate the fractal mean square derivative, fractal mean square integral, and the order of random variables on a fractal curve, extending the concepts of mean square calculus [32]. Additionally, the exact formal solution to the nonlinear Klein-Gordon equation has been established [7], and numerical methods have been applied to solve one-dimensional nonlinear Klein-Gordon equations with quadratic and cubic nonlinearities [8, 56]. The asymptotic behavior of the semilinear Klein-Gordon equation with fractional-order nonlinearity has also been analyzed [13, 14]. The Homotopy Perturbation Method has proven effective in finding approximate analytical solutions to the fractional nonlinear Klein-Gordon equation [19]. The focus of this paper is to apply the Homotopy Perturbation Method to solve the fractal nonlinear Klein-Gordon equation.

2 Basic Tools

2.1 Fundamental Definitions in Fractal Calculus

In this section, we present an overview of fractal calculus as it relates to the Cantor set $F \subset [c, d] \subset \mathbb{R}$, drawing on insights from [21, 50].

Definition 1 The indicator function $\mathbb{I}_F(K)$ for the set F is defined by

$$\mathbb{I}_F(K) = \begin{cases} 1, & \text{if } F \cap K \neq \emptyset, \\ 0, & \text{otherwise,} \end{cases}$$

where $K = [c, d] \subset \mathbb{R}$.

Definition 2 The coarse-grained measure $\mu_\epsilon^\gamma(F, c, d)$ of $F \cap [c, d]$ is given by

$$\mu_\epsilon^\gamma(F, c, d) = \inf_{|\mathcal{P}| \leq \epsilon} \sum_{i=0}^{m-1} \Gamma(\gamma + 1)(z_{i+1} - z_i)^\gamma \mathbb{I}_F([z_i, z_{i+1}]),$$

where $|\mathcal{P}| = \max_{0 \leq i \leq m-1}(z_{i+1} - z_i)$, $\mathcal{P}_{[c,d]} = \{z_0 = c, z_1, \ldots, z_m = d\}$, $0 < \gamma \leq 1$, and $\Gamma(\cdot)$ denotes the Gamma function.

Definition 3 The measure function $\mu^\gamma(F, c, d)$ for F is defined as

$$\mu^\gamma(F, c, d) = \lim_{\epsilon \to 0} \mu_\epsilon^\gamma(F, c, d).$$

Definition 4 The fractal ν-dimension of $F \cap [c, d]$ is defined by

$$\begin{aligned} \dim_\nu(F \cap [c, d]) &= \inf\{\gamma : \mu^\gamma(F, c, d) = 0\} \\ &= \sup\{\gamma : \mu^\gamma(F, c, d) = \infty\}. \end{aligned} \tag{1}$$

Definition 5 The integral staircase function $S_F^\gamma(z)$ is defined as

$$S_F^\gamma(z) = \begin{cases} \mu^\gamma(F, c_0, z), & \text{if } z \geq c_0, \\ -\mu^\gamma(F, z, c_0), & \text{otherwise,} \end{cases}$$

where $c_0 \in \mathbb{R}$ is a fixed constant.

Definition 6 The characteristic function $\chi_F(z)$ on a fractal set F is defined by

$$\chi_F(z) = \begin{cases} \Gamma(\gamma + 1), & \text{if } z \in F, \\ 0, & \text{otherwise.} \end{cases} \tag{2}$$

Here, γ denotes the fractal dimension as defined in (1).

Definition 7 For a function $k : F \to \mathbb{R}$, the $F\_\lim$ of $k(z)$ at a point $z \in F$ is the number ℓ such that for every $\epsilon > 0$, there exists a $\delta > 0$ satisfying

$$y \in F \text{ and } |y - z| < \delta \implies |k(y) - \ell| < \epsilon.$$

If such an ℓ exists, it is denoted by

$$\ell = F\_\lim_{y \to z} k(y).$$

This definition does not involve the function values at y if $y \notin F$.

Definition 8 A function $k : F \to \mathbb{R}$ is said to be F-continuous at $z \in F$ if

$$k(z) = F\_\lim_{y \to z} k(y), \tag{3}$$

whenever the $F\_\lim$ exists.

Definition 9 For a function k on a γ-perfect fractal set F, the F^{γ}-derivative of k at z is defined as

$$D_F^{\gamma} k(z) = \begin{cases} F\_\lim\limits_{y \to z} \frac{k(y)-k(z)}{S_F^{\gamma}(y)-S_F^{\gamma}(z)}, & \text{if } z \in F, \\ 0, & \text{otherwise,} \end{cases}$$

provided that the $F\_\lim$ exists.

Definition 10 The F^{γ}-integral of a bounded function $k(z)$, where $k \in B(F)$ (i.e., k is bounded on F), is defined as

$$\int_c^d k(z)\, d_F^{\gamma} z = \sup_{\mathcal{P}_{[c,d]}} \sum_{i=0}^{m-1} \inf_{z \in F \cap K} k(z)\, (S_F^{\gamma}(z_{i+1}) - S_F^{\gamma}(z_i)) \tag{4}$$

$$= \inf_{\mathcal{P}_{[c,d]}} \sum_{i=0}^{m-1} \sup_{z \in F \cap K} k(z)\, (S_F^{\gamma}(z_{i+1}) - S_F^{\gamma}(z_i)), \tag{5}$$

where $z \in F$, and the infimum or supremum is taken over all partitions $\mathcal{P}_{[c,d]}$ [21, 50].

2.2 *Homotopy Perturbation Method*

The concept of the Homotopy Perturbation Method (HPM) and its application to various differential equations have been explored in [36–40]. We now provide a brief overview of HPM. Consider the nonlinear differential equation:

$$A(v) + B(v) = g(x), \quad x \in \Omega,$$

with the boundary condition:

$$C\left(v, \frac{\partial v}{\partial n}\right) = 0, \quad x \in \partial\Omega,$$

where A is a linear operator, B is a nonlinear operator, $\partial\Omega$ is the boundary of the domain Ω, and $g(x)$ is a known analytic function. Using He's homotopy perturbation technique, we define a homotopy $V(x, q) : \Omega \times [0, 1] \to \mathbb{R}$, which satisfies the following equation:

$$H(V, q) = (1 - q)[A(V) - A(v_0)] + q[A(V) + B(V) - g(x)] = 0,$$

where $q \in [0, 1]$ is the embedding parameter, and v_0 is an initial approximation that satisfies the boundary conditions. As q changes from 0 to 1, $V(x, q)$ deforms from the initial approximation v_0 to the solution $v(x)$.

The fundamental assumption of HPM is that the solution $V(x, q)$ can be expressed as a power series in q:

$$V = V_0 + qV_1 + q^2V_2 + \cdots$$

Therefore, the approximate solution of the original equation is given by

$$v = \lim_{q\to 1} V = V_0 + V_1 + V_2 + \cdots$$

The convergence of this series has been established in [36–40].

3 Fractal Nonlinear Klein-Gordon Equation

The nonlinear Klein-Gordon equation is widely recognized for its numerous applications in physics [53]. In this section, we apply the Homotopy Perturbation Method (HPM) to solve the fractal version of the nonlinear Klein-Gordon equation. Consider the following fractal Klein-Gordon equation with an initial condition:

$$D^{\alpha}_{F,t}v(x,t) = D^{2\beta}_{F,x}v(x,t) + av(x,t) + bv^2(x,t) + cv^3(x,t), \quad v(x,0) = v_0, \quad x, t \in F, \tag{6}$$

where a, b, and c are real constants, α represents the dimension of time, and β represents the dimension of space. To solve Eq. (6) using HPM, we construct the following homotopy equation:

$$\mathcal{H}(V, p) = (1 - p)D^{\alpha}_{F,t}V + p\left(D^{\alpha}_{F,t}V - D^{2\beta}_{F,x}V(x,t) - aV(x,t) - bV^2(x,t) - cV^3(x,t)\right) = 0. \tag{7}$$

Here, $p \in [0, 1]$ is an embedding parameter. When $p = 0$, the homotopy equation corresponds to a simpler form of the original equation, and as $p \to 1$, it gradually transforms into the original fractal Klein-Gordon equation. This approach enables us to obtain an approximate solution by expanding the solution as a series in terms of p. Let the solution of Eq. (7) be expressed as a series:

$$V = V_0 + pV_1 + p^2 V_2 + \cdots \tag{8}$$

Substituting Eq. (8) into Eq. (7) and equating the coefficients of terms with identical powers of p, we obtain the following system of equations: Let $V_0 = v_0$, the initial condition, then the system becomes:
Let p^0, we have

$$D^{\alpha}_{F,t} v_0 = 0, \tag{9}$$

Let p^1, we have

$$D^{\alpha}_{F,t} V_1 - D^{2\beta}_{F,x} v_0 - a v_0 - b v_0^2 - c v_0^3 = 0, \quad V_1(x, 0) = 0, \tag{10}$$

Let p^2, we have

$$D^{\alpha}_{F,t} V_2 - D^{2\beta}_{F,x} V_1 - a V_1 - 2b v_0 V_1 - c\left(v_0 V_1^2 + 2V_1 v_0^2\right) = 0, \quad V_2(x, 0) = 0, \tag{11}$$

Let p^j, we have

$$D^{\alpha}_{F,t} V_j - D^{2\beta}_{F,x} V_{j-1} - a V_{j-1} - b \sum_{i=0}^{j-1} V_i V_{j-i-1} - c \sum_{i=0}^{j-1} \sum_{k=0}^{j-i-1} V_i V_k V_{j-k-i-1} = 0,$$
$$V_j(x, 0) = 0. \tag{12}$$

Assuming $V_0 = v_0 = v(x, 0)$, Eq. (12) leads to the following relations:

$$V_j = \int_0^t \left(D^{2\beta}_{F,x} V_{j-1} + a V_{j-1} + b \sum_{i=0}^{j-1} V_i V_{j-i-1} + c \sum_{i=0}^{j-1} \sum_{k=0}^{j-i-1} V_i V_k V_{j-k-i-1} \right) d^{\alpha}_{F} \tau. \tag{13}$$

Thus, the solution of Eq. (6) can be obtained by setting $p = 1$:

$$v = \lim_{p \to 1} V = V_0 + V_1 + V_2 + \cdots \tag{14}$$

4 Illustrative Examples

In this section, we utilize the Homotopy Perturbation Method (HPM) to solve the fractal linear Klein-Gordon equation, demonstrating the effectiveness of this method in addressing fractal differential equations.

Example 1 Consider the fractal linear Klein-Gordon equation with the initial condition:

$$D^{\alpha}_{F,t}v(x,t) - D^{2\beta}_{F,x}v(x,t) - v(x,t) = 0, \quad t > 0, \quad v(x,0) = 1 + \sin(S^{\alpha}_{F}(x)), \tag{15}$$

where $t, x \in F$, and α and β represent the fractal dimensions of time and space, respectively. Applying the Homotopy Perturbation Method (HPM) to solve this equation, we use the homotopy equation (7) with parameters $a = 1$, $b = 0$, and $c = 0$. This yields the following terms in the series expansion from Eqs. (13) and (14):

$$V_1(x,t) = S^{\alpha}_{F}(t), \tag{16}$$

$$V_2(x,t) = \frac{S^{\alpha}_{F}(t)^2}{2!}, \tag{17}$$

$$\vdots \tag{18}$$

$$V_n(x,t) = \frac{S^{\alpha}_{F}(t)^n}{n!}. \tag{19}$$

Thus, the approximate solution becomes

$$v(x,t) = 1 + \sin(S^{\beta}_{F}(x)) + \sum_{n=1}^{\infty} \frac{S^{\alpha}_{F}(t)^n}{n!} \tag{20}$$

$$\approx 1 + \sin(S^{\beta}_{F}(x)) + \sum_{n=1}^{\infty} \frac{t^{n\alpha}}{n!}. \tag{21}$$

The behavior of the solution $v(x,t)$ is illustrated for various values of α in Fig. 1a, b, c, and d. These plots demonstrate the influence of the fractal dimension of time on the system's dynamics.

Similarly, the impact of varying the spatial dimension β is shown in Fig. 2a through d. As β increases, we observe notable changes in the system's dynamics, highlighting the role of spatial fractal dimensions.

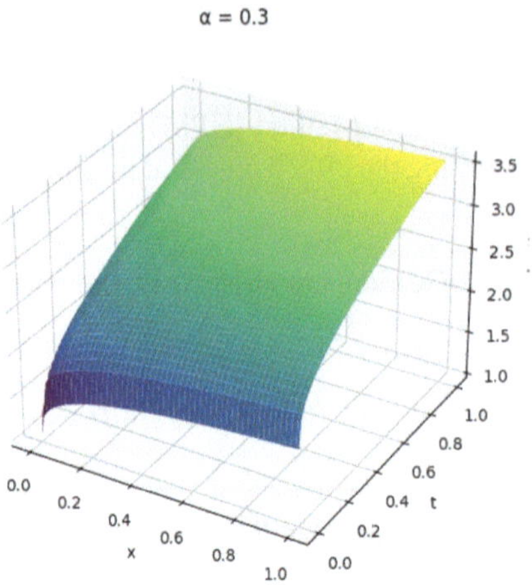

Surface plot of $v(x, t)$ for $\alpha = 0.3$.

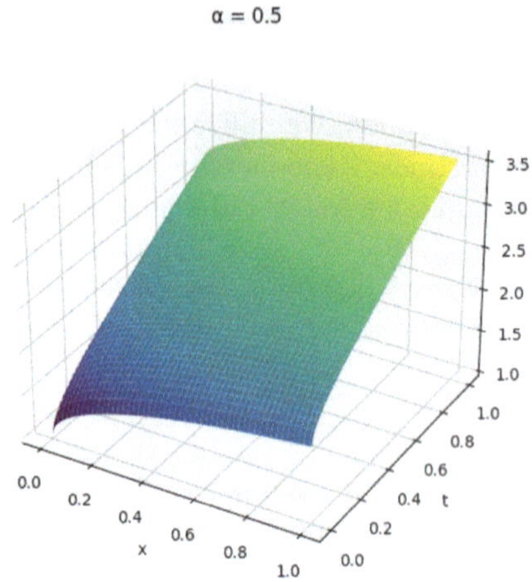

Surface plot of $v(x, t)$ for $\alpha = 0.5$.

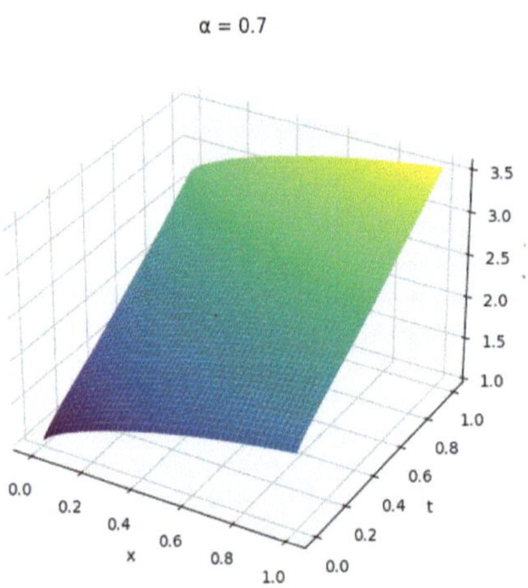

Surface plot of $v(x, t)$ for $\alpha = 0.7$.

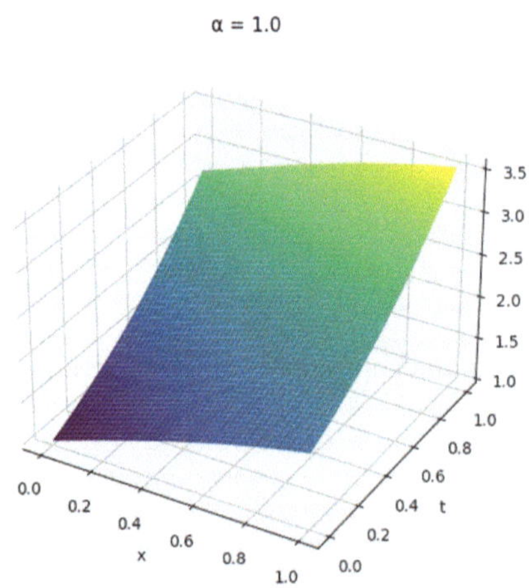

Surface plot of $v(x, t)$ for $\alpha = 1.0$.

Fig. 1 Surface plots of $v(x, t)$ for different values of α (dimension of time)

Example 2 Consider the fractal nonlinear Klein-Gordon equation:

$$D_{F,t}^{\alpha} v(x,t) - D_{F,x}^{2\beta} v(x,t) + v^2(x,t) = 0, \quad t > 0, \quad v(x,0) = 1 + \sin(x), \quad t, x \in F, \tag{22}$$

where α and β denote the fractal dimensions of time and space, respectively.

To solve this equation using the Homotopy Perturbation Method (HPM), we construct the homotopy equation based on Eq. (7), with parameters $a = 0$, $b = -1$, and

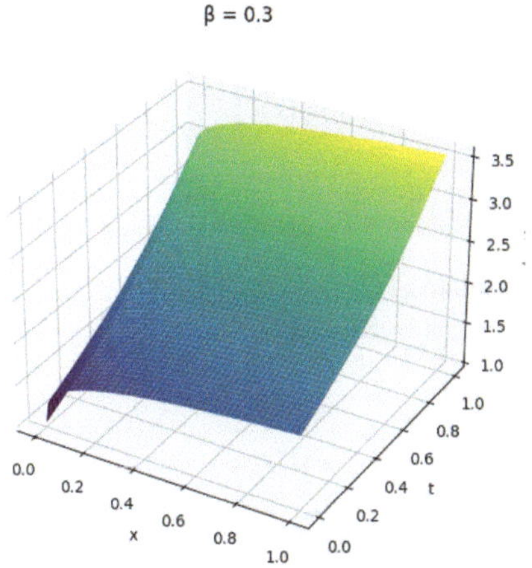

Surface plot of $v(x,t)$ for $\beta = 0.3$.

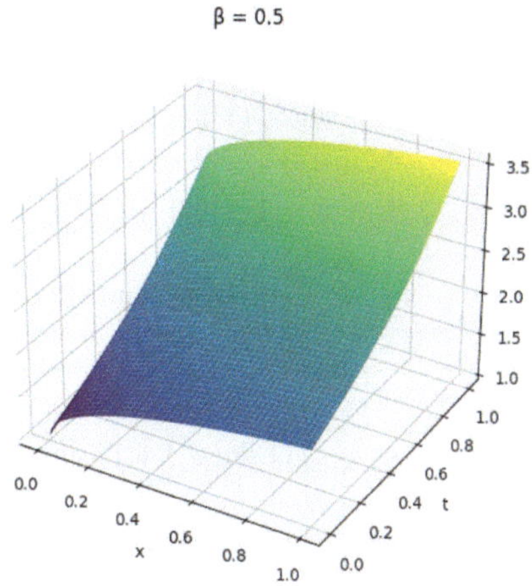

Surface plot of $v(x,t)$ for $\beta = 0.5$.

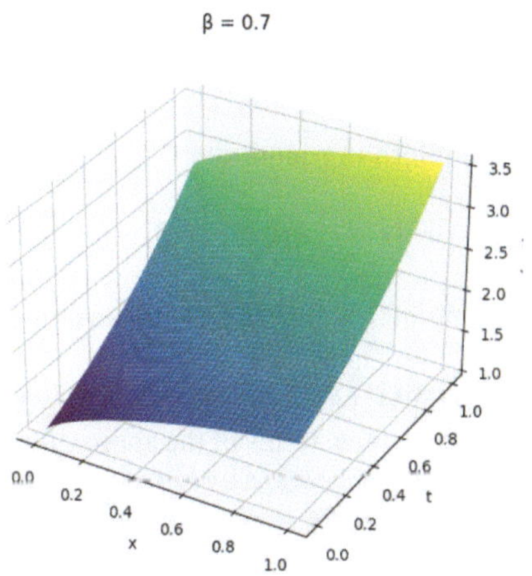

Surface plot of $v(x,t)$ for $\beta = 0.7$.

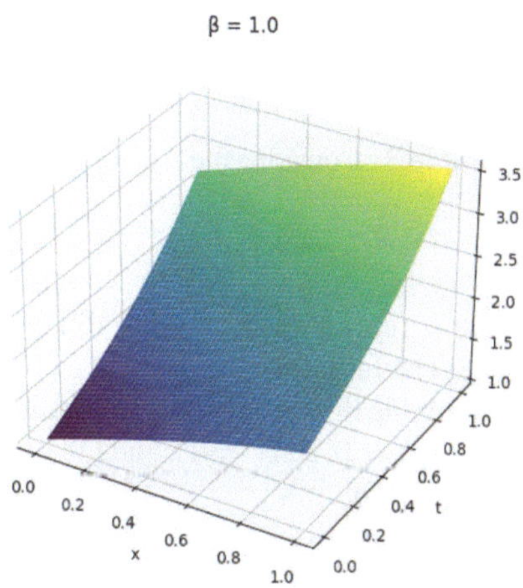

Surface plot of $v(x,t)$ for $\beta = 1.0$.

Fig. 2 Surface plots of $v(x, t)$ for different values of β (dimension of space)

$c = 0$. Using Eqs. (13) and (14), we derive the following terms for the approximate solution:

$$
\begin{aligned}
V_1(x,t) &= -S_F^{\alpha}(t)\left(1 + 3\sin(S_F^{\beta}(x)) + \sin^2(S_F^{\beta}(x))\right),\\
V_2(x,t) &= \frac{S_F^{\alpha}(t)^2}{2!}\left(11\sin(S_F^{\beta}(x)) + 12\sin^2(S_F^{\beta}(x)) + 2\sin^3(S_F^{\beta}(x))\right),\\
V_3(x,t) &= \frac{S_F^{\alpha}(t)^3}{3!}\left(18 - 57\sin(S_F^{\beta}(x)) - 160\sin^2(S_F^{\beta}(x)) - 82\sin^3(S_F^{\beta}(x)) - 10\sin(4S_F^{\beta}(x))\right),\\
&\vdots
\end{aligned}
\tag{23}
$$

Thus, the approximate solution is expressed as

$$\begin{aligned} v(x,t) &= 1+\sin(S_F^{\beta}(x)) - S_F^{\alpha}(t)\left(1+3\sin(S_F^{\beta}(x))+\sin^2(S_F^{\beta}(x))\right)\\ &+\frac{S_F^{\alpha}(t)^2}{2!}\left(11\sin(S_F^{\beta}(x))+12\sin^2(S_F^{\beta}(x))+2\sin^3(S_F^{\beta}(x))\right)\\ &+\frac{S_F^{\alpha}(t)^3}{3!}\left(18-57\sin(S_F^{\beta}(x))-160\sin^2(S_F^{\beta}(x))-82\sin^3(S_F^{\beta}(x))\right.\\ &\left.-10\sin(4S_F^{\beta}(x))\right)+\cdots \approx 1+\sin(x^{\beta})-t^{\alpha}\left(1+3\sin(x^{\beta})+\sin^2(x^{\beta})\right)\\ &+\frac{t^{2\alpha}}{2!}\left(11\sin(x^{\beta})+12\sin^2(x^{\beta})+2\sin^3(x^{\beta})\right)\\ &+\frac{t^{3\alpha}}{3!}\left(18-57\sin(x^{\beta})-160\sin^2(x^{\beta})-82\sin^3(x^{\beta})-10\sin(4x^{\beta})\right)+\cdots \end{aligned} \tag{24}$$

This solution captures the behavior of the fractal nonlinear Klein-Gordon equation for different time and space fractal dimensions α and β, showcasing how the solution evolves over time.

Example 3 Consider the fractal nonlinear Klein-Gordon equation:

$$D_{F,t}^{\alpha}v(x,t) - D_{F,x}^{2\beta}v(x,t) + v(x,t) - v^3(x,t) = 0, \quad t > 0, \tag{25}$$

where α represents the fractal dimension of time and β the fractal dimension of space. The initial condition is given by $v(x,0) = -\text{sech}(S_F^{\beta}(x))$.

Applying the Homotopy Perturbation Method (HPM) based on Eqs. (7), (13) (with $a=-1, b=0$, and $c=1$), and (14), we derive the first few terms of the approximate solution as follows:

$$\begin{aligned} V_1(x,t) &= -S_F^{\alpha}(t)\left(2\,\text{sech}(S_F^{\beta}(x)) - 3\,\text{sech}^3(S_F^{\beta}(x))\right),\\ V_2(x,t) &= -\frac{S_F^{\alpha}(t)^2}{2!}\left(3\,\text{sech}(S_F^{\beta}(x)) - 34\,\text{sech}^3(S_F^{\beta}(x)) - 18\,\text{sech}^5(S_F^{\beta}(x))\right),\\ V_3(x,t) &= -\frac{S_F^{\alpha}(t)^3}{3!}\left(64\,\text{sech}^3(S_F^{\beta}(x)) - 288\,\text{sech}^5(S_F^{\beta}(x)) + 240\,\text{sech}^7(S_F^{\beta}(x))\right),\\ &\vdots \end{aligned} \tag{26}$$

Thus, the approximate solution for $v(x,t)$ is expressed as

$$\begin{aligned}
v(x,t) &= -\operatorname{sech}(S_F^{\beta}(x)) - S_F^{\alpha}(t)\left(2\operatorname{sech}(S_F^{\beta}(x)) - 3\operatorname{sech}^3(S_F^{\beta}(x))\right)\\
&\quad - \frac{S_F^{\alpha}(t)^2}{2!}\left(3\operatorname{sech}(S_F^{\beta}(x)) - 34\operatorname{sech}^3(S_F^{\beta}(x)) - 18\operatorname{sech}^5(S_F^{\beta}(x))\right)\\
&\quad - \frac{S_F^{\alpha}(t)^3}{3!}\left(64\operatorname{sech}^3(S_F^{\beta}(x)) - 288\operatorname{sech}^5(S_F^{\beta}(x)) + 240\operatorname{sech}^7(S_F^{\beta}(x))\right) + \cdots\\
&\approx -\operatorname{sech}(x^{\beta}) - t^{\alpha}\left(2\operatorname{sech}(x^{\beta}) - 3\operatorname{sech}^3(x^{\beta})\right)\\
&\quad - \frac{t^{2\alpha}}{2!}\left(3\operatorname{sech}(x^{\beta}) - 34\operatorname{sech}^3(x^{\beta}) - 18\operatorname{sech}^5(x^{\beta})\right)\\
&\quad - \frac{t^{3\alpha}}{3!}\left(64\operatorname{sech}^3(x^{\beta}) - 288\operatorname{sech}^5(x^{\beta}) + 240\operatorname{sech}^7(x^{\beta})\right) + \cdots
\end{aligned} \tag{27}$$

This series provides an approximation to the solution of the fractal nonlinear Klein-Gordon equation, demonstrating how the solution evolves over time with respect to the fractal dimensions α and β.

5 Conclusion

In this paper, we have successfully solved the fractal nonlinear Klein-Gordon equation using the Homotopy Perturbation Method (HPM), demonstrating how the solution evolves over time in relation to the fractal dimensions of space and time. Several illustrative examples were provided to showcase the details of the solution process and the effectiveness of the method. These examples highlight the applicability of HPM in addressing complex fractal differential equations, offering insights into the behavior of solutions within fractal geometries.

References

1. Banchuin, R.: Noise analysis of electrical circuits on fractal set. Compel- Int. J. Comput. Math. Electr. Electron. Eng. **41**(5), 1464–1490 (2022)
2. Banchuin, R.: Nonlocal fractal calculus based analyses of electrical circuits on fractal set. Compel- Int. J. Comput. Math. Electr. Electron. Eng. **41**(1), 528–549 (2022)
3. Barlow, M.T., Perkins, E.A.: Brownian motion on the sierpinski gasket. Probab. Theory Rel. **79**(4), 543–623 (1988)
4. Bongiorno, D.: Derivatives not first return integrable on a fractal set. Ric. Mat. **67**(2), 597–604 (2018)
5. Bongiorno, D., Corrao, G.: An integral on a complete metric measure space. Real Anal. Exch. **40**(1), 157–178 (2015)
6. Bongiorno, D., Corrao, G.: On the fundamental theorem of calculus for fractal sets. Fractals **23**(02), 1550008 (2015)
7. Burt, P., Reid, J.: Exact solution to a nonlinear Klein-Gordon equation. J. Math. Anal. Appl. **55**(1), 43–45 (1976)

8. Dehghan, M., Shokri, A.: Numerical solution of the nonlinear Klein-Gordon equation using radial basis functions. J. Comput. Appl. Math. **230**(2), 400–410 (2009)
9. Deppman, A., Megías, E., Pasechnik, R.: Fractal derivatives, fractional derivatives and q-deformed calculus. Entropy **25**(7), 1008 (2023)
10. Falconer, K.: Fractal Geometry: Mathematical Foundations and Applications. Wiley, New York (2004)
11. Feder, J.: Fractals. Springer Science & Business Media, New York (2013)
12. Freiberg, U., Zähle, M.: Harmonic calculus on fractals-a measure geometric approach I. Potential Anal. **16**(3), 265–277 (2002)
13. Georgiev, V.: Decay estimates for the Klein ndesh Gordon equation. Comm. Part. Differ. Equ. **17**(7–8), 1111–1139 (1992)
14. Georgiev, V., Lucente, S.: Decay for nonlinear Klein-Gordon equations. Nonlinear Differ. Equ. Appl. **11**, 529–555 (2004)
15. Golmankhaneh, A.K., Ali, K., Yilmazer, R., Kaabar, M.: Local fractal Fourier transform and applications. Comput. Methods Differ. Equ. **10**(3), 595–607 (2021)
16. Golmankhaneh, A.K., Balankin, A.S.: Sub-and super-diffusion on cantor sets: beyond the paradox. Phys. Lett. A **382**(14), 960–967 (2018)
17. Golmankhaneh, A.K., Baleanu, D.: Fractal calculus involving gauge function. Commun. Nonlinear Sci. Numer. Simul. **37**, 125–130 (2016)
18. Golmankhaneh, A.K., Baleanu, D.: Non-local integrals and derivatives on fractal sets with applications. Open Phys. **14**(1), 542–548 (2016)
19. Golmankhaneh, A.K., Golmankhaneh, A.K., Baleanu, D.: On nonlinear fractional klein-Gordon equation. Signal Process. **91**(3), 446–451 (2011)
20. Golmankhaneh, A.K., Tunç, C.: Sumudu transform in fractal calculus. Appl. Math. Comput. **350**, 386–401 (2019)
21. Golmankhaneh, A.K.: Fractal Calculus and its Applications. World Scientific, Singapore (2022)
22. Golmankhaneh, A.K., Cattani, C.: Fractal logistic equation. Fractal Fract. **3**(3), 41 (2019)
23. Golmankhaneh, A.K., Fernandez, A.: Fractal calculus of functions on Cantor tartan spaces. Fractal Fract. **2**(4), 30 (2018)
24. Golmankhaneh, A.K., Fernandez, A., Golmankhaneh, A.K., Baleanu, D.: Diffusion on middle-ξ Cantor sets. Entropy **20**(7), 504 (2018)
25. Golmankhaneh, A.K., Jørgensen, P.E., Schlichtinger, A.M.: Einstein field equations extended to fractal manifolds: a fractal perspective. J. Geom. Phys. **196**, 105081 (2023)
26. Golmankhaneh, A.K., Nia, S.M.: Laplace equations on the fractal cubes and Casimir effect. Eur. Phys. J. Special Topics **230**(21), 3895–3900 (2021)
27. Golmankhaneh, A.K., Tejado, I., Şevli, H., Valdés, J.E.N.: On initial value problems of fractal delay equations. Appl. Math. Comput. **449**, 127980 (2023)
28. Golmankhaneh, A.K., Tunç, C., Şevli, H.: Hyers-Ulam stability on local fractal calculus and radioactive decay. Eur. Phys. J. Special Topics **230**(21), 3889–3894 (2021)
29. Golmankhaneh, A.K., Wanliss, J.: Expansion of the universe on fractal time: a study on the dynamics of cosmic growth. Int. J. Mod. Phys. A **39**(13), 2450059 (2024)
30. Golmankhaneh, A.K., Welch, K.: Equilibrium and non-equilibrium statistical mechanics with generalized fractal derivatives: a review. Mod. Phys. Lett. A **36**(14), 2140002 (2021)
31. Golmankhaneh, A.K., Welch, K., Serpa, C., Jørgensen, P.E.: Non-standard analysis for fractal calculus. J. Anal. **31**, 1895–1916 (2023)
32. Golmankhaneh, A.K., Welch, K., Serpa, C., Stamova, I.: Stochastic processes and mean square calculus on fractal curves. Random Oper. Stoch. Equ. **32**(3), 211–222 (2024)
33. Golmankhaneh, K.A., Welch, K., Serpa, C., Jørgensen, P.E.: Fractal Mellin transform and non-local derivatives. Georgian Math. J. **31**(3), 423–436 (2024)
34. Golmankhaneh, K.A., Welch, K., Serpa, C., Rodríguez-López, R.: Fractal Laplace transform: analyzing fractal curves. J. Anal. **32**(2), 1111–1137 (2024)
35. Gowrisankar, A., Golmankhaneh, A.K., Serpa, C.: Fractal calculus on fractal interpolation functions. Fractal Fract. **5**(4), 157 (2021)

36. He, J.H.: Homotopy perturbation technique. Comput. Methods Appl. Mech. Eng. **178**(3–4), 257–262 (1999)
37. He, J.H.: A coupling method of a homotopy technique and a perturbation technique for non-linear problems. Int. J. Non-linear Mech. **35**(1), 37–43 (2000)
38. He, J.H.: Homotopy perturbation method: a new nonlinear analytical technique. Appl. Math. Comput. **135**(1), 73–79 (2003)
39. He, J.H.: Comparison of homotopy perturbation method and homotopy analysis method. Appl. Math. Comput. **156**(2), 527–539 (2004)
40. He, J.H.: The homotopy perturbation method for nonlinear oscillators with discontinuities. Appl. Math. Comput. **151**(1), 287–292 (2004)
41. Jiang, H., Su, W.: Some fundamental results of calculus on fractal sets. Commun. Nonlinear Sci. Numer. Simul. **3**(1), 22–26 (1998)
42. Jorgensen, P.E.: Analysis and Probability: Wavelets, Signals, Fractals, vol. 234. Springer Science & Business Media, New York (2006)
43. Juraev, D.A., Mammadzada, N.M.: Fractals and its applications. Karshi Multidis. Int. Sci. J. **1**(1), 27–40 (2024)
44. Khalili Golmankhaneh, A., Tunc, C.: Analogues to lie method and Noether's theorem in fractal calculus. Fractal Fract. **3**(2), 25 (2019)
45. Kigami, J.: Analysis on Fractals, vol. 143. Cambridge University Press, Cambridge (2001)
46. Mandelbrot, B.B.: The Fractal Geometry of Nature. WH freeman, New York (1982)
47. Massopust, P.R.: Fractal Functions, Fractal Surfaces, and Wavelets. Academic Press, Massachusetts (2017)
48. Nottale, L.: Fractal Space-Time and Microphysics: Towards a Theory of Scale Relativity. World Scientific, Singapore (1993)
49. Parvate, A., Gangal, A.: Calculus on fractal subsets of real line-II: conjugacy with ordinary calculus. Fractals **19**(03), 271–290 (2011)
50. Parvate, A., Gangal, A.D.: Calculus on fractal subsets of real line-I: formulation. Fractals **17**(01), 53–81 (2009)
51. Pietronero, L., Tosatti, E.: Fractals in Physics. Elsevier, Amsterdam (1986)
52. Satin, S.E., Parvate, A., Gangal, A.: Fokker-Planck equation on fractal curves. Chaos Solit. Fractals **52**, 30–35 (2013)
53. Scott, A.C.: A nonlinear Klein-Gordon equation. Am. J. Phys. **37**(1), 52–61 (1969)
54. Shlesinger, M.F.: Fractal time in condensed matter. Annu. Rev. Phys. Chem. **39**(1), 269–290 (1988)
55. Stillinger, F.H.: Axiomatic basis for spaces with noninteger dimension. J. Math. Phys. **18**(6), 1224–1234 (1977)
56. Strauss, W., Vazquez, L.: Numerical solution of a nonlinear Klein-Gordon equation. J. Comput. Phys. **28**(2), 271–278 (1978)
57. Tarasov, V.E.: Fractional Dynamics. Springer, Berlin Heidelberg, New York (2010)
58. Vrobel, S.: Fractal Time. World Scientific, Singapore (jan (2011)
59. Welch, K.: A Fractal Topology of Time: Deepening into Timelessness. Fox Finding Press, Austin (2020)
60. Withers, W.: Fundamental theorems of calculus for Hausdorff measures on the real line. J. Math. Anal. Appl. **129**(2), 581–595 (1988)

Survey on Different Near Linear Topological Spaces

Gabriela Apreutesei and Anca Croitoru

Abstract In this paper we present some relationships among different types of near linear spaces, such as: spaces with non-negative scalars, weak linear spaces, almost linear spaces, quasilinear spaces, semilinear spaces and conlinear spaces. Some examples, properties and results on their topological structures are also established.

Keywords Space with non-negative scalars · Weak linear space · Almost linear space · Quasilinear space · Semilinear space · Conlinear space

1 Introduction

2 Introduction and Historical Considerations

Let E be a linear space over the scalar field $\mathbb{R}$ and $\mathcal{P}_0(E)$ the family of non-void parts of E. If we consider a family of sets $\mathcal{A} \subseteq \mathcal{P}_0(E)$, $\mathcal{A} \neq \emptyset$, and $\mathcal{A}$ is chosen appropriately, then the operations of addition and multiplication by scalars on E can be transferred in specific operations on $\mathcal{A}$:

"$+$":$\mathcal{A} \times \mathcal{A} \to \mathcal{A}$, $A + B = \{a + b; a \in A, b \in B\}$,

"$\cdot$":$\mathbb{R} \times \mathcal{A} \to \mathcal{A}$, $\lambda \cdot A = \{\lambda a; a \in A\}$.

But the linear space structure of E is not transferred entirely to $\mathcal{A}$. It is easily shown that $(\mathcal{A}, +)$ is not a group because the elements $A \in \mathcal{A}$ do not admit, generally, an opposite element.

Depending on the concrete choice of the family $\mathcal{A}$, the multiplication by scalars on $\mathcal{A}$ verifies only some of the axioms of multiplication by scalars on a linear space, but also other specific properties. Exactly these particular properties led to the definition of different "near linear spaces".

G. Apreutesei (✉) · A. Croitoru
"Alexandru Ioan Cuza" University of Iasi, Iasi, Romania
e-mail: gapreutesei@yahoo.com

A. Croitoru
e-mail: croitoru@uaic.ro

H. Kalita et al. (eds.), *Nonlinear Analysis and Computational Techniques*,
Springer Proceedings in Mathematics & Statistics 501,
https://doi.org/10.1007/978-981-96-6038-4_6

There are also other ways to define addition and multiplication by scalars for a family of sets $\mathcal{A}$. For example, if $\mathcal{A}$ is the family of closed sets of a topological linear space E, the sum in the Minkowski sense is used so that the addition is an internal operation:

$A \bigoplus B = \overline{\{a + b; a \in A, b \in B\}}$, where by $\overline{M}$ we denote the closure of the set $M \subseteq E$ in the topology of E.

Of course, these notions were formalized and usually studied beyond the context from which they appeared.

Also, functions similar to metrics can be defined on $\mathcal{A} \times \mathcal{A}$. These metrics are conceived to be compatible with "near linear" structures. A reference example for such metric is the Pompeiu-Hausdorff metric, whose basic properties can be found in almost all the presented examples:

Let (E, d) be a metric space. For a non-void set $A \subseteq E$, $x \in E$ and $\varepsilon > 0$, denote $d(x, A) = inf\{d(x, a); a \in A\}$ and $S_\varepsilon(A) = \{x \in E; d(x, A) < \varepsilon\}$. On $\mathcal{A} \subseteq \mathcal{P}_0(E)$ it can be defined the extended-valued semi-metric H_d (Pompeiu-Hausdorff semi-metric), namely

$H_d : \mathcal{A} \times \mathcal{A} \to [0, +\infty]$, $H_d(A, B) = \sup\{| d(x, A) - d(x, B) |; x \in E\}$.

Or, equivalently, $H_d(A, B) = \sup\{e(A, B), e(B, A)\}$, where $e(A, B) = \sup\{d(a, B); a \in A\}$ is the *Hausdorff excess of* A *with respect to* B.

Another expression for H_d is $H_d(A, B) = \inf\{\varepsilon \geq 0; A \subseteq S_\varepsilon(B), B \subseteq S_\varepsilon(A)\}$.

If $\mathcal{A} = \mathcal{C}l_0(E)$ (respectively $\mathcal{A} = \mathcal{P}b(E)$) denotes the family of non-empty closed (respectively the family of non-empty bounded) subsets of E, then the restriction of H_d to $\mathcal{C}l_0(E) \times \mathcal{C}l_0(E)$ (respectively to $\mathcal{P}b(E) \times \mathcal{P}b(E)$) is an extended-valued metric (respectively a real valued function).

An important result is Radström's Theorem that was extended by Labuschagne et al. [34] to embeddings of near vector lattices and then applied to different hyperspaces, such as $\mathcal{P}cwk\ (E)$, the family of non-empty convex weakly compact subsets of E, and $\mathcal{P}ck\ (E)$, the family of non-empty convex compact subsets of E, where E is a Banach space.

In some cases (for example when E is a normed linear space) applications can be defined on $\mathcal{A}$ that verify only a part of the axioms of the norm and/or some additional axioms. Thus, these "near metrics" and "near norms" induce topologies on $\mathcal{A}$ that approach topologies of linear normed spaces or, more generally, linear topological spaces.

In this paper we aim to present some of these concepts, specifying many times the context in which appeared. We also expose some significant examples. Of course, this presentation is not an exhaustive one.

Many of the different types of these "near linear spaces" have been introduced by several authors independently under different names. Also, some names are common for different types of "near linear spaces". We present them in the chronological order of their appearance. But we preferred to keep those names that were most used later in the literature.

For example, Borş introduced in 1969 [9] a generalization of linear spaces which he called "quasivector spaces". The name "quasilinear spaces" appeared later in 1985 in Aseev [7], but also in Markov [37], although in different meanings compared to Borş, but also different from each other.

Aseev added to the axioms of linear space an order relation compatible with it and an axiom concerning multiplication by the scalar 0. This space is distinguished from the space with non-negative real scalars, introduced by Climescu in 1965 [17].

Later Markov calls "quasilinear spaces" various generalizations of linear spaces, including the spaces introduced in 1970 by Octav Mayer [39], which we call as "almost linear spaces". Godini also works with these spaces [29–32].

In 1998 Apreutesei introduced in [1] a new notion of "near linear space" called "almost linear space", but which does not coincide with that of Mayer and Godini. Later, for the same definition, the author preferred the name "semilinear space" because it coincides with the notion introduced by Hamel in 2005 [33]. Löhne, in turn, used the name "semilinear space" for a similar structure, but in which the scalars are non-negative ([36], 2003). These spaces also appear in [33] , but called "conlinear spaces".

Most of these algebraic structures were endowed with compatible topologies either by means of metrics (Mayer in [39], Aseev in [7]) or order (pseudo or pre) metrics (Hamel in [33]), either by some "norms" (Godini [29–32], Apreutesei [1], Borş and Croitoru [12, 13, 19]), or by requirements similar to linear topologies (Apreutesei [2]).

The introduction of near linear spaces was motivated by various applications in optimization, functional analysis, interval analysis, computer science (e.g. [3–6, 8, 11–16, 18, 20, 22–26, 37]).

The structure of the paper is the following: Section 1 is dedicated to the introduction and historical notes. Section 2 contains some preliminaries concerning different axioms on the operations of sum and multiplication by scalars of a linear space and on a partial order relation. In Sect. 3, we present definitions, examples, properties and topological considerations of different near linear spaces, such as: spaces with non-negative scalars, weak linear spaces, almost linear spaces, quasilinear spaces, semilinear spaces and conlinear spaces. Some relationships among these spaces are established in Sect. 4 and in the end a Conclusion Sect. 5 follows.

3 Preliminaries

In the sequel we denote $\mathbb{R}_+ = [0, +\infty)$, $\mathbb{R}_- = (-\infty, 0]$, $\mathbb{N} = \{0, 1, 2, \ldots\}$ and $\mathbb{N}^* = \mathbb{N}\backslash\{0\}$.

Now we present some classical notions related to linear spaces, metric spaces and normed linear spaces.

Definition 1 Consider a non-void set X and "+" : $X \times X \to X$ an operation which verifies the following axioms:

$(G1)$ $(x + y) + z = x + (y + z)$ for all $x, y, z \in X$ (*associativity*).

$(G2)$ $x + y = y + x$ for all $x, y \in X$ (*commutativity*).

$(G3)$ There exists $\theta \in X$ (named *neutral element*) such that $x + \theta = x$, for all $x \in X$.

Then $(X, +)$ is called a *commutative monoid.*

Definition 2 Consider the following axioms:

$(G4)$ For any $x \in X$ there exists $-x \in X$ (named the *opposite element* of x) such that $x + (-x) = \theta$.

$\left(G4'\right)$ If $x, y, z \in X$ the equality $x + y = x + z$ implies $y = z$ (*the cancellation law*).

The monoid $(X, +)$ is called a *commutative group* iff the property $(G4)$ or, equivalent, $\left(G4'\right)$ is fulfilled.

Definition 3 Let $\mathbb{R}$ be a field of scalars and the operation "$\cdot$" : $\mathbb{R} \times X \to X$ such that

$(L1)$ $\alpha \cdot (\beta \cdot x) = (\alpha\beta) \cdot x$ for all $\alpha, \beta \in \mathbb{R}$, $x \in X$.

$(L2)$ There exists $1 \in X$ such that $1 \cdot x = x$ for all $x \in X$.

$(L3)$ $\alpha \cdot (x + y) = \alpha \cdot x + \alpha \cdot y$ for all $\alpha \in \mathbb{R}$, $x, y \in X$ (*first distributivity*).

$(L4)$ $(\alpha + \beta) \cdot x = \alpha \cdot x + \beta \cdot x$ for all $\alpha, \beta \in \mathbb{R}$, $x \in X$ (*second distributivity*).

A commutative group $(X, +)$ which satisfies also the axioms $(L1) - (L4)$ is called a *linear space* (abbreviated l.s.).

(In order not to load the writing, we noted with "+" both the addition from X and the one from the field of scalars; the choice comes from the context.)

A useful property refers to the behavior of multiplying the elements of the space X by the scalar 0 from $\mathbb{R}$:

$(L5)$ $0 \cdot x = \theta$ for all $x \in X$.

Definition 4 Let X be a non-void set. A relation "$\leq$" on X is called a *partial order relation* iff the following conditions are met:

$(O1)$ $x \leq x$ for all $x \in X$.

$(O2)$ For all $x, y, z \in X$ such that $x \leq y$ and $y \leq z$ it follows $x \leq z$.

$(O3)$ For all $x, y \in X$ such that $x \leq y$ and $y \leq x$ it follows $x = y$.

Definition 5 By a *partially ordered group* we mean a group $(X, +)$ endowed with a partially order relation "$\leq$" satisfying:

$(OG1)$ $(x \leq y) \Rightarrow (x + z \leq y + z)$ for all $x, y, z \in X$.

$(OG2)$ $(x \leq y) \Rightarrow (z + x \leq z + y)$ for all $x, y, z \in X$.

Definition 6 One says that a partially ordered group $(X, +, \leq)$ satisfies the *Riesz property* iff

(R) For all $x_1, x_2, y_1, y_2 \in X$ with $x_k \geq \theta$, $y_k \geq \theta$, $k = 1, 2$, such that $x_1 + x_2 = y_1 + y_2$ there exist $a, b, c, d \in X$, $a, b, c, d \geq \theta$ with $x_1 = a + b$, $x_2 = c + d$, $y_1 = a + c$, $y_2 = b + d$.

Definition 7 By an *ordered vector space* we mean a real linear space X endowed with a partially order relation "$\leq$" satisfying:

$(OV1)$ $(x \leq y) \Rightarrow (x + z \leq y + z)$ for all $x, y, z \in X$;

$(OV2)$ $(x \leq y) \Rightarrow (\alpha x \leq \alpha y)$ for all $\alpha \in \mathbb{R}_+$, $x, y \in X$.

For ordered vector spaces the following result is known [21, 27, 28]:

Theorem 1 *Let X be an ordered vector space. Then the following conditions are equivalent:*

(i) $(X, +, \leq)$ has the Riesz property.

(ii) For any $x, y_1, y_2 \in X$, $x, y_1, y_2 \geq \theta$ which satisfy the inequality $x \leq y_1 + y_2$, there exist $x_1, x_2 \in X$ such that $\theta \leq x_1 \leq y_1$, $\theta \leq x_1 \leq y_2$ and $x = x_1 + x_2$.

(iii) X has the finite interpolation property: for any $x_1, x_2, y_1, y_2 \in X$ with $x_i \leq y_j$ for $i, j \in \{1, 2\}$, there exists $z \in X$ such that $x_i \leq z + y_j$ for all $i, j \in \{1, 2\}$.

Now we recall the definitions of a metric and a norm.

Definition 8 Consider X a non-void set. A function $d : X \times X \to \mathbb{R}_+$ is called a *metric* on X iff:

$(M1)$ $d(x, y) = 0$ if and only if $x = y$;

$(M2)$ $d(x, y) = d(y, x)$ for all $x, y \in X$;

$(M3)$ $d(x, y) \leq d(x, z) + d(z, y)$ for all $x, y, z \in X$.

The couple (X, d) is called a metric space.

Definition 9 If d verifies $(M2)$, $(M3)$ and

$\left(M1'\right)$ $d(x, x) = 0$ for all $x \in X$,

d is said to be a *premetric*.

Definition 10 If X is a l.s. over $\mathbb{R}$, an application $\|\cdot\| : X \to \mathbb{R}_+$ is called a *norm* on X iff:

$(N1)$ $\|x\| = 0$ if and only if $x = \theta$;

$(N2)$ $\|\lambda \cdot x\| = |\lambda| \, \|x\|$ for all $\lambda \in \mathbb{R}$, $x \in X$;

$(N3)$ $\|x + y\| \leq \|x\| + \|y\|$ for all $x, y \in X$.

The pair $(X, \|\cdot\|)$ is called a *normed linear space* (abbreviated, n.l.s.).

If X is an ordered vector space and $\|\cdot\| : X \to \mathbb{R}_+$ is an application, we also consider the following axiom:

$(N4)$ For all $x, y \in X$ the inequality $x \leq y$ implies $\|x\| \leq \|y\|$.

As we will see in the definitions of the next section, weak linear spaces, almost linear spaces, semilinear spaces and conlinear spaces are endowed with only two operations (addition and multiplication by scalars). Spaces with non-negative scalars and quasilinear spaces consist of two operations (addition and multiplication by scalars) and a partial order relation.

4 Near Linear Topological Spaces

4.1 Spaces with Non-negative Scalars

The first notion of a near linear space was introduced by Climescu [17] in 1965 extracted from the study of several examples.

Definition 11 [Climescu [17], 1965] Let $S \subseteq \mathbb{R}_+$ be a nonempty set which is closed under addition and multiplication of $\mathbb{R}_+$. A nonempty set X is called a space with non-negative scalars of S if it is endowed with two operations "+": $X \times X \to X$, "·": $S \times X \to X$ and a relation "≤" that satisfy the following properties:

(i) $(G1)$, $(G2)$, $(G3)$ (that is $(X, +)$ is a commutative monoid);

(ii) $(O1)$, $(O2)$, $(O3)$ (that is "≤" is a partial ordered relation on X);

(iii) (R) (Riesz property);

$(P1)$ For all $x_1, x_2, y_1, y_2 \in X$ such that $x_1 \le y_1$, $x_2 \le y_2$ it results $x_1 + x_2 \le y_1 + y_2$.

$(P2)$ $(\alpha + \beta) \cdot x \le \alpha \cdot x + \beta \cdot x$ for all $\alpha, \beta \in S, x \in X$.

$(P3)$ $\alpha \cdot (x + y) \le \alpha \cdot x + \alpha \cdot y$ for all $\alpha \in S, x, y \in X$.

X will also be denoted by (X, S).

S could be:

- $[a, +\infty)$ or $(a, +\infty)$, where $a = 0$ or $a \ge 1$;
- $\mathbb{N}_j$, $j \in \mathbb{N}$, where $\mathbb{N}_j = \{j, j+1, j+2, ...\}$;
- $\mathbb{Q}_+$ or $\mathbb{Q}_+^* = \mathbb{Q}_+ \setminus \{0\}$.

Example 1 ([17]) I. Consider $X = \left\{(a, b) \in \mathbb{R}^2; 0 \le a \le b\right\}$ and $S = \mathbb{R}_+$.

For all $\lambda \in S$, $x, y \in X$, $x = (a, b)$, $y = (c, d)$, define

(i) $x + y = (a + c, b + d)$;

(ii) $\lambda \cdot x = (\lambda a, \lambda b)$;

(iii) $x \le y$ iff $a \le c$ and $b \le d$.

Then (X, S) is a space with non-negative scalars of $\mathbb{R}_+$.

II. Denote $X = \{f; f : [a, b] \to \mathbb{R}, f$ continuous and increasing$\}$ with usual addition and multiplication by scalars from $S = [0, +\infty)$. The order relation is: $f \le g \Leftrightarrow \{f(x) \le g(x)$, for all $x \in [a, b]\}$. Then (X, S) is a space with non-negative scalars with neutral element θ, where $\theta(x) = 0$ for all $x \in [a, b]$.

The notion of space with non-negative scalars appeared in an algebraic context. In [3] we defined and studied a norm on a space with non-negative scalars.

Definition 12 Let (X, S) be a space with non-negative scalars (of S). An application $\|\cdot\|_p : X \to \mathbb{R}_+$ is called a norm with non-negative scalars iff it satisfies the axioms $(N1)$, $(N3)$, $(N4)$ and

$(PN2)$ $\|\alpha \cdot x\|_p \le \alpha \|x\|_p$ for all $\alpha \in S, x \in X$.

Example 2 Let (X, S) be the space from Example 13-I. The following applications are norms with non-negative scalars:

I. $\|x\|_p = \sqrt{a^2 + b^2}$ for all $x = (a, b) \in X$;

II. $\|x\|_p = b$, for all $x = (a, b) \in X$.

4.2 Weak Linear Spaces

In 1969, Borş [9] defined weak linear spaces, initially named quasivector spaces. After Aseev [7] introduced in 1985 the concept of quasilinear spaces, other authors have studied these spaces and the prefix quasi was consecrated in the literature. So in [13] quasivector spaces were renamed weak linear spaces. They were studied for instance in [10–13].

Borş arrived at this definition by studying various algebraic structures on the Cartesian product between an algebra $\mathcal{A}$ of parts of a nonempty set and a linear space (respectively group, ring, Hilbert space) X such that the addition on $\mathcal{A} \times X$ admits a symmetric element for each $(A, x) \in \mathcal{A} \times X$ (see Example 4 below).

Definition 13 [Borş [9], 1969] Let $(\Gamma, +, \cdot)$ be a field with distinct 0 and 1. A non-empty set X is called a weak linear space over Γ (shortly Γ-w.l.s.) if it is endowed with two mappings "+":$X \times X \to X$ and "·":$\Gamma \times X \to X$, satisfying:

(i) $(G1)$, $(G2)$, $(G3)$, $(G4)$ (that is $(X, +)$ is a commutative group);

(ii) $(L1)$, $(L2)$, $(L3)$;

(W) $(\alpha + \beta + \gamma)x = \alpha x + \beta x + \gamma x$ for all $\alpha, \beta, \gamma \in \Gamma$, $x \in X$.

If $\Gamma = \mathbb{R}$, then X is called a real weak linear spaces (shortly, w.l.s.).

Definition 14 Let X be a Γ-w.l.s. A non-empty subset $Y \subseteq X$ is called a Γ-weak linear subspace of X iff Y is a Γ-w.l.s. with respect to the restrictions on Y of the operations of X.

Remark 1 [9] Consider X be a Γ-w.l.s.

(i) There exists $x \in X$ such that $0 \cdot x \neq \theta$. For instance, in Example 4, consider a non-empty set $A \in \mathcal{A}$. Then $0 \cdot (A, \theta) = (A, \theta) \neq e = (\varnothing, \theta)$.

(ii) $0 \cdot x + 0 \cdot x = \theta$ and $\alpha \cdot \theta = \theta$ for all $\alpha \in \Gamma$.

(iii) There exist $\alpha, \beta \in \Gamma$, $x \in X$ such that $(\alpha + \beta) \cdot x \neq \alpha \cdot x + \beta \cdot x$.

(iv) $(\alpha \cdot x = \theta) \Rightarrow (x = \theta)$ for all $\alpha \in \Gamma \backslash \{0\}$, $x \in X$.

(v) Condition (W) is equivalent with (W'): $(\alpha + \beta) \cdot x = \alpha \cdot x + \beta \cdot x + 0 \cdot x$ for all $\alpha, \beta \in \Gamma$, $x \in X$.

(vi) The relation from (v) can be generalized as follows:

For all $\alpha_1, \ldots, \alpha_n \in \Gamma$ (where $n \in N^*$), $x \in X$, the following formula holds:

$$(\alpha_1 + \ldots + \ldots \alpha_n) \cdot x = \begin{cases} \alpha_1 \cdot x + \ldots + \alpha_n \cdot x, & n \text{ odd} \\ \alpha_1 \cdot x + \ldots + \alpha_n \cdot x + 0 \cdot x, & n \text{ even} \end{cases}.$$

(vii) Denote $X^0 = \{0 \cdot x | x \in X\}$, $X_0 = \{x + 0 \cdot x | x \in X\}$. Then X_0 is a linear subspace of X, X^0 is a weak linear subspace of X and $X_0 \cap X^0 = \{\theta\}$.

Denote $-x = (-1) \cdot x$ for all $x \in X$.

In the sequel we define a weak norm, a distance and a semi-metric on w.l.s., offering the possibility of a topological study of these spaces.

This weak norm has weaker properties than those of a norm, but heaving a good behavior relative to θ, to $-x$ and to addition:

Definition 15 [13] An application on a w.l.s. X, $\|\cdot\|_w : X \to \mathbb{R}_+$, is called a weak norm iff:

$(N1)$ and $(N3)$ hold and

$(WN2)$ $\|-x\|_w = \|x\|_w$ for all $x \in X$.

Basically, the second property of norm is verified for $\lambda = -1$.

Definition 16 The pair $(X, \|\cdot\|_w)$ is named *normed weak linear space (abbreviated n.w.l.s.)*.

If $(X, \|\cdot\|_w)$ is a n.w.l.s. then $d : X \times X \to \mathbb{R}_+$, $d(x, y) = \|x - y\|_w$ is a metric (called the metric induced by the weak norm). This metric verifies the following properties of compatibility with the weak structure:

(i) $|d(x, \theta) - d(y, \theta)| \leq d(x, y)$ for all $x, y \in X$;

(ii) $d(x + z, y + z) = d(x, y)$ for all $x, y, z \in X$.

Example 3 Consider $k \in \mathbb{N}^*$ and n.l.s. $\left(X_j, \|\cdot\|_j\right)$, $j \in \{1, 2, ..., k\}$. If $X = X_1 \times X_2 \times .. \times X_k$ we define

$\|\cdot\|_w : X \to \mathbb{R}$,

$$\|(x_1, x_2, ..., x_k)\|_w = \sum_{j=1}^{k} \frac{\|x_j\|_j}{1 + \|x_j\|_j}.$$

Then $(X, \|\cdot\|_w)$ is a n.w.l.s..

Example 4 ([9]) Let T be a nonempty set and $\mathcal{P}(T)$ the family of all subsets of T. Consider $\mathcal{A} \subseteq \mathcal{P}(T)$ an algebra of subsets of T (i.e. for every $A, B \in \mathcal{A}$ we have $A \cup B \in \mathcal{A}$, $A \backslash B \in \mathcal{A}$ and $T \in \mathcal{A}$). If X is a w.l.s. (on the scalar field Γ) with neutral element θ, then $U = \mathcal{A} \times X$ is a w.l.s. with respect to operations:

$(A, x) + (B, y) = ((A \backslash B) \cup (B \backslash A), x + y)$;

$\alpha \cdot (A, x) = (A, \alpha \cdot x)$ for all $\alpha \in \Gamma$, $A \in \mathcal{A}$, $x \in X$.

The neutral element with respect to sum is $e = (\varnothing, \theta)$ and the opposite of $u = (A, x) \in U$ is $-u = (A, -x)$.

In the particular case: T is a finite set, $\mathcal{A} = \mathcal{P}(T)$ and $(X, \|\cdot\|_w)$ is a n.l.s. over $\mathbb{R}$, then $\|\cdot\|_w : U \to \mathbb{R}$, $\|(A, x)\|_w = card\ A + \|x\|$ is a weak norm on U, where *card* A denote the cardinality of A.

In the sequel we define another metric on a w.l.s. via positively homogeneous operators.

Definition 17 Let X be an arbitrary non-empty set.

(i) Denote by $F(X, \mathbb{R})$ the space of all real functions defined on X. A non-empty subset $E \subseteq F(X, \mathbb{R})$ is called total over X iff for every $x, y \in X$ and every $h \in E$, it holds:

$h(x) = h(y) \Longrightarrow x = y$.

(ii) If X is a w.l.s., a function $h \in F(X, \mathbb{R})$ is called positively homogeneous (shortly, p.h.) iff for every $x \in X$ and $\alpha \in \mathbb{R}_+$, $h(\alpha \cdot x) = \alpha h(x)$.

(iii) Suppose $(X, \|\cdot\|_w)$ is a n.w.l.s.. A function $h : X \to \mathbb{R}$ is called $\|\cdot\|_w$-bounded iff there exists $\alpha \in \mathbb{R}_+$ such that $|h(x)| \leq \alpha \|x\|_w$ for all $x \in X$.

For a $\|\cdot\|_w$-bounded function $h : X \to \mathbb{R}$ denote
$l(h) = \inf\{\alpha \in \mathbb{R}_+; |h(x)| \le \alpha \|x\|_w \text{ for all } x \in X\}$.
Consider

$$HB(X, \mathbb{R}) = \{h \in F(X, \mathbb{R}); h \text{ is p.h., } \|\cdot\|_w\text{-bounded and } l(h) < +\infty\}$$

and $S = \{h \in HB(X, \mathbb{R}); l(h) \le 1\}$.

Remark 2 We observe that for every $\|\cdot\|_w$-bounded function $h : X \to \mathbb{R}$ and every $\beta \in \mathbb{R}_+$, $l(\beta h) = \beta l(h)$.

Theorem 2 *Let $(X, \|\cdot\|_w)$ be a n.w.l.s., d be the distance induced by $\|\cdot\|_w$ and the application $s : X \times X \to \mathbb{R}_+$, given by $s(x, y) = \sup_{h \in S} |h(x) - h(y)|$. Then:*
(i) s is a semi-metric on X.
(ii) If $HB(X, \mathbb{R})$ is total over X, then s is a distance on X.
(iii) If we replace $HB(X, \mathbb{R})$ by the space
$LB(X, \mathbb{R}) = \{h \in F(X, \mathbb{R}); h \text{ is linear, } \|\cdot\|_w\text{-bounded and } l(h) < +\infty\}$,
then $s \le d$.

Proof (i) It results by the definition of s.
(ii) We prove that for every $x, y \in X$, $s(x, y) = 0 \Longrightarrow x = y$. If $s(x, y) = 0$ then $h(x) = h(y)$ for $h \in S$. Let g be in $HB(X, \mathbb{R})$ such that $l(g) > 1$. Consider $h = g/l(g)$. It results $l(h) = 1$ and $h \in S$, which implies $h(x) = h(y)$. It follows $g(x) = g(y)$. So $h(x) = h(y)$ for every $h \in HB(X, \mathbb{R})$. But $HB(X, \mathbb{R})$ is total over X, which leads to $x = y$.
(iii) Consider $h \in S$ and $x, y \in X$. Now, the following inequalities hold:
$|h(x) - h(y)| = |h(x - y)| \le l(h) \|x - y\|_w \le \|x - y\|_w = d(x, y)$,
which implies $s(x, y) \le d(x, y)$.

4.3 Almost Linear Spaces

In 1970, Mayer [39] defined an abstract space of the "near linear space" type, called "quasilinear space", but which is different from the one defined by Aseev [7]. Later this space received the name "almost linear space."

The purpose of this type of space was to examine rounding errors in numerical calculation using interval analysis. He also introduced some metrics compatible with almost linear structure of this space to deduce convergence tests.

Definition 18 [Mayer [39], 1970] A nonempty set X is called an almost linear space (shortly, a.l.s.) iff it is endowed with two mappings "+":$X \times X \to X$ and "$\cdot$":$\mathbb{R} \times X \to X$ satisfying:
(i) $(M1)$, $(M2)$, $(M3)$ (so $(X, +)$ is a commutative monoid);
(ii) $(L1)$, $(L2)$, $(L3)$, $(L5)$;
(iii) $(AL4)$: $(\alpha + \beta) \cdot x = \alpha \cdot x + \beta \cdot x$ for all $\alpha, \beta \in \mathbb{R}_+$, $x \in X$.

Axiom $(AL4)$ is a restriction of $(L4)$ for non-negative real scalars. In fact, $(AL4)$ is true for real scalars of the same sign.

Definition 19 Let X be an a.l.s.. A non-empty subset $Y \subseteq X$ is called an almost linear subspace of X iff Y is an a.l.s. with respect to the restrictions on Y of the operations of X.

Other calculation rules in an a.l.s. are given below:

Remark 3 [30] Let X be an a.l.s..

(i) $\alpha \cdot \theta = \theta$ for all $\alpha \in \mathbb{R}$.

(ii) We denote $(-1) \cdot x$ by $-x$ and $x + (-y)$ by $x - y$. We observe that $x - x$ need not be equal to θ since an element of X does not have an inverse element. Denote $V_X = \{x \in X; x - x = \theta\}$ and $W_X = \{x \in X; x = -x\}$.

(iv) $(\alpha + \beta) \cdot x = \alpha \cdot x + \beta \cdot x$ for all $\alpha, \beta \in \mathbb{R}$, $x \in V_X$.

(v) V_X is a linear subspace of X, $W_X = \{x - x; x \in X\}$ and W_X is an almost linear subspace of X, and $V_X \cap W_X = \{\theta\}$.

(vi) The following conditions are equivalent:

(a) X is a linear space.

(b) $V_X = X$.

(c) $W_X = \{\theta\}$.

Example 5 ([40]) Let $(E, \leq)$ be an ordered linear space (with neutral element θ). An interval from E is a set of form $X = [\underline{x}, \overline{x}] = \{a \in E; \underline{x} \leq a \leq \overline{x}\}$, where $\underline{x}, \overline{x} \in E$.

Let $\mathcal{I}$ be the family of all the intervals with elements from E. Then $(\mathcal{I}, +, \cdot)$ is an a.l.s. if we consider on $\mathcal{I}$ the usual addition and multiplication with scalars.

A compatible metric on $\mathcal{I}$ is defined as follows:

$d : \mathcal{I} \times \mathcal{I} \to \mathbb{R}_+, d(X, Y) = \max\{|\underline{x} - \underline{y}|, |\overline{x} - \overline{y}|\}$, where $X = [\underline{x}, \overline{x}]$ and $Y = [\underline{y}, \overline{y}]$.

The width of an interval X is $l(X) = \overline{x} - \underline{x}$.

This metric verifies:

1. $d(X, Y) \leq l(Y)$ for all $X, Y \in \mathcal{I}$, $X \subseteq Y$;
2. $d(X + Z, Y + Z) = d(X, Y)$ for all $X, Y, Z \in \mathcal{I}$.

The notion of a normed almost linear space was introduced by Godini [30] in 1985 and then studied for example in [29–32, 35].

The context that led Godini to study the concept of normed almost linear spaces was to obtain the natural framework for finding the points of best approximation on n.l.s..

More exactly, consider a real normed linear space $(E, \|\cdot\|)$, $\varnothing \neq G \subseteq E$ and let $\mathcal{A}$ be the family of all bounded, convex, nonempty subsets of E. The study of the set-values application P_G is considered, where $P_G : E \to \mathcal{A}$, $P_G(x) = \left\{g_0 \in G; \|x - g_0\| = \inf_{g \in G} \|x - g\|\right\}$.

In order to apply to this multifunction P_G the optimization results similar to those in the classical context of n.l.s., the author introduced a "near norm" on $\mathcal{A}$:

Definition 20 An almost norm on an a.l.s. X is a functional $\|\cdot\|_a : X \to \mathbb{R}_+$ having the properties $(N1)$, $(N2)$, $(N3)$ and

$(AN4)$ $\|x-y\|_a \le \|x-z\|_a + \|z-y\|_a$ for all $x, y, z \in X$.

Example 6 ([29]) Let $\mathcal{A}$ be the family of all bounded, convex, nonempty subsets of a real n.l.s. E. Then $X = (\mathcal{A}, +, \cdot)$ is an a.l.s. with neutral element $\theta = \{0\}$.

If we define $\|\cdot\|_a : \mathcal{A} \to \mathbb{R}$, $\|A\|_a = \sup\limits_{x\in A} \|x\|$, then $(\mathcal{A}, \|\cdot\|_a)$ is a n.a.l.s..

4.4 *Quasilinear Spaces*

Quasilinear spaces and normed quasilinear spaces were introduced by Aseev (see [7]) in 1985 and then studied for instance in [41, 42].

Definition 21 [Aseev [7], 1985] A nonempty set X is called a quasilinear space (shortly, q.l.s.) iff a relation "$\le$" and two operations "$+$":$X \times X \to X$, "$\cdot$":$\mathbb{R} \times X \to X$ are defined on it and satisfy the following:

(i) $(G1)$, $(G2)$, $(G3)$ with neutral element θ;

(ii) $(O1)$, $(O2)$, $(O3)$;

(iii) $(L1)$, $(L2)$, $(L3)$, $(L4)$, $(L5)$;

(iv) $(OV3)$ If $x, y, z, t \in X$ such that $x \le y$, $z \le t$ then $x + z \le y + t$;

(v) $(OV4)$ $(x \le y) \Rightarrow (\alpha x \le \alpha y)$ for all $\alpha \in \mathbb{R}$, $x, y \in X$.

Remark 4 [7] Let X be a q.l.s..

(i) An element $x' \in X$ is called an inverse of $x \in X$ if $x + x' = \theta$. If $x \in X$ has an inverse x', then x' is unique.

(ii) If every element $x \in X$ has an inverse $x' \in X$, then X is a real linear space.

Definition 22 Let X be a q.l.s.. An application $\|\cdot\|_q : X \to \mathbb{R}_+$ is called a quasilinear norm iff the following axioms hold:

$(QN1)$ $\|x\|_q > 0$ if $x \ne \theta$;

$(N2)$ and $(N3)$.

$(QN4)$ If $x, y \in X$ such that $x \le y$ then $\|x\|_q \le \|y\|_q$.

$(QN5)$ If for any $\varepsilon > 0$ there exists an element $x_\varepsilon \in X$ such that $x \le y + x_\varepsilon$ and $\|x_\varepsilon\|_q < \varepsilon$ then $x \le y$.

The pair $\left(X, \|\cdot\|_q\right)$ is called a quasilinear normed space (shortly, q.l.n.s.).

Definition 23 Let $\left(X, \|\cdot\|_q\right)$ be a q.l.n.s.. Let us define on X a generalization of the Pompeiu-Hausdorff metric (see Sect. 1) $d : X \times X \to \mathbb{R}_+$ by equality

$$d(x, y) = \inf\left\{\varepsilon \ge 0; x \le y + a_1^\varepsilon, y \le x + a_2^\varepsilon, a_j^\varepsilon \in X, \left\|a_j^\varepsilon\right\|_q \le \varepsilon, j = 1, 2\right\}.$$

We notice that

(i) $d(x, y) \le \|x - y\|_q$ for all $x, y \in X$;

(ii) $d(\alpha x, \alpha y) = |\alpha| \, d(x, y)$ for all $\alpha \in \mathbb{R}_+$ and $x, y \in X$;

(iii) $d(x + y, z + t) \le d(x, z) + d(y, t)$ for all $x, y, z, t \in X$;

(iv) $d(x, \theta) = \|x\|_q$ for all $x \in X$.

Example 7 Let X be a real linear normed space. Let us denote by $\mathcal{C}cb\,(X)$ space of non-empty convex closed bounded subsets of X. The algebraic operation sum is defined in Minkowski sense by expression $A \bigoplus B = \overline{\{a + b; a \in A, b \in B\}}$ and multiplication by a real number is the usual one. We will assume that the partial order on $\mathcal{C}cb\,(X)$ is given by inclusion. The norm on the space $\mathcal{C}cb\,(X)$ is define by equality $\|A\|_q = \sup\limits_{a \in A} \|a\|$. Then $\mathcal{C}cb\,(X)$ is a quasilinear normalized space. In this case, the Hausdorff metric is defined in the usual way.

Example 8 Let S be a compact topological space, X be a complete normed quasilinear space. Let us denote by $C(S, X)$ the space of all continuous mappings $f : S \to X$. We assume that $f_1 \leq f_2$ if $f_1(s) \leq f_2(s)$ for any $s \in S$. We define algebraic operations of sum and multiplication by the real numbers as follows:

$(f_1 + f_2)(s) = f_1(s) + f_2(s)$, $(\alpha f) = \alpha \cdot f(s)$.

We define the norm in the space $C(S, X)$ by the equality

$\|f\|_q = \max\limits_{s \in S} \|f(s)\|$.

Then $\left(C(S, X), \|\cdot\|_q\right)$ is a q.l.n.s..

4.5 Semilinear Spaces

Semilinear spaces (initially called almost linear spaces) were defined by Apreutesei ([1], 1998 and [2], 2002) and then studied for example in [3–6, 33].

The context in which this notion appears was that of the families $\mathcal{A}$ of non-void subsets of a n.l.s. To study the continuity of some applications between families $\mathcal{A}$, the author endowed $\mathcal{A}$ with usual multiplication with scalars and the (Minkowski) addition and obtained a "near topological" structure on $\mathcal{A}$, similar to that of a topological linear space.

Definition 24 [Apreutesei [1],1998 and [2], 2002] A nonempty set X is called a semilinear space (shortly, s.l.s.) if two operations "+":$X \times X \to X$, "·":$\mathbb{R} \times X \to X$ are defined and satisfy the following:

(i) $(G1)$, $(G2)$, $(G3)$ with neutral element θ;

(ii) $(L1)$, $(L2)$, $(L3)$, $(L5)$.

Remark 5 ([33]) Let $(X, +, \cdot)$ be a semilinear space. Then:

(i) If $\lambda \in \mathbb{R}$, then $\lambda \cdot \theta = \theta$.

(ii) If $\lambda \in \mathbb{R}\setminus\{0\}$, $x_1, x_2 \in X$ and $\lambda \cdot x_1 = \lambda \cdot x_2$, then $x_1 = x_2$.

(iii) If $x \in X\setminus\{\theta\}$ and $\lambda \in \mathbb{R}\setminus\{0\}$, then $\lambda \cdot x \neq \theta$.

Definition 25 (i) Let $(X, +\cdot)$ be a s.l.s.. A topology τ on X is called a semilinear topology if the operations "+" and "·" are continuous with respect to τ. The structure $(X, +, \cdot, \tau)$ is called semilinear topological space (s.l.t.s.).

(ii) If $(X, +, \cdot, \tau)$ is a s.l.t.s. and $\mathcal{V}$ is a fundamental system of τ-neighbourhood of the origin $\theta \in X$, then the family

$\mathcal{U}(x) = \{U \subseteq X$; there exists $V \in \mathcal{V}$ such that $x + V \subseteq U\}$

generates a new topology σ on X, called the translated topology of τ.

Remark 6 (i) The translated topology of a semilinear topology τ is coarser than τ.

(ii) There are translated topologies σ that are not semilinear.

(iii) The advantage of σ is given by an easier use of neighbourhoods of any element $x \in X$, by the form $x + V$, where $V \in \mathcal{V}(\theta)$, $\mathcal{V}(\theta)$ is a fundamental neighbourhoods system of the origin θ such that V is absorbent, balanced etc.

(iv) A lot of properties from topological linear spaces are adequately transposed to s.l.t.s. by σ, like T1 separation, metrizability, totally boundness and compactness (see [5]). But other properties, such as Cauchy property, completeness and others, behave differently [4].

Example 9 Consider a n.l.s. E. Now we recall that the Hausdorff topology on $\mathcal{P}b(E)$ is the supremum of two topologies: $\tau_H = \tau_H^- \vee \tau_H^+$ (the coarsest topology that contains both τ_H^- and τ_H^+), where a basic neighbourhoods of an arbitrary set $A_0 \in \mathcal{P}b(E)$ is respectively:

$U_-(A_0, \varepsilon) = \{A \in \mathcal{P}b(E)\,;\ A_0 \subset S_\varepsilon(A)\}$ in τ_H^-, and

$U_+(A_0, \varepsilon) = \{A \in \mathcal{P}b(E)\,;\ A \subset S_\varepsilon(A_0)\}$ in τ_H^+, for all $\varepsilon > 0$.

If we endowed $(\mathcal{P}b(E), +, \cdot)$ with some hyperspacial topologies, like Hausdorff topology τ_H (or lower Hausdorff τ_H^- or upper Hausdorff τ_H^+ topology), we obtain a s.l.t.s..

Consider $\|\cdot\|_s : \mathcal{P}b(E) \to \mathbb{R}_+$, $\|A\|_s = \sup\limits_{a \in A} \|a\|$. This function verifies the axioms of a norm (but it is not a norm because $(\mathcal{P}b(E), +, \cdot)$ is not a linear space). The convergence of a sequence $(A_n)_{n\in\mathbb{N}} \subseteq \mathcal{P}b(E)$ to A in $(\mathcal{P}b(E), \tau_H)$ implies the convergence of $(\|A_n\|_s)_{n\in\mathbb{N}}$ to $\|A\|_s$.

The same type of "near linear" space was introduced in 2005 by Hamel [33] who offered various generalizations of Ekeland's variational principle and equivalent formulations for functions with values in ordered groupes or in s.l.s. ([33], Corollaries 15–18).

4.6 *Conlinear Spaces*

These spaces were introduced by Löhne [36] in 2003 (named semilinear spaces) and then studied for instance in [33] under the name of *conlinear spaces*.

Definition 26 [Löhne [36], 2003] A non-empty set X is called a conlinear space (shortly, c.l.s.) iff it is endowed with two operations "+":$X \times X \to X$, "·":$R_+ \times X \to X$ satisfying the following:

(i) $(G1)$, $(G2)$, $(G3)$ with neutral element θ;

(ii) $(SL1)$: $\alpha \cdot (\beta \cdot x) = (\alpha\beta) \cdot x$ for all $\alpha, \beta \in \mathbb{R}_+$, $x \in X$;

(iii) $(L2)$, $(L5)$;

(iv) $(SL3)$: $\alpha \cdot (x + y) = \alpha \cdot x + \alpha \cdot y$ for all $\alpha \in \mathbb{R}_+$, $x, y \in X$;

(iv) $(AL4)$.

Remark 7 (i) $(SL3)$ is the restriction of $(L3)$ to non-negative scalars.
(ii) Suppose X is a c.l.s.. Then θ is unique and $\alpha \cdot \theta = \theta$ for all $\alpha \in \mathbb{R}_+$.

Example 10 [36] Consider $X = \mathbb{R} \cup \{+\infty\}$ together with the extended operations:

$$x + (+\infty) = (+\infty) + x = +\infty, \forall x \in \mathbb{R} \cup \{+\infty\},$$

$$\alpha \cdot (+\infty) = \begin{cases} +\infty, & \alpha > 0 \\ 0, & \alpha = 0. \end{cases}.$$

Then X is a c.l.s..

Example 11 If X is a c.l.s. with neutral element θ, then every convex cone $C \subseteq X$ with $\theta \in C$ is a c.l.s..

Remark 8 Using the properties of c.l.s. Löhne arrives interesting properties for convex multifunctions. For example, consider E a l.s., $C \subseteq E$ a convex set, V a separated topological linear space, $\mathcal{Cl}(V)$ the family of closed subsets of V, $f : C \to \mathcal{Cl}(V)$ a convex function. Denote by $core\,(A)$ the algebraic interior of the set $A \subseteq E$ and $dom\ f = \{x \in C; f(x) \neq \varnothing\}$. If there exists a point $x_0 \in core\,(dom\ f)$ such that $f(x_0)$ is compact then f is compact-valued in any point of $dom\ f$.

5 Considerations on Different Types of Near Linear Spaces

In this section some considerations, examples and relationships among the spaces presented in Sect. 3 are given.

Remark 9 I. Every s.l.s. is c.l.s. since s.l.s. involve all real scalars and c.l.s. use only non-negative scalars.

The spaces with non-negative scalars appeared as a result of some examples in which the multiplication by real scalars does not work. However, we can pose the theoretical problem of extending the multiplication by scalars of $S \subseteq \mathbb{R}_+$ to one by scalars of $S \subseteq \mathbb{R}$ so that the operation remains internal. In this hypothesis, the c.l.s. of Example 10 can not be extended to a s.l.s..

II. Every q.l.s. is s.l.s.. The converse is not true. For example, let E be a n.l.s.. Then $\mathcal{P}b(E)$ is s.l.s., but not q.l.s. since it does not verify $(L4)$.

Remark 10 There is no relationship between w.l.s. and a.l.s..

I. Let U be the w.l.s. as in Example 4.

The axiom $(L5)$ of a.l.s. is not accomplished and so U is not an a.l.s..

II. Let be $X = \mathbb{R}_+$ and the operations defined for every $x, y \in X$ and every $\alpha \in \mathbb{R}$ by

$$x + y = \max\{x, y\},$$

$$\alpha x = \begin{cases} x, & \alpha \neq 0 \\ 0, & \alpha = 0. \end{cases}$$

Then X is an a.l.s. with $\theta = 0$, but it is not a w.l.s. since $(G4)$ is false.

Remark 11 There is no relationship between w.l.s. and s.l.s..

I. Let U be the w.l.s. as in Example 4. Then U is not s.l.s. since $(L5)$ is not valid. Indeed, let be $A \in \mathcal{A}$, $A \neq \varnothing$ and $x \in X$. Then $0 \cdot (A, x) = (A, 0) \neq (\varnothing, 0) = \theta$.

II. Let X be a real linear space and $U = \mathcal{P}_0(X)$ the family of all nonempty subsets of X such that $card\ U \leq 2$. We consider U together with the usual set operations, where $card\ U$ means the cardinality of U.

Then U is a s.l.s. with $\theta = \{0\}$. Now, considering $A = \{x, y\}$ such that $x \neq y$, it results $A + B \neq \theta$ for every $B \in U$, so $(G4)$ is not satisfied. Consequently, U is not a w.l.s..

Remark 12 There is no relationship between w.l.s. and q.l.s..

I. Let T be a real normed space and $X = \mathcal{P}bc_0(T)$ the family of all non-empty closed bounded subsets of T endowed with the order relation given by the usual inclusion "$\subseteq$" and the operations defined for every A, $B \in X$ and every $\alpha \in \mathbb{R}$ by

$$A \bigoplus B = \overline{\{a + b; a \in A, b \in B\}}, \alpha A = \{\alpha a, a \in A\},$$

where $\overline{E}$ is the closure of $E \subseteq T$ relative to the topology induced by the norm of T.

Then X is q.l.s. that is not a w.l.s. since $(G4)$ is not satisfied.

II. Let U be the w.l.s. like in Example 4 and let "$\leq$" be the relation defined for every $(A, x), (B, y) \in U$ by $(A, x) \leq (B, y)$ iff $A \subseteq B$.

Then U is not a q.l.s. since $(L5)$ is not accomplished (see Remark 11-I).

Remark 13 I. If X is an a.l.s., then X is a s.l.s..

II. The reverse of the above proposition is not true. Indeed, consider $X = \mathbb{R}$ and U the s.l.s. of Remark 11-II. Now, taking $A = \{0, 1\}$, $\alpha = 2$ and $\beta = 3$, we have $(\alpha + \beta)A = \{0, 5\}$, $\alpha A = \beta A = \{0, 2, 3, 5\}$ and $(\alpha + \beta)A \neq \alpha A + \beta A$. So U is not an a.l.s. since $(AL4)$ is not true.

Remark 14 There is no relationship between spaces with non-negative scalars and q.l.s..

I. Let $X = Pbc_0(\mathbb{R})$ be the q.l.s. of Remark 12-I. Suppose X satisfies the Riesz property (R). Considering $A_1 = \{0, 1\}$, $A_2 = \{0, 1, 2\}$, $B_1 = \{0, 1\}$, $B_2 = \{0, 2\}$, it results $A_1 + A_2 = B_1 + B_2$. Then there exist $C, D, E, F \geq \theta$ such that

(1) $A_1 = C + D$,

(2) $A_2 = E + F$,

(3) $B_1 = C + E$,

(4) $B_2 = D + F$.

From (1) we have two cases:

(i) $C = \theta$, $D = \{0, 1\}$. By (3) it follows $B_1 = \{0, 1\}$ and by (2) we obtain $F = \{0, 1\}$. Now, (4) implies $\{0, 2\} = B_2 = D + F = \{0, 1\} + \{0, 1\} = \{0, 1, 2\}$, which is false.

(ii) $C = \{0, 1\}$, $D = \theta$. By (4) it follows $F = \{0, 2\}$ and by (3) we obtain $E = \theta$. Now, (2) implies $\{0, 1, 2\} = A_2 = E + F = \theta + \{0, 2\}$, which is a contradiction.

Consequently, X does not have the Riesz property (R). So X is not a space with non-negative scalars of $S = \mathbb{R}_+$.

II. Let $\mathbb{C} = \{x + iy; x, y \in \mathbb{R}\}$ be endowed with the usual addition and multiplication by real scalars. Let be $\mathbb{C}^+ = \{z = x + iy; z \in \mathbb{C}, x = y = 0 \text{ or } x > 0, y > 0\}$ and the order relation " $\leq$" defined for every $z, t \in \mathbb{C}$ by

$$z \in t \text{ iff } t - z \in \mathbb{C}^+.$$

Then $\mathbb{C}$ is a space with non-negative scalars of $\mathbb{R}_+$, but $\mathbb{C}$ is not a q.l.s. since $(OV4)$ is not satisfied.

Remark 15 Denote by PLS, WLS, ALS, QLS, SLS and CLS the sets of non-negative linear space, w.l.s., q.l.s., a.l.s., s.l.s. and c.l.s., respectively. Among these six sets there are only the following inclusions:

$SLS \subseteq CLS$,

$QLS \subseteq SLS$,

$ALS \subseteq SLS$, but these six sets are not disjoint.

Let C_1 and C_2 be any of the six sets above.

Suppose $X \in C_1 \cap C_2$ and τ_1 a topology on X compatible with the near linear space structure in C_1. Because $X \in C_2$, it can be endowed with another topology τ_2 compatible with the near linear structure in C_2. Obviously, τ_1 is compatible with the algebraic structure of C_2.

There are two situations: $\tau_1 = \tau_2$ or $\tau_1 \neq \tau_2$. In the last one, τ_1 and τ_2 can compare or not.

6 Conclusion

Motivated by different applications in many domains (e.g. optimization, interval analysis, computer science), various "near linear spaces" were introduced in the literature. These spaces satisfy only a part of the axioms of a linear space.

We treated properties of some near linear spaces, such as: spaces with non-negative scalars, weak linear spaces, almost linear spaces, quasilinear spaces, semilinear spaces and conlinear spaces. Some examples, relationships among these spaces and their topological structures are also established.

As future research:

- We intend to study some variational principles and possible equivalent formulations in some of the near linear spaces studied in this paper.

- We will introduce a near operatorial norm on the studied near linear spaces, similarly to the w.l.s. (Definition 23 and Theorem 2).

- If X is a near linear space and M is a non-empty set, then the set $\mathcal{F}(M, X)$ of all functions $f: M \to X$, with the usual operations, is also a near linear space of the same type as X. By choosing M and X appropriately, we will study various topologies and properties that can be highlighted on $\mathcal{F}(M, X)$. Such a study is made for example, in [6].

Acknowledgements Authors are very grateful to unknown Referees for their valuable suggestions in improvement of this paper.
The second author expresses her thanks to Professor Irina Condurachi and Mr. Gabriel Atodiresei for their help in participating at ICNACT–2024.

References

1. Apreutesei, G.: Hausdorff topology and some operations with subsets. An. Şt. Univ. "Alexandru Ioan Cuza" Iaşi, XLIV, 445–454 (1998)
2. Apreutesei, G.: The hyperspecial topologies and almost linear spaces. An. Şt. Univ. "Alexandru Ioan Cuza" Iaşi, XLVIII, 3–18 (2002)
3. Apreutesei, G., Croitoru, A.: Normed spaces with non-negative scalars, Submitted for publication
4. Apreutesei, G.: Cauchy nets and convergent nets on semilinear topology spaces. Topology Appl. **159**, 2922–2931 (2012)
5. Apreutesei, G., Croitoru, A., Mastorakis, N.E.: Totally boundedness and compactness on semilinear topological spaces. In: Recent Researches in Automatic Control, Systems Science and Communications, pp. 32–35. ISBN: 978-1-61804-103-6 (2012)
6. Apreutesei, G., Croitoru, A.: Semi-linearity on spaces of set-valued functions. J. Nonlinear Convex Anal. **24**(9), 1901–1917 (2023)
7. Aseev, S.M.: Quasilinear operators and their applications in the theory of multivalued mappings (Russian transl.). Tr. MIAN SSSR, Nauka Moskva, 167, 25–52; Proc. Steklov Inst. Math. 2 (1986), 23–52 (1985)
8. Boccuto, A., Candeloro, D., Sambucini, A.R.: Henstock multivalued integrability in Banach lattices with respect to pointwise non atomic measures. Atti Accad. Naz. Lincei Rend. Lincei Mat. Appl. **26**(4), 363–383 (2015). https://doi.org/10.4171/RLM/710
9. Borş, D.M.: O generalizare a spaţiilor vectoriale (A generalization of vector spaces). Bull. Inst. Polit. Iaşi, XV(XIX), s.I, Fasc. 3-4, 17-20 (in Romanian) (1969)
10. Borş, D.M.: Quasivector spaces over fields. Bull. Inst. Polit. Iaşi, XVII (XXI), s. I, Fasc. 1-2, 47-52 (1971)
11. Borş, D.M., Ciobanu, L., Gontineac, M., Grămadă, V.: Quasi-linear functions. Bull. Inst. Polit. Iaşi, XLIII (XLVII), Fasc. 3-4, 21-26 (1997)
12. Borş, D.M., Croitoru, A., Mastorakis, N.E.: Considerations on weak linear spaces. In: Proceedings of the 2-nd International Conference on Mathematics and Computers in Science and Industry (MCSI), Sliema, Malta, Aug 17–19, pp. 158–161, 2015 IEEE. https://doi.org/10.1109/MCSI.2015.26
13. Borş, D.M., Croitoru, A.: Weak-linear spaces. ROMAI J. **17**(2), 1–9 (2021)
14. Candeloro, D., Di Piazza, L., Musial, K., Sambucini, A.R.: Gauge integrals and selections of weakly compact valued multifunctions. J.M.A.A. **441**(1), 293–308 (2016). https://doi.org/10.1016/j.jmaa.2016.04.009

15. Candeloro, D., Di Piazza, L., Musial, K., Sambucini, A.R.: Relations among gauge and Pettis integrals for multifunctions with weakly compact convex values. Annali di Matematica **197**(1), 171–183 (2018). https://doi.org/10.1007/s10231-017-0674-z
16. Candeloro, D., Croitoru, A., Gavrilut, A., Sambucini, A.R.: A multivalued version of the Radon-Nikodym theorem, via the single-valued Gould integral. Aust. J. Math. Anal. Appl. **15**(2), art. 9, pp. 1–16 (2018)
17. Climescu, A.: Les espaces à scalaire ≥ 0. Bull. Inst. Polit. Iaşi, XI, (XV), fasc. 3-4, 1-6 (1965)
18. Costarelli, D., Croitoru, A., Gavrilut, A., Iosif, A., Sambucini, A.R.: The Riemann-Lebesgue integral of interval-valued multifunctions. Mathematics **8**, 2250, 117; https://doi.org/10.3390/math8122250(2020)
19. Croitoru, A., Borş, D.M.: Remarks on weak linear spaces. Bulletin of the Polytechnic Institute of Iasi, Tomul LXI (LXV). Fasc. 2, 67–74 (2015)
20. Croitoru, A., Apreutesei, G.: C-Metric spaces. ROMAI J. **12**(No. 2), 39–47 (2016)
21. Dăneţ, N., Dăneţ, R.M.: Extension theorems and the Riesz decomposition property. Positivity **7**, 87–93 (2003)
22. Di Piazza, L., Musial, K., Marraffa, V., Sambucini, A.R.: Convergence for varying measures. J. Math. Anal. Appl. **518**(2), 126782 (2023). https://doi.org/10.1016/j.jmaa.2022.126782
23. Di Piazza, L., Marraffa, V., Musial, K., Sambucini, A.R.: Convergence for varying measures in the topological case. Annali di Matematica **203**, 71–86 (2024). https://doi.org/10.1007/s10231-023-01353-8
24. Du, W.S., Turinici, M.: Variational principles and equivalent formulations in almost metric spaces. Appl. Math. Sci. **11**(36), 1755–1764. https://doi.org/10.12988/ams.2017.76190 (2017)
25. Duffin, R.J., Karlovitz, L.A.: Formulation of linear programs in analysis. I. Approximation theory, SIAM J. Appl. Math. **16**, 662–675 (1968)
26. Fraigniaud, P., Lebhar, E., Viennot, L.: The inframetric model for the internet. In: IEEE INFOCOM 2008-The 27th Conference on Computer Communications, pp. 1085–1093. IEEE (2008)
27. Fuchs, L.: Riesz groups. Annali della Scuola Normale Superiore di Pisa - Scienze Fisiche e Matematiche, Serie 3, Volume 19 no. 1, pp. 1–34 (1965)
28. Fuchs, L.: Riesz vector spaces and Riesz algebras. Séminaire Dubreil. Algèbre et théorie des nombres, Volume 19 no. 2, Talk no. 23-24, 9 p (1965–1966)
29. Godini, G.: An approach to generalizing Banach spaces: normed almost linear spaces. Rendiconti del Circolo Matematico di Palermo, Serie II, Supplemento No. 5, pp. [33]–50 (1984)
30. Godini, G.: A framework for best simultaneous appoximation: normed almost linear spaces, INCREST, Preprint Series in Mathematics, No. 30(1983), J. Approx. Theory **43**, 338–358 (1985)
31. Godini, G.: Operators in normed almost linear spaces. In: Proceedings of the 14th Winter School on Abstract Analysis, Circolo Matematico di Palermo 309-328 (1987)
32. Godini, G.: On normed almost linear spaces. Math. Ann. **279**, 449–455 (1988)
33. Hamel, A.: Variational principles on metric and uniform spaces, Habilitation Thesis, Martin-Luther University, Haale (2005)
34. Labuschagne, C.C.A., Pinchuck, A.L., van Alten, C.J.: A vector lattice version of Radstrom's embedding theorem. Quaest. Math. **30**(3), 285–308 (2007)
35. Lee, S.H., Jun, K.W.: A metric on normed almost linear spaces. Bull. Korean Math. Soc. **36**, 379–388 (1999)
36. Löhne, A.: On Convex Functions with Values in Semi-linear Spaces, Report of the Institute of Optimization and Stochastics 07. Martin-Luther-University Halle-Wittenberg, Department of Mathematics and Computer Science (2003)
37. Marraffa, V., Sambucini, A.R.: Vitali theorems for varying measures. Symmetry **16**(8), 972 (2024). https://doi.org/10.3390/sym16080972
38. Markov, S.: Quasilinear spaces and their relation to vector spaces. Eletronic J. Math. Comput. **2–1**, 1–21 (2005)

39. Mayer, O.: Algebraische und metrische Strukturen in der Intervallrechnung und einige Anwendungen. Computing **5**, 144–162 (1970)
40. Moore, R.E., Baker Kearfott, R., Cloud, M.J.: Intoduction in interval analysis. SIAM. ISBN 978-0-898716-69-6 (2009)
41. Rojas-Medar, M., Jiménez-Gamero, M.D., Chalco-Cano, Y., Viera-Brandao, A.J.: Fuzzy quasilinear spaces and applications. Fuzzy Sets Syst. **152**(2), 173–190 (2005)
42. Talo, O., Başar, F.: Quasilinearity on the classical sets of sequences of Fuzzy numbers and some related results. Taiwanese J. Math. **14**(5), 1799–1819 (2010)

Cayley Transform of Operators on the Hilbert Space

Pabitra Kumar Jena and Badal Sahoo

Abstract In this note, we study some algebraic properties such as unitariness, normality, additive inverse law, forward order law, backward order law and absorption law for Cayley transform of operators and operator matrices on the Hilbert space.

Keywords Cayley transform · Hilbert space · Operator matrices

1 Introduction

If $\mathcal{H}$ is a Hilbert space, then $\mathcal{B}(\mathcal{H})$ denotes the class of bounded linear operators on $\mathcal{H}$. If $(I + \Omega)$ is invertible for some $\Omega \in \mathcal{B}(\mathcal{H})$, then the Cayley transform of Ω is denoted as $Cl(\Omega)$ and is defined by $Cl(\Omega) = (I + \Omega)^{-1}(I - \Omega)$. In the Hilbert space context, Cayley transform is an operator between two Hilbert spaces. Cayley introduced this notion in [1]. In particular, Cayley transform is considered as an extension of the conformal mapping $f(z) = \frac{1-z}{1+z}$ from the complex plane to itself, later extended to the matrix classes. In this regard, Stein [17] and Taussky [18] considered Cayley transform indirectly providing connections between stable matrices whose eigenvalues lie in the open left half-plane to the convergent matrices whose eigenvalues lie in the open unit disc.

In recent years, many researchers have studied the Cayley transform for different matrix classes. The convergence of a square matrix is interlinked with the existence of Cayley transform of a stable matrix (see [10]), where stable matrices are those matrices whose all characteristic values have negative real parts. In 2002, the Cayley transform $Cl(\Omega)$ is studied for positive definite matrix, an inverse

The first author is supported by Seed Fund Grant, 2023–24 under R and D cell, Berhampur University, letter No. 12605/R and DC/BU/23.

P. Kumar Jena
Berhampur University, Bhanja Bihar, Brahmapur, India

B. Sahoo (✉)
Government Science College, Chhatrapur, India
e-mail: badalsahoo206@gmail.com

H. Kalita et al. (eds.), *Nonlinear Analysis and Computational Techniques*,
Springer Proceedings in Mathematics & Statistics 501,
https://doi.org/10.1007/978-981-96-6038-4_7

$M\ matrix,\ P\ matrix\ and\ M\ matrix$ [7]. In 2024, Mondal et al. [15] investigated Cayley transform for almost skew-Hermitian matrices, positive definite matrices, semi-positive matrices, as well as subclasses of the $M\ matrices,\ H\ matrices$ and its inverses.

Gavriljuk [9] provided solutions to an IVP for the first order differential equation using Cayley transform for some unbounded self-adjoint and positive operators. In this paper, we restrain ourself to study the algebraic properties of Cayley transform only for linear bounded operators on the Hilbert space.

Now, we unroll this paper as follows. In Sect. 1, we discuss about literature review. In Sect. 2, we add some notations and definitions which play crucial role in successive section. In Sect. 3, we study some algebraic properties of Cayley transform for the class of idempotent operators, normal operators, nilpotent operators, positive operators and some other closed range operators. These results are obtained using the matrix representations of operator Ω under certain Hilbert space decompositions provided that $I + \Omega$ is invertible.

2 Preliminaries

If $\Omega \in \mathcal{B}(\mathcal{H})$, then $Ran(\Omega)$ and $Nul(\Omega)$ denote the range space and null space of Ω respectively. An operator $\Omega \in \mathcal{B}(\mathcal{H})$ is called an idempotent if $\Omega^2 = \Omega$ or $Ran(I - \Omega) \subseteq Nul(\Omega)$. Notice that $I - \Omega$ is also an idempotent if Ω is an idempotent. In theorem (1), we prove that Cayley transform of idempotent operator is also an idempotent. An operator Ω is called normal if $\Omega\Omega^* = \Omega^*\Omega$. In addition, Ω will be unitary if it is isometry i.e. $\Omega\Omega^* = I$ as well as co-isometry i.e. $\Omega^*\Omega = I$. We also establish that if sum of an idempotent operator Ω and its adjoint equals to their product, then $Cl(\Omega)$ is unitary.

If $\Omega, \Psi \in \mathcal{B}(\mathcal{H})$, then $[\Omega, \Psi]$ represents commutator operator which is defined by $[\Omega, \Psi] = \Omega\Psi - \Psi\Omega$. An operator $\Omega \in \mathcal{B}(\mathcal{H})$ is said to be $\theta\ operator$ if $[\Omega^*\Omega, \Omega^* + \Omega] = 0$ and quasi-normal if $[\Omega^*\Omega, \Omega] = 0$. In theorem (1), we show that idempotent operators are $\theta\ operator$ if $\Omega\Omega^*\Omega$ is self-adjoint and quasi-normal if $\Omega\Omega^*\Omega = \Omega\Omega^*$. Nilpotent operator is one which satisfy $\Omega^k = 0$ for some $k \in \mathbb{N}$. The least value of k for which $\Omega^k = 0$ is called nilpotency for the operator Ω. The Cayley transform and their algebraic properties for the nilpotent operators are characterised in theorem (3). An operator Ω is positive if $(\Omega x, x) \geq 0\ \forall x \in \mathcal{H}$, where (,) stands for inner product on $\mathcal{H}$. The algebraic properties like normality, unitariness, quasi-normality of Cayley transform are also studied for the class of closed range operators, positive operators and some operator matrices. These properties are validated through an example of idempotent operator on $l^2(\mathcal{H})$-space. If $\Omega^*\Omega \geq \Omega\Omega^*$ for some $\Omega \in \mathcal{B}(\mathcal{H})$, then we say Ω is hyponormal and if $\Omega(\Omega^n\Omega^*) = (\Omega^*\Omega^n)\Omega$, we say Ω is quasi n-normal.

If $\Omega, \Psi \in \mathcal{B}(\mathcal{H})$, then $\Omega^{-1} + \Psi^{-1} = \Omega^{-1}(\Omega + \Psi)\Psi^{-1}$ is called absorption law of operators. In addition, $(\Omega\Psi)^{-1} = \Psi^{-1}\Omega^{-1}$ and $(\Omega\Psi)^{-1} = \Omega^{-1}\Psi^{-1}$ are called reverse and forward order laws for the operators Ω, Ψ respectively. These laws are discussed for Cayley transform of nilpotent operators, zero operators, hyponormal operators and compact operators in theorem (6). Absorption law, reverse and forward order law for generalized inverses of operators and ring elements are found in [3, 5, 8, 12, 14, 16]

Throughout this paper, for any operator $\Omega \in \mathcal{B}(\mathcal{H})$, we consider the matrix representation of Ω under the decomposition $\mathcal{H} = \mathcal{H}_1 \oplus \mathcal{H}_2$ as

$$\Omega = \begin{bmatrix} \Omega_1 & \Omega_2 \\ \Omega_3 & \Omega_4 \end{bmatrix} : \begin{bmatrix} \mathcal{H}_1 \\ \mathcal{H}_2 \end{bmatrix} \to \begin{bmatrix} \mathcal{H}_1 \\ \mathcal{H}_2 \end{bmatrix},$$

where

$$\Omega_1 : \mathcal{H}_1 \to \mathcal{H}_1,\ \Omega_2 : \mathcal{H}_2 \to \mathcal{H}_1,$$

$$\Omega_3 : \mathcal{H}_1 \to \mathcal{H}_2,\ \Omega_4 : \mathcal{H}_2 \to \mathcal{H}_2.$$

Also, the ideas of matrix representations of operators are found in [6].

3 Main Results

3.1 Cayley Transform of Operators and Operator Matrices

Theorem 1 *Let $\Omega \in \mathcal{B}(\mathcal{H})$ be an idempotent operator, then*

(a) $Cl(\Omega)$ is self-adjoint iff Ω is self-adjoint.
(b) $Cl(\Omega)$ is normal iff Ω is normal.
(c) $Cl(\Omega)$ is unitary iff $\Omega\Omega^* = \Omega^*\Omega = \Omega + \Omega^*$.
(d) $Cl(\Omega)$ is θ operator iff $\Omega\Omega^*\Omega$ is self-adjoint.
(e) $Cl(\Omega)$ is quasinormal iff $\Omega\Omega^*\Omega = \Omega\Omega^*$.
(f) $[Cl(\Omega)]^n = Cl(\Omega)\ \forall n \in \mathbb{N}$.
(g) $Cl(\Omega)$ is quasi n-normal iff Ω is normal.
(h) $Cl(\Omega)$ is hyponormal iff Ω is hyponormal.
(i) $Cl(\Omega)$ is isometry iff $\Omega^*\Omega = \Omega + \Omega^*$ and co-isometry iff $\Omega\Omega^* = \Omega + \Omega^*$.
(j) $Cl^n(\Omega) = \begin{cases} \Omega, & \text{if } n \text{ even} \\ I - \Omega, & n \text{ odd} \end{cases},\ \forall n \in \mathbb{N}$.

Proof Since Ω is idempotent, then notice that $Nul(I - \Omega) = Ran(\Omega)$ is closed. So $\mathcal{H} = Ran(\Omega) \oplus Nul(\Omega^*)$. Then from [11], we have Ω has the following representation,

$$\Omega = \begin{bmatrix} I & \Omega_2 \\ 0 & 0 \end{bmatrix} : \begin{bmatrix} Ran(\Omega) \\ Nul(\Omega^*) \end{bmatrix} \to \begin{bmatrix} Ran(\Omega) \\ Nul(\Omega^*) \end{bmatrix}.$$

Thus, $I + \Omega = \begin{bmatrix} 2I & \Omega_2 \\ 0 & I \end{bmatrix}$ implies $(I + \Omega)^{-1} = \begin{bmatrix} \frac{I}{2} & \frac{-\Omega_2}{2} \\ 0 & I \end{bmatrix}$.

Now, $Cl(\Omega) = (I + \Omega)^{-1}(I - \Omega) = \begin{bmatrix} 0 & -\Omega_2 \\ 0 & I \end{bmatrix} = I - \Omega$.

To prove (a), one can easily show that $Cl(\Omega)$ is self-adjoint iff Ω is.

For (b), Suppose $Cl(\Omega)$ is normal, then

$$\begin{aligned} Cl(\Omega)Cl(\Omega)^* = Cl(\Omega)^*Cl(\Omega) &\Leftrightarrow (I-\Omega)(I-\Omega^*) = (I-\Omega^*)(I-\Omega) \\ &\Leftrightarrow I-\Omega-\Omega^*+\Omega\Omega^* = I-\Omega^*-\Omega-\Omega^*\Omega \\ &\Leftrightarrow \Omega\; is\; normal. \end{aligned}$$

(c) $Cl(\Omega)$ is unitary iff $Cl(\Omega)^*Cl(\Omega) = I = Cl(\Omega)Cl(\Omega)^*$ which implies $I - \Omega - \Omega^* + \Omega\Omega^* = I$ and $I - \Omega^* - \Omega + \Omega^*\Omega = I$ iff $\Omega\Omega^* = \Omega + \Omega^*$ and $\Omega^*\Omega = \Omega + \Omega^*$. Hence the result follows.

To prove (d),

$$\begin{aligned} Cl(\Omega)^*Cl(\Omega)[Cl(\Omega)^* + Cl(\Omega)] &= (I-\Omega^*)(I-\Omega)(2I-\Omega^*-\Omega) \\ &= 2I - 2\Omega - 2\Omega^* + \Omega\Omega^* + 2\Omega^*\Omega - \Omega^*\Omega\Omega^*. \end{aligned}$$

Again,

$$\begin{aligned} [Cl(\Omega)^* + Cl(\Omega)]Cl(\Omega)^*Cl(\Omega) &= (2I-\Omega^*-\Omega)(I-\Omega^*)(I-\Omega) \\ &= 2I - 2\Omega - 2\Omega^* + \Omega\Omega^* + 2\Omega^*\Omega - \Omega\Omega^*\Omega. \end{aligned}$$

To prove, $Cl(\Omega)$ to be a θ *operator*, we have to show that $[Cl(\Omega)^*Cl(\Omega), Cl(\Omega)^* + Cl(\Omega)] = 0$ iff $\Omega^*\Omega\Omega^* = \Omega\Omega^*\Omega$.

Hence, $Cl(\Omega)$ is θ *operator* iff $\Omega\Omega^*\Omega$ is self-adjoint.

For (e), $Cl(\Omega)$ is quasinormal iff $[Cl(\Omega)^*Cl(\Omega), Cl(\Omega)] = 0$.

Now,

$$\begin{aligned} (Cl(\Omega)^*Cl(\Omega))Cl(\Omega) &= ((I-\Omega^*)(I-\Omega))(I-\Omega) \\ &= (I-\Omega-\Omega^*+\Omega^*\Omega)(I-\Omega) = I-\Omega-\Omega^*+\Omega^*\Omega. \end{aligned}$$

Similarly,

$$\begin{aligned} Cl(\Omega)(Cl(\Omega))^*Cl(\Omega)) &= (I-\Omega)((I-\Omega^*))(I-\Omega)) \\ &= (I-\Omega)(I-\Omega-\Omega^*+\Omega^*\Omega) < \\ &= I-\Omega-\Omega^*+\Omega^*\Omega+\Omega\Omega^*-\Omega\Omega^*\Omega. \end{aligned}$$

Notice that $[Cl(\Omega)^*Cl(\Omega), Cl(\Omega)] = 0$ iff $\Omega\Omega^*\Omega = \Omega\Omega^*$. Hence the result follows.

(f) Here, $Cl(\Omega) = I - \Omega,\ (Cl(\Omega))^2 = Cl(\Omega),\ (Cl(\Omega))^3 = Cl(\Omega), \ldots$. Inductively, we get the result.

(g) Now, for $n \in \mathbb{N},\ Cl(\Omega)[Cl(\Omega)^n Cl(\Omega)^*] = Cl(\Omega)[Cl(\Omega)C(\Omega^*)] = I - \Omega - \Omega^* + \Omega\Omega^*$.

Similarly, $[Cl(\Omega)^*Cl(\Omega)^n]Cl(\Omega) = I - \Omega - \Omega^* + \Omega^*\Omega$.

So, $Cl(\Omega)$ is quasi n-normal iff $Cl(\Omega)[Cl(\Omega)^n Cl(\Omega)^*] = Cl(\Omega)^* Cl(\Omega)^n] Cl(\Omega)$ iff Ω is normal.

(h) For hyponormality of $Cl(\Omega)$, we have $Cl(\Omega)^* Cl(\Omega) \geq Cl(\Omega) Cl(\Omega)^*$ iff $\Omega^*\Omega \geq \Omega\Omega^*$. Hence the result.

(i) Notice that, the proof of (i) directly follows from (c).

For (j), We have $Cl(\Omega) = I - \Omega$, then for $n = 2k, k \in \mathbb{N}$, $Cl^n(\Omega) = Cl(I - \Omega) = \Omega$. Similarly, for $n = 2k + 1, k \in \mathbb{N}$, $Cl^n(\Omega) = Cl(\Omega) = I - \Omega$.

Example 1 Define $\Omega : l^2 \to l^2$ by

$\Omega(x_1, x_2, x_3, \ldots) = (b_1x_1, b_2x_2, b_3x_3, \ldots)$, where $b_i = 0 \, or \, 1$. Here $\Omega^2 = \Omega$ i.e. Ω is idempotent.

$Cl(\Omega) = I - \Omega = ((1 - b_1)x_1, (1 - b_2)x_2, \ldots)$. Direct computation shows that Ω and $Cl(\Omega)$ are self-adjoint. Thus $Cl(\Omega)$ is normal.

Also, $Cl(\Omega)$ is unitary iff $\Omega = 0$. If $Cl(\Omega)^2 = I$, then $Cl(\Omega) = I$ implies $\Omega = 0$. The converse is obvious.

In the next theorem, we study algebraic properties for Cayley transform of nilpotent operators. If $a \in \mathbb{N}$ and $a = 1, 2, 3, \ldots, n - 1$, then we write $a \in [n - 1]$.

Theorem 2 *If $\Omega \in \mathcal{B}(\mathcal{H})$ be a nilpotent operator with nilpotency n and $\|\Omega\| \leq 1$, then*

(a) $Cl(\Omega) = I - 2 \sum_{a\in[n-1]} (-1)^{a+1}\Omega^a$, *where* $a \in \mathbb{N}$.
(b) $Cl(\Omega)$ *is self adjoint iff* Ω *is self-adjoint.*
(c) $Cl(\Omega)$ *is normal if* Ω *is normal.*
(d) $Cl(\Omega)$ *is quasinormal if* Ω *is normal.*
(e) $Cl(\Omega)$ *is unitary if* Ω *is normal and* $\sum_{a\in[n-1]} (-1)^{a+1}(\Omega^a + \Omega^{*a})$

$= 2 \sum_{b\in[n-1]} \sum_{c\in[n-1]} (-1)^{b+c+2}\Omega^b(\Omega^*)^c$, *where* $a, b, c \in \mathbb{N}$.

Proof (a) As $\|\Omega\| \leq 1$ and $\Omega^n = 0$, then

$$\begin{aligned}(I + \Omega)^{-1} &= I - \Omega + \Omega^2 - \Omega^3 + \cdots + (-1)^n\Omega^{n-1} \\ &= I - \sum_{a\in[n-1]} (-1)^{a+1}\Omega^a.\end{aligned}$$

Now,

$$\begin{aligned}Cl(\Omega) = (I + \Omega)^{-1}(I - \Omega) &= (I - 2 \sum_{a\in[n-1]} (-1)^{a+1}\Omega^a)(I - \Omega) \\ &= I - 2\Omega + 2\Omega^2 - 2\Omega^3 + \cdots + 2(-1)^n\Omega^{n-1} \\ &= I - 2 \sum_{a\in[n-1]} (-1)^{a+1}\Omega^a.\end{aligned}$$

To prove (b), notice that $Cl(\Omega) = Cl(\Omega)^*$ iff $I - 2 \sum_{a\in[n-1]} (-1)^{a+1}\Omega^a = I - 2 \sum_{a\in[n-1]} (-1)^{a+1}(\Omega^*)^a$ iff $\Omega^* = \Omega$.

(c) Now,

$$\begin{aligned} Cl(\Omega)Cl(\Omega)^* &= (I - 2\sum_{a\in[n-1]} (-1)^{a+1}\Omega^a)(I - 2\sum_{a\in[n-1]} (-1)^{a+1}(\Omega^*)^a) \\ &= I - 2\Big[\sum_{a\in[n-1]} (-1)^{a+1}(\Omega^a + \Omega^{*a})\Big] + 4\sum_{b\in[n-1]}\sum_{c\in[n-1]} (-1)^{b+c+2}\Omega^b(\Omega^*)^c. \end{aligned}$$

Similarly,

$$\begin{aligned} Cl(\Omega)^*Cl(\Omega) &= (I - 2\sum_{a\in[n-1]} (-1)^{a+1}(\Omega^*)^a)(I - 2\sum_{a\in[n-1]} (-1)^{a+1}\Omega^a) \\ &= I - 2\Big[\sum_{a\in[n-1]} (-1)^{a+1}(\Omega^a + \Omega^{*a})\Big] + 4\sum_{b\in[n-1]}\sum_{c\in[n-1]} (-1)^{b+c+2}\Omega^{*b}\Omega^c. \end{aligned}$$

Notice that $Cl(\Omega)$ is normal if Ω is normal.

(d) For $Cl(\Omega)$ to be quasinormal, we have to calculate $[Cl(\Omega)^*Cl(\Omega), Cl(\Omega)]$. Now,

$$\begin{aligned} &(Cl(\Omega)^*Cl(\Omega))Cl(\Omega) \\ &= \Big[I - 2\Big[\sum_{a\in[n-1]} (-1)^{a+1}(\Omega^a + \Omega^{*a})\Big] + 4\sum_{b\in[n-1]}\sum_{c\in[n-1]} (-1)^{b+c+2}\Omega^{*b}\Omega^c\Big] \\ &\quad \Big[I - 2\sum_{d\in[n-1]} (-1)^{d+1}\Omega^d\Big] \\ &= I - 2\Big[\sum_{a\in[n-1]} (-1)^{a+1}(\Omega^a + \Omega^{*a})\Big] + 4\sum_{b\in[n-1]}\sum_{c\in[n-1]} (-1)^{b+c+2}(\Omega^*)^b\Omega^c \\ &\quad - 2\sum_{d\in[n-1]} (-1)^{d+1}\Omega^d \\ &\quad + 4\Big[\sum_{a\in[n-1]} (-1)^{a+1}(\Omega^a + \Omega^{*a})\sum_{d\in[n-1]} (-1)^{d+1}\Omega^d\Big] - 8\sum_{b\in[n-1]}\sum_{c\in[n-1]}\sum_{d\in[n-1]} \\ &\quad (-1)^{b+c+d+3}(\Omega^*)^b\Omega^c\Omega^d \\ &= I - 2\Big[\sum_{a\in[n-1]} (-1)^{a+1}(2\Omega^a + \Omega^{*a})\Big] + 4\sum_{b\in[n-1]}\sum_{c\in[n-1]} (-1)^{b+c+2}(\Omega^*)^b\Omega^c \\ &\quad + 4\Big[\sum_{a\in[n-1]}\sum_{d\in[n-1]} (-1)^{a+d+1}(\Omega^a + (\Omega^*)^a)\Omega^d\Big] - 8\sum_{b\in[n-1]}\sum_{c\in[n-1]}\sum_{d\in[n-1]} \\ &\quad (-1)^{b+c+d+3}(\Omega^*)^b\Omega^{c+d}. \end{aligned}$$

Similarly,

$$(Cl(\Omega)^*Cl(\Omega))Cl(\Omega)$$
$$= I - 2\Big[\sum_{a\in[n-1]}(-1)^{a+1}(2\Omega^a + \Omega^{*a})\Big] + 4\sum_{b\in[n-1]}\sum_{c\in[n-1]}(-1)^{b+c+2}(\Omega^*)^b\Omega^c$$
$$+ 4\Big[\sum_{a\in[n-1]}\sum_{d\in[n-1]}(-1)^{a+d+1}\Omega^d(\Omega^a + (\Omega^*)^a)\Big] - 8\sum_{b\in[n-1]}\sum_{c\in[n-1]}\sum_{d\in[n-1]}$$
$$(-1)^{b+c+d+3}\Omega^d(\Omega^*)^b\Omega^c.$$

Since Ω is normal, then $(Cl(\Omega)^*Cl(\Omega))Cl(\Omega) = (Cl(\Omega)^*Cl(\Omega))Cl(\Omega)$ implies $Cl(\Omega)$ is quasinormal.

To prove (e), we have to show that $Cl(\Omega)Cl(\Omega)^* = I = Cl(\Omega)^*Cl(\Omega)$. Since Ω is normal, then from (b), $Cl(\Omega)Cl(\Omega)^* = Cl(\Omega)^*Cl(\Omega)$.

Again, as $\sum_{a\in[n-1]}(-1)^{a+1}(\Omega^a + \Omega^{*a}) = 2\sum_{b\in[n-1]}\sum_{c\in[n-1]}(-1)^{b+c+2}\Omega^b(\Omega^*)^c$, then $Cl(\Omega)$ is unitary.

Theorem 3 *If* $\Omega \in \mathcal{B}(\mathcal{H}\oplus\mathcal{H})$ *with* $\Omega = \begin{bmatrix} 0 & \Omega_2 \\ \Omega_3 & 0 \end{bmatrix}$ *and* $\Omega_2\Omega_3 = \Omega_3\Omega_2 = -I$, *then*

(a) $Cl(\Omega)$ *is self adjoint if* $\Omega\Omega^* = -I = \Omega^*\Omega$.
(b) $Cl(\Omega)$ *normal if* Ω *is normal.*
(c) $Cl(\Omega)$ *is unitary if* Ω *is normal and* $\Omega + \Omega^* = 0$.
(d) $Cl(\Omega)$ *is quasinormal if* Ω *is normal.*

Proof First we have to calculate $Cl(\Omega)$.

Since $\Omega = \begin{bmatrix} 0 & \Omega_2 \\ \Omega_3 & 0 \end{bmatrix}$, then $I + \Omega = \begin{bmatrix} I & \Omega_2 \\ \Omega_3 & I \end{bmatrix}$.

Again, as $\Omega_2\Omega_3 = \Omega_3\Omega_2 = -I$, then $(I+\Omega)^{-1} = \begin{bmatrix} I & \Omega_2 \\ \Omega_3 & I \end{bmatrix}^{-1} = \frac{1}{2}\begin{bmatrix} I & \Omega_3^{-1} \\ \Omega_2^{-1} & I \end{bmatrix}$.

Thus $Cl(\Omega) = (I+\Omega)^{-1}(I-\Omega) = \frac{1}{2}\begin{bmatrix} 0 & \Omega_3^{-1} - \Omega_2 \\ \Omega_2^{-1} - \Omega_3 & I \end{bmatrix} = \frac{1}{2}(\Omega^{-1} - \Omega)$.

(a) As $\Omega\Omega^* = -I = \Omega^*\Omega$, then it is easy to notice $Cl(\Omega)^* = Cl(\Omega)$.

To prove (b), we have to calculate

$Cl(\Omega)Cl(\Omega)^* = \frac{1}{4}[(\Omega^*\Omega)^{-1} - \Omega^{-1}\Omega^* - \Omega(\Omega^*)^{-1} + \Omega\Omega^*]$

Similarly, $Cl(\Omega)Cl(\Omega)^* = \frac{1}{4}[(\Omega\Omega^*)^{-1} - \Omega^*\Omega^{-1} - (\Omega^*)^{-1}\Omega + \Omega^*\Omega]$.

Since Ω is normal, then $(\Omega^*)^{-1}\Omega = \Omega(\Omega^*)^{-1}$. Hence $Cl(\Omega)$ is normal.

(c) From (b), we have, $Cl(\Omega)^*Cl(\Omega) = Cl(\Omega)Cl(\Omega)^*$ if Ω is normal. As $\Omega + \Omega^* = 0$, then direct computation shows that $Cl(\Omega)$ is unitary.

(d) Now,

$$
\begin{aligned}
(Cl(\Omega)^*Cl(\Omega))Cl(\Omega) &= \frac{1}{8}((\Omega^{*-1} - \Omega^*)(\Omega^{-1} - \Omega))(\Omega^{-1} - \Omega) \\
&= \frac{1}{8}[(\Omega^*\Omega)^{-1} - \Omega^{-1}\Omega^* - \Omega(\Omega^*)^{-1} + \Omega\Omega^*](\Omega^{-1} - \Omega) \\
&= \frac{1}{8}[\Omega^{*-1}\Omega^{-2} - 2\Omega^{*-1} + \Omega^{*-1}\Omega^2 - \Omega^*\Omega^2 + \Omega^*].
\end{aligned}
$$

Again, direct computation gives

$$
\begin{aligned}
&Cl(\Omega)(Cl(\Omega)^*Cl(\Omega)) \\
&= \frac{1}{8}[\Omega^{-1}\Omega^{*-1}\Omega^{-1} - \Omega^{-1}\Omega^*\Omega^{-1} - \Omega^{-1}\Omega^{*-1}\Omega + \Omega^{-1}\Omega^*\Omega \\
&- \Omega\Omega^{*-1}\Omega^{-1} + \Omega\Omega^*\Omega^{-1} + \Omega\Omega^{*-1}\Omega - \Omega\Omega^*\Omega].
\end{aligned}
$$

As Ω is normal, then $(Cl(\Omega)^*Cl(\Omega))Cl(\Omega) = Cl(\Omega)(Cl(\Omega)^*Cl(\Omega))$ i.e. $Cl(\Omega)$ is quasinormal.

Let $\Omega \in \mathcal{B}(\mathcal{H})$, and the matrix representation of Ω with respect to Hilbert space decomposition $\mathcal{H} = K_1 \oplus K_1^\perp$ is
$\Omega = \begin{bmatrix} \Omega_1 & \Omega_2 \\ \Omega_3 & \Omega_4 \end{bmatrix} : \begin{bmatrix} K_1 \\ K_1^\perp \end{bmatrix} \to \begin{bmatrix} K_1 \\ K_1^\perp \end{bmatrix}$. If M be the subspace of $\mathcal{H}$ which reduces Ω i.e. $\Omega(M) \subseteq M$, then we get the following results.

Theorem 4 *Let $\Omega \in \mathcal{B}(\mathcal{H})$ and M reduces Ω. If $\Omega_1 \geq 0$, $\Omega_4 \geq 0$, then*

(a) $Cl(\Omega) = \begin{bmatrix} Cl(\Omega_1) & D \\ 0 & Cl(\Omega_4) \end{bmatrix}$,
where $D = (I + \Omega_1)^{-1}\Omega_2(I - Cl(\Omega_4))$.

(b) $Cl(\Omega)$ *is self-adjoint, normal and unitary if at least one of the following cases holds.*

(i) $Cl(\Omega_1), Cl(\Omega_4)$ *are self adjoint and* $\Omega_2 = 0$.
(ii) $Cl(\Omega_1)$ *is self adjoint and* $\Omega_4 = 0$.
(iii) $Cl(\Omega_1)$ *is self adjoint,* $\Omega_2 = 0$ *and* $\Omega_4 = 0$.

Proof (a) Since the subspace M reduces Ω, then $\Omega = \begin{bmatrix} \Omega_1 & \Omega_2 \\ 0 & \Omega_4 \end{bmatrix}$ under the decomposition $\mathcal{H} = M \oplus M^\perp$. As $\Omega_1 \geq 0$ and $\Omega_4 \geq 0$, then $(I + \Omega)^{-1} = \begin{bmatrix} (I + \Omega_1)^{-1} & -(I + \Omega_1)^{-1}\Omega_2(I + \Omega_4)^{-1} \\ 0 & (I + \Omega_4)^{-1} \end{bmatrix}$.

Now, the Cayley transform of Ω is

$$Cl(\Omega) = (I+\Omega)^{-1}(I-\Omega) = \begin{bmatrix} (I+\Omega_1)^{-1} & -(I+\Omega_1)^{-1}\Omega_2(I+\Omega_4)^{-1} \\ 0 & (I+\Omega_4)^{-1} \end{bmatrix} \begin{bmatrix} I-\Omega_1 & \Omega_2 \\ 0 & I-\Omega_4 \end{bmatrix}$$
$$= \begin{bmatrix} C(\Omega_1) & (I+\Omega_1)^{-1}\Omega_2 - (I+\Omega_1)^{-1}\Omega_2 C(\Omega_4) \\ 0 & I-\Omega_4 \end{bmatrix}$$
$$= \begin{bmatrix} Cl(\Omega_1) & D \\ 0 & Cl(\Omega_4) \end{bmatrix},$$

where $D = (I+\Omega_1)^{-1}\Omega_2(I - Cl(\Omega_4))$. Here $Cl(\Omega)^* = \begin{bmatrix} Cl(\Omega_1)^* & 0 \\ D^* & Cl(\Omega_4)^* \end{bmatrix}$.

(b) From case (i), as $Cl(\Omega_1) = Cl(\Omega_1)^*, Cl(\Omega_4) = Cl(\Omega_4)^*$ and $\Omega_2(I - Cl(\Omega_4)) = 0$, then $D = 0$. Hence $Cl(\Omega)$ is self-adjoint.

From case (ii), $Cl(\Omega_1) = Cl(\Omega_1)^*$ and $\Omega_4 = 0$ means $I - Cl(\Omega_4) = 0$ implies $D = 0$. Thus $Cl(\Omega)$ is self-adjoint. From case (i) and (ii), we get case (iii) which implies $Cl(\Omega)$ is self-adjoint.

Now, we have $Cl(\Omega)^*Cl(\Omega) = \begin{bmatrix} Cl(\Omega_1)^*Cl(\Omega_1) & Cl(\Omega_1)^*D \\ D^*Cl(\Omega_1) & D^*D + Cl(\Omega_4)^*Cl(\Omega_4) \end{bmatrix}$.

Again, $Cl(\Omega)Cl(\Omega)^* = \begin{bmatrix} Cl(\Omega_1)Cl(\Omega_1)^* + DD^* & DC_l(\Omega_4)^* \\ Cl(\Omega_4)D^* & Cl(\Omega_4)Cl(\Omega_4)^* \end{bmatrix}$.

To prove $Cl(\Omega)$ is normal, we have to show that

$$Cl(\Omega_1)^*Cl(\Omega_1) = Cl(\Omega_1)Cl(\Omega_1)^* + DD^*,$$
$$Cl(\Omega_1)^*D = DC_l(\Omega_4)^* \text{ and}$$
$$D^*D + Cl(\Omega_4)^*Cl(\Omega_4) = Cl(\Omega_4)Cl(\Omega_4)^*.$$

Hence any of the cases (i),(ii) and (iii) implies these equations hold. Thus $Cl(\Omega)$ is normal.

Finally, to prove $Cl(\Omega)$ is unitary, we will have $Cl(\Omega_1)^*Cl(\Omega_1) = I, Cl(\Omega_1)^*D = 0$. But, later equality forces D to be zero. Then $\Omega_2(I - Cl(\Omega_4)) = 0$.

Hence any of the given cases (i),(ii) and (iii) implies $Cl(\Omega)$ is unitary.

Theorem 5 *Let $\Omega \in \mathcal{B}(\mathcal{H})$ be any closed range operator, then*

(a) $\Omega = \begin{bmatrix} \Omega_1 & \Omega_2 \\ 0 & 0 \end{bmatrix} : \begin{bmatrix} Ran(\Omega) \\ Nul(\Omega^*) \end{bmatrix} \to \begin{bmatrix} Ran(\Omega) \\ Nul(\Omega^*) \end{bmatrix}$.

(b) If $\Omega_1 \geq 0$, then $Cl(\Omega) = \begin{bmatrix} Cl(\Omega_1) & 0 \\ 0 & I \end{bmatrix}$.

(c) $Cl(\Omega)$ is normal iff $Cl(\Omega_1)$ is normal.

(d) $Cl(\Omega)$ is unitary iff $Cl(\Omega_1)$ is unitary.

(e) $Cl(\Omega)$ is quasinormal, quasi n-normal and hyponormal iff $Cl(\Omega_1)$ is so respectively.

(f) $Cl(\Omega)$ is θ $operator$ iff $Cl(\Omega_1)$ is so.

Proof (a) Since Ω is closed range, then $\mathcal{H} = Ran(\Omega) \oplus Nul(\Omega^*)$. Then (a) follows from [5].

(b) As $\Omega = \begin{bmatrix} \Omega_1 & \Omega_2 \\ 0 & 0 \end{bmatrix}$, then $I + \Omega = \begin{bmatrix} I + \Omega_1 & \Omega_2 \\ 0 & I \end{bmatrix}$.

Since $\Omega_1 \geq 0$, then $I + \Omega_1$ is invertible. Thus $(I + \Omega)^{-1} = \begin{bmatrix} (I + \Omega_1)^{-1} & -(I + \Omega_2)^{-1}\Omega_2 \\ 0 & I \end{bmatrix}$.

Now, $Cl(\Omega) = \begin{bmatrix} (I + \Omega_1)^{-1} & -(I + \Omega_2)^{-1}\Omega_2 \\ 0 & I \end{bmatrix} \begin{bmatrix} I - \Omega_1 & \Omega_2 \\ 0 & I \end{bmatrix} = \begin{bmatrix} Cl(\Omega_1) & 0 \\ 0 & I \end{bmatrix}$.

For (c) and (d), observe that $Cl(\Omega)^* = \begin{bmatrix} Cl(\Omega_1)^* & 0 \\ 0 & I \end{bmatrix}$. Hence the normality and unitariness of $Cl(\Omega)$ can be verified directly.

(e) Now,

$(Cl(\Omega)^* Cl(\Omega))Cl(\Omega) = \begin{bmatrix} Cl(\Omega_1)^* C(\Omega_1)^2 & 0 \\ 0 & I \end{bmatrix}$ and

$Cl(\Omega)(Cl(\Omega)^* Cl(\Omega)) = \begin{bmatrix} Cl(\Omega_1)C(\Omega_1)^* Cl(\Omega_1) & 0 \\ 0 & I \end{bmatrix}$.

Hence, $Cl(\Omega)$ is quasinormal iff $[Cl(\Omega)^* Cl(\Omega), Cl(\Omega)] = 0$ iff $[Cl(\Omega_1)^* Cl(\Omega_1), Cl(\Omega_1)] = 0$.

Quasi-n-normality and Hyponormality of $Cl(\Omega)$ can be computed in similar way.

(f) Now,

$$(Cl(\Omega)^* Cl(\Omega))(Cl(\Omega)^* + Cl(\Omega)) = \begin{bmatrix} Cl(\Omega_1)^* & 0 \\ 0 & I \end{bmatrix} \begin{bmatrix} Cl(\Omega_1) & 0 \\ 0 & I \end{bmatrix} \begin{bmatrix} (Cl(\Omega_1) + Cl(\Omega_1)^*) & 0 \\ 0 & I \end{bmatrix}$$
$$= \begin{bmatrix} Cl(\Omega_1)^* Cl(\Omega_1)(Cl(\Omega_1)^* + Cl(\Omega_1)) & 0 \\ 0 & I \end{bmatrix}.$$

Similarly, $(Cl(\Omega)^* + Cl(\Omega))(Cl(\Omega)^* Cl(\Omega)) = \begin{bmatrix} (Cl(\Omega_1)^* + Cl(\Omega_1))Cl(\Omega_1)^* Cl(\Omega_1) & 0 \\ 0 & I \end{bmatrix}$.

Hence $Cl(\Omega)$ is θ *operator* iff $Cl(\Omega_1)$ is.

Corollary 1 *If $\Omega \in \mathcal{B}(\mathcal{H})$ be the closed range positive operator, then* $Cl(\Omega) = \begin{bmatrix} Cl(\Omega_1) & 0 \\ 0 & I \end{bmatrix}$.

Proof As $\Omega \geq 0$, then $\Omega^* = \Omega$ and from the theorem (5), $\mathcal{H} = Ran(\Omega) \oplus Nul(\Omega)$. Thus, $\Omega_2 = 0$ and

$\Omega = \begin{bmatrix} \Omega_1 & 0 \\ 0 & 0 \end{bmatrix}$. Notice that Ω_1 is positive and invertible. Hence the result follows from the theorem (5).

Now, we give two different operators whose Cayley transforms exists. In addition, these operators satisfy absorption law, reverse and forward order laws.

Theorem 6 *If Ω is a nilpotent operator with nilpotency 2 and Ψ is the zero operator, then*

(a) $Cl(\Omega)^{-1}(Cl(\Omega) + Cl(\Psi))Cl(\Psi)^{-1} = Cl(\Omega)^{-1} + Cl(\Psi)^{-1}$

(b) $(Cl(\Omega)Cl(\Psi))^{-1} = Cl(\Psi)^{-1}Cl(\Omega)^{-1}$

(c) $(Cl(\Omega)Cl(\Psi))^{-1} = Cl(\Omega)^{-1}Cl(\Psi)^{-1}$

Proof For (a), As $\Omega^2 = 0$, then from theorem (2), we have $Cl(\Omega) = I - 2\Omega$ implies $Cl(\Omega)^{-1} = I + 2\Omega$.

Now, $\Psi = 0$ implies $Cl(\Psi) = I$. Thus, $Cl(\Omega)^{-1}(Cl(\Omega) + Cl(\Psi))Cl(\Psi)^{-1} = 2(I + 2\Omega)(I - \Omega) = 2(I + \Omega)$.

Similarly, $Cl(\Omega)^{-1} + Cl(\Psi)^{-1} = 2(I + \Omega)$. Hence the result follows.

The proof of (b) and (c) are obvious.

Now, We proceed to examine the consequences of the theorem on the behavior of compact operators.

Corollary 2 *If $\Omega, \Psi \in B(\mathcal{H})$ be two the compact operators with $\Omega^4 \simeq \Omega^2 \oplus \Omega^2$ and $\Psi^2 \simeq \Psi \oplus \Psi$, then*
(a) $Cl(\Omega)^{-1}(Cl(\Omega) + Cl(\Psi))Cl(\Psi)^{-1} = Cl(\Omega)^{-1} + Cl(\Psi)^{-1}$
(b)$(Cl(\Omega)Cl(\Psi))^{-1} = Cl(\Psi)^{-1}Cl(\Omega)^{-1}$
(c)$(Cl(\Omega)Cl(\Psi))^{-1} = Cl(\Omega)^{-1}Cl(\Psi)^{-1}$

Proof Since $\Omega, \Psi \in \mathcal{B}(\mathcal{H})$ be two the compact operators with $\Omega^4 \simeq \Omega^2 \oplus \Omega^2$ and $\Psi^2 \simeq \Psi \oplus \Psi$, then from [2], we have $\Omega^2 = 0$ and $\Psi = 0$. Hence the result follows from the theorem (6).

Now, we impose certain conditions on compact and hyponormal operators to assure the desired outcomes of the theorem.

Corollary 3 *Let $\Omega, \Psi \in B(\mathcal{H})$. If Ω be the compact operator with $\Omega^2 \simeq \Omega \oplus \Omega$ and Ψ be the hyponormal operator with $|\Psi|^2 + \alpha(BE - EB) \leq 0$ for some $E \in B(\mathcal{H})$, $\alpha \in \mathbb{C}$, then*
(a) $Cl(\Omega)^{-1}(Cl(\Omega) + Cl(\Psi))Cl(\Psi)^{-1} = Cl(\Omega)^{-1} + Cl(\Psi)^{-1}$
(b)$(Cl(\Omega)Cl(\Psi))^{-1} = Cl(\Psi)^{-1}Cl(\Omega)^{-1}$
(c)$(Cl(\Omega)Cl(\Psi))^{-1} = Cl(\Omega)^{-1}Cl(\Psi)^{-1}$

Proof As Ω be the compact operator with $\Omega^2 \simeq \Omega \oplus \Omega$, then from [2], $\Omega^2 = 0$. Again, Ψ be the hyponormal operator with $|\Psi|^2 + \alpha(BE - EB) \leq 0$ implies $\Psi = 0$ which follows from [13]. Hence the result follows.

4 Conclusion

In this work, we study the algebraic properties of Cayley transform of some closed range operators which have matrix representations with respect to the Hilbert space decompositions.

Acknowledgements The authors thank the referees for providing helpful suggestions that improved the paper.

References

1. Cayley, A.: About the algebraic structure of the orthogonal group and the other classical groups in a field of characteristic zero or a prime characteristic. J. Reine Angew. Math. **32**, 6 (1846)
2. Conway, J.B., Prajitur, G., Martínez, A.R.: Powers and direct sums. J. Math. Anal. Appl. **413**, 880–889 (2014). https://doi.org/10.1016/j.jmaa.2013.12.007
3. Deng, C.Y.: Reverse order law for the group inverses. J. Math. Anal. Appl. **382**, 663–671 (2011). https://doi.org/10.1016/j.jmaa.2011.04.085
4. Dincic, N.C., Djordjevic, D.S.: Identities concerning the reverse order law for the Moore-Penrose inverse. Appl. Math. Comput. **220**, 439–445 (2013). https://doi.org/10.1016/j.amc.2013.05.076
5. Djordjevic, D.S., Dincic, N.C.: Reverse order law for Moore-Penrose inverse. J. Math. Anal. Appl. **361**, 252–261 (2010). https://doi.org/10.1016/j.jmaa.2009.08.056
6. Douglas, R.G.: Banach Algebra Techniques in Operator Theory, 2nd edn. Springer Publications, USA (2010)
7. Fallat, S.M., Tsatsomeros, M.J.: On the Cayley transform of positivity classes of matrices. Electron. J. Linear Algebra **9**, 190–196 (2002). https://doi.org/10.13001/1081-3810.1086
8. Gao, Y., Chen, J., Wang, L., Zou, H.: Absorption laws and reverse order laws for generalized core inverses. Commun. Algebra **49**, 3241–3254 (2015). https://doi.org/10.1080/00927872.2021.1892712
9. Gavrilyuk, I.P., Makarov, V.L.: The Cayley transform and the solution of an initial value problem for a first order differential equation with an unbounded operator coefficient in Hilbert space. Numer. Func. Anal. Opt. **15**, 583–598 (1994). https://doi.org/10.1080/01630569408816582
10. Haynes, T.: Stable matrices, the Cayley transform, and convergent matrices. Int. J. Math. Math. Sci. **14**, 77–82 (1991). https://doi.org/10.1155/S0161171291000078
11. Jena, P.K., Sahoo, B.: Core inverse of operators in Hilbert spaces. Oper. Matrices Accepted (2024)
12. Jin, H., Benitez, J.: The absorption laws for the generalized inverses in rings. Electron. J. Linear Algebra **30**, 827–842 (2015). https://doi.org/10.13001/1081-3810.3092
13. Kittaneh, F.: Some characterizations of self-adjoint operators. Acta Sci. Math. **47**, 441–444 (1984)
14. Liu, X., Jin, H., Cvetkovic-Ilic, D.S.: The absorption laws for the generalized inverses. Appl. Math. Comput. **219**, 2053–2059 (2012). https://doi.org/10.13001/1081-3810.3092
15. Mondal, S., Sivakumar, K.C., Tsatsomeros, M.: The Cayley transform of prevalent matrix classes. Linear Algebra Appl. **681**, 1–20 (2024). https://doi.org/10.1016/j.laa.2023.10.017
16. Padhy, C., Jena, P.K., Paikray, S.K.: Aluthge transform of operators on the Bergman space. Arab. J. Math., **9**, 645–651 (2020). https://doi.org/10.1007/s40065-019-00272-y
17. Stein, P.: On the ranges of two functions of positive definite matrices. J. Algebra **2**, 350–353 (1965)
18. Taussky, O.: Matrices C with $C^n \to 0$. J. Algebra **1**, 5–10 (1964)

Solvability of Hybrid Differential Equation in Hölder Space

Manalisha Bhujel and Bipan Hazarika

Abstract The aim of the present paper is to study about solvability of hybrid differential equation. We use fixed point theorem namely Schauder fixed point theorem to establish the existence result. The investigations of the paper are placed in the space of continuous and tempered functions on closed interval $[0, T]$. We provide one example to verify the applicability of our result.

Keywords Relative compactness · Hölder function space · Fixed point theorem (FPT) · Hybrid differential equation (HDE)

1 Introduction

The study of the dynamical system involving non-linear property can be analyzed with the help of perturbation technique. This method is useful for analyzing a non-linear differential or integral equation representing a dynamical system. Whenever the solving of a differential or integral equation becomes difficult, perturbation method can be applied along with the other existing methods to unveil different aspects of the analysis. The broad classification of perturbed differential equation is as follows: (i) Perturbed differential equation of first kind and (ii) Perturbed differential equation of second kind. A differential equation involving perturbation of free unknown functions is referred as perturbated differential equation of first kind and that of involving perturbation of unknown functions under the derivatives is referred as perturbed differential equation of second kind. On the other hand, the perturbation of terms in a non-linear equation can be classified into two categories. The perturbation of the

The research of the first author (Manalisha Bhujel) is supported by SERB, DST, New Delhi, India under the INSPIRE code DST/INSPIRE Fellowship/2019/IF190364.

M. Bhujel (✉) · B. Hazarika
Department of Mathematics, Gauhati University, Guwahati, Assam, India
e-mail: bhujelmanalisha321@gmail.com

B. Hazarika
e-mail: bh_gu@gauhati.ac.in

H. Kalita et al. (eds.), *Nonlinear Analysis and Computational Techniques*,
Springer Proceedings in Mathematics & Statistics 501,
https://doi.org/10.1007/978-981-96-6038-4_8

terms involving addition or subtraction is referred as linear perturbation and that of involving multiplication or division by a term is referred as quadratic perturbation. In the study of Dhage [7], a broad classification of different types of perturbations of differential equation has been mentioned. Such perturbed differential equation have been referred as hybrid differential equations. The study of HDE and integral equation was initiated by Dhage [6] in 1988.

In 2010, Dhage and Lakshmikantham [8] discussed the first order ordinary HDE with quadratic perturbation of first kind

$$\begin{cases} \frac{d}{d\eta_1}\left[\frac{x(\eta_1)}{\zeta(\eta_1,x(\eta_1))}\right] = \psi(\eta_1, x(\eta_1)); \\ x(t_0) = x_0 \end{cases} \tag{1}$$

for $\eta_1 \in J = [t_0, t_0 + a)$, $t_0, a \in \mathbb{R}$ with $a > 0$. Here $\zeta \in C(J \times \mathbb{R}, \mathbb{R} - \{0\})$ and $\psi \in C(J \times \mathbb{R}, \mathbb{R})$.

In 2015, Dhage and Dhage [10] investigated about approximations of the positive solutions for a periodic boundary value problem of the type

$$\begin{cases} \frac{d}{d\eta_1}\left[\frac{x(\eta_1)}{\zeta(\eta_1,x(\eta_1))}\right] + \lambda\left[\frac{x(\eta_1)}{\zeta(\eta_1,x(\eta_1))}\right] = \psi(\eta_1, x(\eta_1)) \\ x(0) = x(T), \end{cases}$$

for $t \in J = [0, T]$, $T \in \mathbb{R}$. Here $\zeta \in C(J \times \mathbb{R}, \mathbb{R} - \{0\})$ and $\psi \in C(J \times \mathbb{R}, \mathbb{R})$. In the same year, Dhage [11] initiated some basic results on Causal Nonlinear Differential Equation of the type

$$\begin{cases} \frac{d}{d\eta_1}\left[\frac{x(\eta_1)-\psi(\eta_1,x(\eta_1))}{\zeta(\eta_1,x(\eta_1))}\right] = Q(x(\eta_1)) \\ x(t_0) = x_0, \end{cases}$$

for $\eta_1 \in J = [t_0, t_0 + a)$, $t_0, a \in \mathbb{R}$ with $a > 0$. Here $\zeta \in C(J \times \mathbb{R}, \mathbb{R} - \{0\})$ and $\psi \in C(J \times \mathbb{R}, \mathbb{R})$. He studied the existence result, strict and non-strict inequalities and existence of the maximal and minimal solutions of the above mentioned equation.

All above mentioned research work is proved under mixed Lipschitz and Caratheodory condition via a FPT in Banach algebra due to Dhage [5]. There have been many works on the theory of HDE, readers can see [9, 12, 14, 15].

Inspired by above works, we extended the concept of HDE to different continuous function space namely, Hölder space. The fixed-point theorems used for hybrid differential equations with perturbation of first or second type are those based on the composition of the solution as sum or product of two operators such as the Dhage Fixed point theorem. In this paper, we thought of using Schauder fixed point theorem as an existence result for which we will have a single operator. To the best of our knowledge, up to now this is the first research work on HDE in Hölder space.

2 Preliminaries

In this section we introduce the space of functions satisfying the Hölder condition. Let $C[a_1, a_2]$, $C_\omega[a_1, a_2]$ be the space of all continuous functions on $[a_1, a_2]$ and the space of functions with tempered increments respectively (see [3]). For a fixed $0 < \alpha \leq 1$, we denote $H_\alpha[a_1, a_2]$ the space of real function x illustrated on $[a_1, a_2]$ for which there exists the least possible constant $H_x > 0$ such that the following condition satisfy:

$$|x(\eta_1) - x(\eta_2)| \leq H_x|\eta_1 - \eta_2|^\alpha,$$

for all $\eta_1, \eta_2 \in [a_1, a_2]$. This condition is called Hölder condition. It is easily seen that $H_\alpha[a_1, a_2]$ is a linear subspace of $C[a_1, a_2]$. More precisely, for $H_x > 0$, we put:

$$H_x = \sup\left\{\frac{|x(\eta_1) - x(\eta_2)|}{|\eta_1 - \eta_2|^\alpha} : \eta_1, \eta_2 \in [a_1, a_2], \eta_1 \neq \eta_2\right\}.$$

Let us recall that, the norm in the space $H_\alpha[a_1, a_2]$ can defined as follows:

$$\|x\|_\alpha = |x(a_1)| + H_x.$$

Further, the space $(H_\alpha[a_1, a_2], ||.||_\alpha)$ with $0 < \alpha \leq 1$ is a Banach space with the above mentioned norm [3]. Moreover, the following inequality holds [3]

$$\|x\|_\infty \leq \max\{1, (a_2 - a_1)^\alpha\}\|x\|_\alpha$$
$$\|x\|_\alpha \leq \max\{1, (a_2 - a_1)^{\beta-\alpha}\}\|x\|_\beta,$$

where $0 < \alpha < \beta \leq 1$.

Finally we recall a criterion for relative compactness which is mentioned in Example 6 of [3]. It is worth to mention that the below mentioned sufficient condition will play a vital role in this research work.

Theorem 1 ([3]) *A bounded subset X in $H_\beta[a_1, a_2]$ is relatively compact in $H_\beta[a_1, a_2]$.*

Theorem 2 ([4]) *Let $B_{r_0} = \{x \in H_\alpha[a_1, a_2] : ||x||_\alpha \leq r_0\}$ is a relatively compact and closed in $H_\alpha[a_1, a_2]$, then B_{r_0} is a compact subset in the space $H_\alpha[a_1, a_2]$.*

Lemma 1 ([8]) *The function $x \to \frac{x}{\zeta(\eta_1, x)}$ is increasing in $\mathbb{R}$ a.e. for $\eta_1 \in J$. Then for any $h \in L^1(J, \mathbb{R})$, the function $x \in AC(J, \mathbb{R})$ is a solution of the HDE*

$$\begin{cases} \frac{d}{d\eta_1}\left[\frac{x(\eta_1)}{\zeta(\eta_1, x(\eta_1))}\right] = \psi(\eta_1, x(\eta_1)) \\ x(0) = x_0, \end{cases}$$

if and only if x satisfies the hybrid integral equation (HIE)

$$x(\eta_1) = \zeta(\eta_1, x(\eta_1)) \left\{ \frac{x_0}{\zeta(0, x_0)} + \int_0^{\eta_1} \psi(s, x(s))ds \right\}, \eta_1 \in J.$$

3 Main Results

The aim is to enquire the existence of solution of HDE of the type

$$\begin{cases} \frac{d}{d\eta_1}\left[\frac{x(\eta_1)}{\zeta(\eta_1, x(\eta_1))}\right] = \psi(\eta_1, x(\eta_1)) \\ x(0) = x_0, \end{cases} \tag{2}$$

for $\eta_1 \in [0, T]$. Where $\zeta \in H_\beta([0, T] \times \mathbb{R}, \mathbb{R} - \{0\})$ and $\psi \in \mathcal{C}([0, T] \times \mathbb{R}, \mathbb{R})$.

Following are the assumption we will use through this section:

(i) The function $x \rightarrow \frac{x}{\zeta(\eta_1, x)}$ is increasing in $\mathbb{R}$ almost everywhere for $\eta_1 \in [0, T]$.

(ii) $\zeta : [0, T] \times \mathbb{R} \rightarrow \mathbb{R} - \{0\}$ and $\psi : [0, T] \times \mathbb{R} \rightarrow \mathbb{R}$ are continuous and non-decreasing function and there exist constants $\mathcal{P}_n > 0$ for $n = 1, 2, 3$ such that the inequalities

$$|\zeta(\eta_1, x(\eta_1)) - \zeta(\eta_2, y(\eta_2))| \leq \mathcal{P}_1|\eta_1 - \eta_2| + \mathcal{P}_2|x - y|$$

and

$$|\psi(\eta_1, x) - \psi(\eta_1, y)| \leq \mathcal{P}_3|x - y|$$

are satisfied.

Let us denote

$$\bar{\zeta} = \sup\{\zeta(\eta_1, 0) : \eta_1 \in [0, T]\},$$
$$\Lambda = \frac{x_0}{\zeta(0, x_0)}.$$

(iii) There exists a function $h \in L^1([0, 1], \mathbb{R})$ such that

$$|\psi(\eta_1, x)| \leq h(\eta_1)$$

for all $x \in \mathbb{R}$.

(iv) There exists a number r_0 which is a solution of the inequality

$$x_0 + [\Lambda + ||h||_{L_1}][\mathcal{P}_1 + \mathcal{P}_2 r_0] + [\mathcal{P}_2 \max\{1, T^\beta\}r_0 + \bar{\zeta}]||h||_{L_1} < r_0.$$

Theorem 3 *Assume that hypotheses $(i) - (iv)$ are satisfied. Then there exists atleast one solution of Eq. (2) in the Hölder space $H_\alpha[0, T],\ 0 < \alpha < \beta \leq 1$.*

Proof By using Lemma 1, we will get the following equivalent integral equation of Eq. (2)

$$x(\eta_1) = \zeta(\eta_1, x(\eta_1)) \left\{ \frac{x_0}{\zeta(0, x_0)} + \int_0^{\eta_1} \psi(s, x(s))ds \right\}. \tag{3}$$

Let us define the operator Θ on $[0, T]$ for an arbitrarily fixed function $x \in H_\alpha[0, T]$ as

$$\Theta x(\eta_1) = \zeta(\eta_1, x(\eta_1)) \left\{ \frac{x_0}{\zeta(0, x_0)} + \int_0^{\eta_1} \psi(s, x(s))ds \right\}.$$

Choosing $\eta_1, \eta_2 \in [0, T]$ arbitrarily with $\eta_1 \neq \eta_2$.

To show that $\Theta x \in H_\alpha[0, T]$. From the imposed assumption, we obtain

$$\begin{aligned}
&|\Theta x(\eta_1) - \Theta x(\eta_2)| \\
&= \Bigg| \zeta(\eta_1, x(\eta_1)) \left\{ \frac{x_0}{\zeta(0, x_0)} + \int_0^{\eta_1} \psi(s, x(s))ds \right\} - \zeta(\eta_2, x(\eta_2)) \\
&\qquad \left\{ \frac{x_0}{\zeta(0, x_0)} + \int_0^{\eta_2} \psi(s, x(s))ds \right\} \Bigg| \\
&\le \Bigg| \zeta(\eta_1, x(\eta_1)) \left\{ \frac{x_0}{\zeta(0, x_0)} + \int_0^{\eta_1} \psi(s, x(s))ds \right\} - \zeta(\eta_2, x(\eta_1)) \\
&\quad \left\{ \frac{x_0}{\zeta(0, x_0)} + \int_0^{\eta_1} \psi(s, x(s))ds \right\} \Bigg| \\
&\quad + \Bigg| \zeta(\eta_2, x(\eta_1)) \left\{ \frac{x_0}{\zeta(0, x_0)} + \int_0^{\eta_1} \psi(s, x(s))ds \right\} - \zeta(\eta_2, x(\eta_2)) \\
&\quad \left\{ \frac{x_0}{\zeta(0, x_0)} + \int_0^{\eta_1} \psi(s, x(s))ds \right\} \Bigg| \\
&\quad + \Bigg| \zeta(\eta_2, x(\eta_2)) \left\{ \frac{x_0}{\zeta(0, x_0)} + \int_0^{\eta_1} \psi(s, x(s))ds \right\} - \zeta(\eta_2, x(\eta_2)) \\
&\quad \left\{ \frac{x_0}{\zeta(0, x_0)} + \int_0^{\eta_2} \psi(s, x(s))ds \right\} \Bigg| \\
&\le |\zeta(\eta_1, x(\eta_1)) - \zeta(\eta_2, x(\eta_1))| \left| \frac{x_0}{\zeta(0, x_0)} + \int_0^{\eta_1} \psi(s, x(s))ds \right| \\
&\quad + |\zeta(\eta_2, x(\eta_1)) - \zeta(\eta_2, x(\eta_2))| \left| \frac{x_0}{\zeta(0, x_0)} + \int_0^{\eta_1} \psi(s, x(s))ds \right| \\
&\quad + |\zeta(\eta_2, x(\eta_2)) - \zeta(\eta_2, 0) + \zeta(\eta_2, 0)| \left| \int_{\eta_1}^{\eta_2} \psi(s, x(s))ds \right|
\end{aligned}$$

Therefore,

$$
\begin{aligned}
\frac{|\Theta x(\eta_1)-\Theta x(\eta_2)|}{|\eta_1-\eta_2|^{\beta}} \leq & \mathcal{P}_1|\eta_1-\eta_2|^{1-\beta}\left|\frac{x_0}{\zeta(0,x_0)}+\int_0^{\eta_1}\psi(s,x(s))ds\right| \\
& +\frac{\mathcal{P}_2|x(\eta_1)-x(\eta_2)|}{|\eta_1-\eta_2|^{\beta}}\left|\frac{x_0}{\zeta(0,x_0)}+\int_0^{\eta_1}\psi(s,x(s))ds\right| \\
& +\frac{\mathcal{P}_2|x(\eta_2)|+\bar{\zeta}}{|\eta_1-\eta_2|^{\beta}}\left|\int_{\eta_1}^{\eta_2}\psi(s,x(s))ds\right| \\
\leq & \mathcal{P}_1|\eta_1-\eta_2|^{1-\beta}|\Lambda+||h||_{L_1}|+\mathcal{P}_2||x||_{\beta}[\Lambda+||h||_{L_1}] \\
& +[\mathcal{P}_2\max\{1,T^{\beta}\}||x||_{\beta}+\bar{\zeta}]|\eta_1-\eta_2|^{-\beta}||h||_{L_1} \\
\leq & \mathcal{P}_1[\Lambda+||h||_{L_1}]+\mathcal{P}_2||x||_{\beta}[\Lambda+||h||_{L_1}]+[\mathcal{P}_2\max\{1,T^{\beta}\} \\
& \quad ||x||_{\beta}+\bar{\zeta}]||h||_{L_1} \\
\leq & [\Lambda+||h||_{L_1}][\mathcal{P}_1+\mathcal{P}_2||x||_{\beta}]+[\mathcal{P}_2\max\{1,T^{\beta}\}||x||_{\beta}+\bar{\zeta}]||h||_{L_1}.
\end{aligned} \tag{4}
$$

In the similar way

$$
|\Theta x(0)| = x_0. \tag{5}
$$

Combining estimates (4) and (5) we get

$$
\begin{aligned}
|\Theta x(0)|+\frac{|\Theta x(\eta_1)-\Theta x(\eta_2)|}{|\eta_1-\eta_2|^{\beta}} &\leq x_0+[\Lambda+||h||_{L_1}][\mathcal{P}_1 \\
&+\mathcal{P}_2||x||_{\beta}]+[\mathcal{P}_2\max\{1,T^{\beta}\}||x||_{\beta}+\bar{\zeta}]||h||_{L_1} \\
\Longrightarrow ||\Theta x||_{\beta} &\leq x_0+[\Lambda+||h||_{L_1}][\mathcal{P}_1+\mathcal{P}_2||x||_{\beta}]+[\mathcal{P}_2\max\{1,T^{\beta}\}||x||_{\beta}+\bar{\zeta}]||h||_{L_1}.
\end{aligned}
$$

According to assumption (iv) the above inequality

$$
||\Theta x||_{\beta} \leq x_0+[\Lambda+||h||_{L_1}][\mathcal{P}_1+\mathcal{P}_2 r_0]+[\mathcal{P}_2\max\{1,T^{\beta}\}r_0+\bar{\zeta}]||h||_{L_1} < r_0
$$

holds. Thus, the operator Θ is well defined. Additionally, the operator Θ transforms the ball B_{r_0} with radius r_0 into itself. The ball B_{r_0} is compact in $H_{\alpha}[0,T]$ according to Theorems 1 and 2. Which implies the compactness of B_{r_0} in $H_{\alpha}[0,T]$. Next to show the continuity of the operator Θ on B_{r_0}. Suppose that $y \in B_{r_0}$ such that $||x-y||_{\alpha} < \delta$, where $x \in B_{r_0}$ arbitrarily fixed and $\delta > 0$. By using our assumption, we obtain

$$
\begin{aligned}
&\frac{[(\Theta x)(\eta_1)-(\Theta y)(\eta_1)]-[(\Theta x)(\eta_2)-(\Theta y)(\eta_2)]}{|\eta_1-\eta_2|^{\alpha}} \\
=&\Bigg|\Bigg[\zeta(\eta_1,x(\eta_1))\left\{\frac{x_0}{\zeta(0,x_0)}+\int_0^{\eta_1}\psi(s,x(s))ds\right\}-\zeta(\eta_1,y(\eta_1))\left\{\frac{x_0}{\zeta(0,x_0)}+\int_0^{\eta_1}\psi(s,y(s))ds\right\}\Bigg] \\
&-\Bigg[\zeta(\eta_2,x(\eta_2))\left\{\frac{x_0}{\zeta(0,x_0)}+\int_0^{\eta_2}\psi(s,x(s))ds\right\}-\zeta(\eta_2,y(\eta_2))\left\{\frac{x_0}{\zeta(0,x_0)}+\int_0^{\eta_2}\psi(s,y(s))ds\right\}\Bigg]\Bigg||\eta_1-\eta_2|^{-\alpha} \\
\leq&\Bigg|\zeta(\eta_1,x(\eta_1))\left\{\frac{x_0}{\zeta(0,x_0)}+\int_0^{\eta_1}\psi(s,x(s))ds\right\}-\zeta(\eta_2,x(\eta_2))\left\{\frac{x_0}{\zeta(0,x_0)}+\int_0^{\eta_1}\psi(s,x(s))ds\right\} \\
&+\zeta(\eta_2,x(\eta_2))\left\{\frac{x_0}{\zeta(0,x_0)}+\int_0^{\eta_1}\psi(s,x(s))ds\right\}-\zeta(\eta_1,y(\eta_1))\left\{\frac{x_0}{\zeta(0,x_0)}+\int_0^{\eta_1}\psi(s,y(s))ds\right\} \\
&-\Bigg[\zeta(\eta_2,x(\eta_2))\left\{\frac{x_0}{\zeta(0,x_0)}+\int_0^{\eta_2}\psi(s,x(s))ds\right\}-\zeta(\eta_2,y(\eta_2))\left\{\frac{x_0}{\zeta(0,x_0)}+\int_0^{\eta_2}\psi(s,y(s))ds\right\}\Bigg]\Bigg||\eta_1-\eta_2|^{-\alpha}
\end{aligned}
$$

$$
\begin{aligned}
\leq & \bigg|\zeta(\eta_1, x(\eta_1))\left\{\frac{x_0}{\zeta(0,x_0)} + \int_0^{\eta_1}\psi(s,x(s))ds\right\} - \zeta(\eta_2, x(\eta_2))\left\{\frac{x_0}{\zeta(0,x_0)} + \int_0^{\eta_1}\psi(s,x(s))ds\right\}\bigg| \\
& + \bigg|\zeta(\eta_2, x(\eta_2))\left\{\frac{x_0}{\zeta(0,x_0)} + \int_0^{\eta_1}\psi(s,x(s))ds\right\} - \zeta(\eta_1, y(\eta_1))\left\{\frac{x_0}{\zeta(0,x_0)} + \int_0^{\eta_1}\psi(s,y(s))ds\right\} \\
& + \zeta(\eta_2, y(\eta_2))\left\{\frac{x_0}{\zeta(0,x_0)} + \int_0^{\eta_1}\psi(s,y(s))ds\right\} - \zeta(\eta_2, y(\eta_2))\left\{\frac{x_0}{\zeta(0,x_0)} + \int_0^{\eta_1}\psi(s,y(s))ds\right\} \\
& - \left[\zeta(\eta_2, x(\eta_2))\left\{\frac{x_0}{\zeta(0,x_0)} + \int_0^{\eta_2}\psi(s,x(s))ds\right\} - \zeta(\eta_2, y(\eta_2))\left\{\frac{x_0}{\zeta(0,x_0)} + \int_0^{\eta_2}\psi(s,y(s))ds\right\}\right]\bigg| \\
\leq & \bigg|\zeta(\eta_1, x(\eta_1))\left\{\frac{x_0}{\zeta(0,x_0)} + \int_0^{\eta_1}\psi(s,x(s))ds\right\} - \zeta(\eta_2, x(\eta_2))\left\{\frac{x_0}{\zeta(0,x_0)} + \int_0^{\eta_1}\psi(s,x(s))ds\right\}\bigg| \\
& + \bigg|\zeta(\eta_2, y(\eta_2))\left\{\frac{x_0}{\zeta(0,x_0)} + \int_0^{\eta_1}\psi(s,y(s))ds\right\} - \zeta(\eta_1, y(\eta_1))\left\{\frac{x_0}{\zeta(0,x_0)} + \int_0^{\eta_1}\psi(s,y(s))ds\right\}\bigg| \\
& + \bigg|\zeta(\eta_2, x(\eta_2))\left\{\frac{x_0}{\zeta(0,x_0)} + \int_0^{\eta_1}\psi(s,x(s))ds\right\} - \zeta(\eta_2, y(\eta_2))\left\{\frac{x_0}{\zeta(0,x_0)} + \int_0^{\eta_1}\psi(s,y(s))ds\right\} \\
& - \left[\zeta(\eta_2, x(\eta_2))\left\{\frac{x_0}{\zeta(0,x_0)} + \int_0^{\eta_2}\psi(s,x(s))ds\right\} - \zeta(\eta_2, y(\eta_2))\left\{\frac{x_0}{\zeta(0,x_0)} + \int_0^{\eta_2}\psi(s,y(s))ds\right\}\right]\bigg|
\end{aligned}
$$

$$
\begin{aligned}
\leq & \bigg|\zeta(\eta_1, x(\eta_1))\left\{\frac{x_0}{\zeta(0,x_0)} + \int_0^{\eta_1}\psi(s,x(s))ds\right\} - \zeta(\eta_2, x(\eta_2))\left\{\frac{x_0}{\zeta(0,x_0)} + \int_0^{\eta_1}\psi(s,x(s))ds\right\}\bigg| \\
& + \bigg|\zeta(\eta_2, y(\eta_2))\left\{\frac{x_0}{\zeta(0,x_0)} + \int_0^{\eta_1}\psi(s,y(s))ds\right\} - \zeta(\eta_1, y(\eta_1))\left\{\frac{x_0}{\zeta(0,x_0)} + \int_0^{\eta_1}\psi(s,y(s))ds\right\}\bigg| \\
& + \bigg|\left[\zeta(\eta_2, x(\eta_2))\left\{\frac{x_0}{\zeta(0,x_0)} + \int_0^{\eta_1}\psi(s,x(s))ds\right\} - \zeta(\eta_2, x(\eta_2))\left\{\frac{x_0}{\zeta(0,x_0)} + \int_0^{\eta_2}\psi(s,x(s))ds\right\}\right] \\
& + \left[\zeta(\eta_2, y(\eta_2))\left\{\frac{x_0}{\zeta(0,x_0)} + \int_0^{\eta_2}\psi(s,y(s))ds\right\} - \zeta(\eta_2, y(\eta_2))\left\{\frac{x_0}{\zeta(0,x_0)} + \int_0^{\eta_1}\psi(s,y(s))ds\right\}\right]\bigg| \\
\leq & \bigg|\zeta(\eta_1, x(\eta_1))\left\{\frac{x_0}{\zeta(0,x_0)} + \int_0^{\eta_1}\psi(s,x(s))ds\right\} - \zeta(\eta_2, x(\eta_2))\left\{\frac{x_0}{\zeta(0,x_0)} + \int_0^{\eta_1}\psi(s,x(s))ds\right\}\bigg| \\
& + \bigg|\zeta(\eta_2, y(\eta_2))\left\{\frac{x_0}{\zeta(0,x_0)} + \int_0^{\eta_1}\psi(s,y(s))ds\right\} - \zeta(\eta_1, y(\eta_1))\left\{\frac{x_0}{\zeta(0,x_0)} + \int_0^{\eta_1}\psi(s,y(s))ds\right\}\bigg| \\
& + \bigg|\zeta(\eta_2, x(\eta_2))\left[\int_0^{\eta_1}\psi(s,x(s))ds - \int_0^{\eta_2}\psi(s,x(s))ds\right] \\
& + \zeta(\eta_2, y(\eta_2))\left[\int_0^{\eta_2}\psi(s,y(s))ds - \int_0^{\eta_1}\psi(s,y(s))ds\right]\bigg| \\
\leq & \bigg|\zeta(\eta_1, x(\eta_1))\left\{\frac{x_0}{\zeta(0,x_0)} + \int_0^{\eta_1}\psi(s,x(s))ds\right\} - \zeta(\eta_2, x(\eta_2))\left\{\frac{x_0}{\zeta(0,x_0)} + \int_0^{\eta_1}\psi(s,x(s))ds\right\}\bigg| \\
& + \bigg|\zeta(\eta_2, y(\eta_2))\left\{\frac{x_0}{\zeta(0,x_0)} + \int_0^{\eta_1}\psi(s,y(s))ds\right\} - \zeta(\eta_1, y(\eta_1))\left\{\frac{x_0}{\zeta(0,x_0)} + \int_0^{\eta_1}\psi(s,y(s))ds\right\}\bigg| \\
& + \bigg|\zeta(\eta_2, x(\eta_2))\int_{\eta_2}^{\eta_1}\psi(s,x(s))ds + \zeta(\eta_2, y(\eta_2))\int_{\eta_1}^{\eta_2}\psi(s,y(s))ds\bigg| \\
\leq & \bigg|\zeta(\eta_1, x(\eta_1))\left\{\frac{x_0}{\zeta(0,x_0)} + \int_0^{\eta_1}\psi(s,x(s))ds\right\} - \zeta(\eta_2, x(\eta_2))\left\{\frac{x_0}{\zeta(0,x_0)} + \int_0^{\eta_1}\psi(s,x(s))ds\right\}\bigg| \\
& + \bigg|\zeta(\eta_2, y(\eta_2))\left\{\frac{x_0}{\zeta(0,x_0)} + \int_0^{\eta_1}\psi(s,y(s))ds\right\} - \zeta(\eta_1, y(\eta_1))\left\{\frac{x_0}{\zeta(0,x_0)} + \int_0^{\eta_1}\psi(s,y(s))ds\right\}\bigg| \\
& + \bigg|-\zeta(\eta_2, x(\eta_2))\int_{\eta_1}^{\eta_2}\psi(s,x(s))ds + \zeta(\eta_2, y(\eta_2))\int_{\eta_1}^{\eta_2}\psi(s,x(s))ds \\
& - \zeta(\eta_2, y(\eta_2))\int_{\eta_1}^{\eta_2}\psi(s,x(s))ds + \zeta(\eta_2, y(\eta_2))\int_{\eta_1}^{\eta_2}\psi(s,y(s))ds\bigg|
\end{aligned}
$$

$$
\frac{[(\Theta x)(\eta_1) - (\Theta y)(\eta_1)] - [(\Theta x)(\eta_2) - (\Theta y)(\eta_2)]}{|\eta_1 - \eta_2|^{\alpha}}
$$

$$
\leq \frac{|\zeta(\eta_1, x(\eta_1)) - \zeta(\eta_2, x(\eta_2))|}{|\eta_1 - \eta_2|^{\alpha}}\left|\frac{x_0}{\zeta(0,x_0)} + \int_0^{\eta_1}\psi(s,x(s))ds\right|
$$

$$+\frac{|\zeta(\eta_2,y(\eta_2))-\zeta(\eta_1,y(\eta_1))|}{|\eta_1-\eta_2|^\alpha}\left|\frac{x_0}{\zeta(0,x_0)}+\int_0^{\eta_1}\psi(s,y(s))ds\right|$$
$$+\frac{|\zeta(\eta_2,y(\eta_2))-\zeta(\eta_2,x(\eta_2))|}{|\eta_1-\eta_2|^\alpha}\left|\int_{\eta_1}^{\eta_2}\psi(s,x(s))ds\right|+\frac{|\zeta(\eta_2,y(\eta_2))|}{|\eta_1-\eta_2|^\alpha}\left|\int_{\eta_1}^{\eta_2}(\psi(s,y(s))-\psi(s,x(s)))ds\right|$$

Now,

$$\begin{aligned}
&\le\frac{\mathcal{P}_1|\eta_1-\eta_2|+\mathcal{P}_2|x(\eta_1)-x(\eta_2)|}{|\eta_1-\eta_2|^\alpha}\left[\Lambda+\int_0^{\eta_1}h(s)|ds|\right]\\
&+\frac{\mathcal{P}_1|\eta_2-\eta_1|+\mathcal{P}_2|y(\eta_2)-y(\eta_1)|}{|\eta_1-\eta_2|^\alpha}\left[\Lambda+\int_0^{\eta_1}h(s)|ds|\right]\\
&+\frac{\mathcal{P}_2|y(\eta_2)-x(\eta_2)|}{|\eta_1-\eta_2|^\alpha}\int_{\eta_1}^{\eta_2}h(s)|ds|+\frac{|\zeta(\eta_2,y(\eta_2))-\zeta(\eta_2,0)+\zeta(\eta_2,0)|}{|\eta_1-\eta_2|^\alpha}\int_{\eta_1}^{\eta_2}\mathcal{P}_3|y(s)-x(s)||ds|\\
&\le[\mathcal{P}_1|\eta_1-\eta_2|^{1-\alpha}+\mathcal{P}_2||x||_\alpha][\Lambda+||h||_{L_1}]+[\mathcal{P}_1|\eta_1-\eta_2|^{1-\alpha}+\mathcal{P}_2||y||_\alpha][\Lambda+||h||_{L_1}]\\
&+\mathcal{P}_2\frac{|y(\eta_2)-x(\eta_2)-[y(0)-x(0)]|}{|\eta_1-\eta_2|^\alpha}\int_{\eta_1}^{\eta_2}h(s)|ds|+\mathcal{P}_2\frac{|y(0)-x(0)|}{|\eta_1-\eta_2|^\alpha}\int_{\eta_1}^{\eta_2}h(s)|ds|\\
&+\frac{\mathcal{P}_2|y(\eta_2)|+\bar{\zeta}}{|\eta_1-\eta_2|^\alpha}\mathcal{P}_3\delta|\eta_2-\eta_1|\\
&\le[\mathcal{P}_1|\eta_1-\eta_2|^{1-\alpha}+\mathcal{P}_2||x||_\alpha][\Lambda+||h||_{L_1}]+[\mathcal{P}_1|\eta_1-\eta_2|^{1-\alpha}+\mathcal{P}_2(||x||_\alpha+\delta)][\Lambda+||h||_{L_1}] \qquad (6)\\
&+\mathcal{P}_2\sup_{\eta_1,\eta_2\in[0,T]}\frac{|y(\eta_2)-x(\eta_2)-[y(\eta_1)-x(\eta_1)]|}{|\eta_1-\eta_2|^\alpha}||h||_{L_1}\\
&+\mathcal{P}_2\delta||h||_{L_1}|\eta_1-\eta_2|^{-\alpha}+\mathcal{P}_3\delta[\mathcal{P}_2||y||_\infty+\bar{\zeta}]|\eta_1-\eta_2|^{1-\alpha}\\
&\le[2(\mathcal{P}_1+\mathcal{P}_2||x||_\alpha)+\mathcal{P}_2\delta][\Lambda+||h||_{L_1}]\\
&+\mathcal{P}_2\sup_{\eta_1,\eta_2\in[0,T]}\frac{|y(\eta_2)-x(\eta_2)-[y(\eta_1)-x(\eta_1)]|}{|\eta_1-\eta_2|^\alpha}||h||_{L_1}\\
&+\mathcal{P}_2\delta||h||_{L_1}+\mathcal{P}_3\delta[\mathcal{P}_2(||x||_\alpha+\delta)+\bar{\zeta}]\\
&\le[2(\mathcal{P}_1+\mathcal{P}_2||x||_\alpha)+\mathcal{P}_2\delta][\Lambda+||h||_{L_1}]+\mathcal{P}_2||x-y||_\alpha||h||_{L_1}\\
&+\mathcal{P}_2\delta||h||_{L_1}+\mathcal{P}_3\delta[\mathcal{P}_2(||x||_\alpha+\delta)+\bar{\zeta}]\\
&\le[2(\mathcal{P}_1+\mathcal{P}_2||x||_\alpha)+\mathcal{P}_2\delta][\Lambda+||h||_{L_1}]+2\mathcal{P}_2\delta||h||_{L_1}+\mathcal{P}_3\delta[\mathcal{P}_2(||x||_\alpha+\delta)+\bar{\zeta}]
\end{aligned}$$

In the similar way

$$|(\Theta x)(0)-(\Theta y)(0)|=0. \qquad (7)$$

Now, combining (6) and (7), we obtain

$$\begin{aligned}||\Theta x-\Theta y||_\alpha\le&[2(\mathcal{P}_1+\mathcal{P}_2||x||_\alpha)+\mathcal{P}_2\delta][\Lambda+||h||_{L_1}]+2\mathcal{P}_2\delta||h||_{L_1}\\&+\mathcal{P}_3\delta[\mathcal{P}_2(||x||_\alpha+\delta)+\bar{\zeta}]<\epsilon.\end{aligned}$$

From the above inequality and assumption (ii), we infer that the continuity of the operator Θ at $x\in B_{r_0}(\subset H_\alpha[0,T])$. Thus, by the Schauder FPT [13], there exists atleast one fixed point $x\in B_{r_0}$ of the operator Θ. This completes the proof.

We provide an example which illustrates the applicability of our results.

Example 1 Consider the following HDE of the type

$$\frac{d}{d\eta_1}\left[\frac{x(\eta_1)}{\frac{1}{8}\{e^{-\eta_1}+\sin|x(\eta_1)|+32\}}\right]=\frac{\sin(1+|x(\eta_1))}{100(1+|x(\eta_1)|)} \tag{8}$$
$$x(0)=1, \quad for\ \eta_1\in[0,1].$$

Comparing Eq. (8) with Eq. (2), we have

$$\begin{aligned}\zeta(\eta_1,x(\eta_1)) =&\frac{1}{8}\{e^{-\eta_1}+\sin|x(\eta_1)|+32\},\\ \psi(\eta_1,x(\eta_1)) =&\frac{1}{100}\frac{\sin(1+|x(\eta_1)|)}{1+|x(\eta_1)|}.\end{aligned}$$

Now,

$$\begin{aligned}|\zeta(\eta_1,x(\eta_1))-\zeta(\eta_2,y(\eta_2))| =&\frac{1}{8}|e^{-\eta_1}+\sin|x(\eta_1)|+32-\{e^{-\eta_2}+\sin|y(\eta_2)|+32\}|\\ \leq&\frac{1}{8}|e^{-\eta_1}-e^{-\eta_2}|+\frac{1}{8}|\sin|x(\eta_1)|-\sin|y(\eta_2)||\\ \leq&\frac{1}{8}|\eta_1-\eta_2|+\frac{1}{8}|\sin|x(\eta_1)|-\sin|x(\eta_2)||+\frac{1}{8}|\sin|x(\eta_2)|-\sin|y(\eta_2)||\\ \leq&\frac{1}{4}|\eta_1-\eta_2|+\frac{1}{8}|x-y|.\end{aligned}$$

Hence, we obtain $\mathcal{P}_1=\frac{1}{4}$, $\mathcal{P}_2=\frac{1}{8}$.

$$\begin{aligned}|\psi(\eta_1,x(\eta_1))-\psi(\eta_1,y(\eta_1))| =&\frac{1}{100}\left|\frac{\sin(1+|x(\eta_1)|)}{1+|x(\eta_1)|}-\frac{\sin(1+|y(\eta_1)|)}{1+|y(\eta_1)|}\right|\\ \leq&\frac{1}{100}\left|\frac{\sin(1+|x(\eta_1)|)}{1+|x(\eta_1)|}-\frac{\sin(1+|y(\eta_1)|)}{1+|x(\eta_1)|}\right|\\ &+\left|\frac{\sin(1+|y(\eta_1)|)}{1+|x(\eta_1)|}-\frac{\sin(1+|y(\eta_1)|)}{1+|y(\eta_1)|}\right|\\ \leq&\frac{1}{100}\left|\frac{x(\eta_1)-y(\eta_1)}{1+|x(\eta_1)|}\right|\\ &+\left|\left(\frac{\sin(1+y(\eta_1))|y(\eta_1)-x(\eta_1)|}{(1+|x(\eta_1)|)(1+|y(\eta_1)|)}\right)\right|\\ \leq&\frac{1}{100}\frac{|x-y|}{1+|x|}\left(1+\frac{|\sin(1+y|}{1+|y|}\right)\\ \leq&\frac{2}{100}\frac{|x-y|}{1+|x|}\\ \leq&\frac{1}{50}|x-y|.\end{aligned}$$

Therefore, assumption (ii) is satisfied. Moreover,

$$|\psi(\eta_1, x(\eta_1))| \leq \left|\frac{\sin(1 + |x(\eta_1)|)}{100(1 + |x(\eta_1)|)}\right| \leq \frac{1}{100}.$$

Therefore, $\mathcal{P}_3 = \frac{1}{50}$ and $h(\eta_1) = \frac{1}{100}$. So,

$$\|h\|_{L^1} = \frac{1}{100}\int_0^1 d\eta_1 = \frac{1}{100}.$$

Assumption (iii) is satisfied.
Also assumption (i) is satisfied as the map $x \to \frac{x(\eta_1)}{\frac{1}{8}\{e^{-\eta_1} + \sin|x(\eta_1)| + 32\}}$ is increasing in $\mathbb{R}$ almost everywhere for $\eta_1 \in [0, 1]$.

$$\bar{\zeta} = \sup\{|\zeta(\eta_1, 0)| : \eta_1 \in [0, 1]\} = \sup_{\eta_1 \in [0,1]} \{\frac{1}{8}(e^{-\eta_1} + 32)\} = 4.125.$$

$$\Lambda = \frac{x_0}{\zeta(0, x_0)} = 0.2364\,.$$

The inequality of assumption (iv) has the form

$$x_0 + [\Lambda + ||h||_{L_1}][\mathcal{P}_1 + \mathcal{P}_2 r_0] + [\mathcal{P}_2 \max\{1, T^\beta\}r_0 + \bar{\zeta}]||h||_{L_1} \leq r_0$$
$$\implies 1 + (0.2364 + 0.01)(\frac{1}{4} + \frac{1}{8}r_0) + (\frac{1}{8}r_0 + 4.125)0.01 \leq r_0$$
$$\implies 1.1029 + 0.0321 r_0 \leq r_0.$$

Clearly, the above inequality is true for $r = 1.2$. Therefore, using Theorem 3, we assert that Eq. (8) has atleast one solution for $r = 1.2$ in the space $H_\alpha[0, 1]$.

4 Conclusion

In this work, we discussed the existence of a solution of HDE in the space $H_\alpha[0, T]$ defined on $[0, T]$. Quadratic perturbation of a differential equation is widely known as HDE. These type of equations are highly recommended as they include many dynamic systems as special case that is open to research. HDE are usually handled by using hybrid fixed point theorems, which are combination of Schauder FPT and Banach Contraction principle. In this research work, we tackled HDE by using Schauder Fixed point theorem only. We used Schauder fixed point point theorem to prove existence result for the solution. To show the efficiency of theoretical approach, we provide an example.

References

1. Agarwal, R.P., Meehan, M., O'Regan, D.: Fixed point theory and applications. Cambridge University Press (2021)
2. Banach, S.: Sur less opérations dans les ensembles abstraits ets leur applications aux equation integrates. Fund. Math. **3**, 133–181 (1922)
3. Banaś, J., Nalepa, R.: On the space of functions with growths tempered by a modulus of continuity and its applications. J. Funct. Spaces Appl. **2013**, 820437 (2013)
4. Caballero, M.J., Darwish, M.A., Sadarangani, K.: Solvability of a quadratic integral equation of Fredholm type in Hölder spaces. Electron. J. Diff. Equ. **31**, 1–10 (2014)
5. Dhage, B.C.: A fixed point theorem in Banach Algebras with applications to functional integral equations. Kyungpook Math. J. **44**, 145–155 (2004)
6. Dhage, B.C.: On some variants of Schauder fixed point principle and applications to nonlinear integral equations. J. Math. Phys. Sci. **22**, 603–611 (1988)
7. Dhage, B.C.: Quadratic perturbations of periodic boundary value problems of second order ordinary differential equations. Diff. Eq. Appl. **2**(4), 465–486 (2010)
8. Dhage, B.C., Lakshmikantham, V.: Basic results on hybrid differential equation. Nonlinear Anal. Hybrid **4**, 414–424 (2010)
9. Dhage, B.C., Jadhav, N.S.: Basic results in the the theory of hybrid differential equations with linear perturbations of second type. Tamkang J. Math. **44**(2), 171–186 (2013)
10. Dhage, B.C., Dhage, S.B.: Approximating positive solutions of PBVPs of nonlinear first order ordinary quadratic differential equations. Aplied Math. Lett. **46**, 133–142 (2015)
11. Dhage, B.C.: Basic results in the theory of hybrid casual nonlinear differential equations. KYUNGPOOK Math. J. **55**, 1069–1088 (2015)
12. Lu, H., Sun, S., Yang, D., Teng, H.: Theory of fractional hybrid differential equations with linear perturbations of second type. Bound. Value Probl. **2013**(23) (2013)
13. Schauder, J.: Der: Fixpunktsatz in Funktionalriiumen. Studia Math. **2**(1), 171–180 (1930). https://doi.org/10.4060/sm-2-1-171-180
14. Sun, S., Zhao, Y., Han, Z., Li, Y.: The existence of solutions for boundary value problem of fractional hybrid differential equations. Commun. Nonlinear Sci. Numer. Simul. **17**, 4961–4967 (2012)
15. Zhao, Y., Suna, S., Hana, Z., Li, Q.: Theory of fractional hybrid differential equations. Comput. Math. Appl. **62**(3), 1312–1324 (2011)

The Exponential Stability of Nonlinear Shallow Water Equations Within a Trapezoidal Channel

Seydou Sore, Babacar Mbaye Ndiaye, and Yacouba Simporé

Abstract This research focuses on the exponential stability of the nonlinear Saint-Venant equations within a trapezoidal channel. We examine a general framework that incorporates arbitrary friction, spatially varying slopes, and channel geometry variations, resulting in non-uniform steady states. To analyze stability, we construct an explicit quadratic Lyapunov function as a weighting function for small perturbations around the steady state. It is then demonstrated that, with a well-designed boundary feedback control, the nonlinear Saint-Venant equations governing a trapezoidal channel achieve local exponential stability in the H^2-norm. Finally, we explicitly define this control and validate our findings through numerical simulations.

Keywords The inhomogeneous equations of Saint-Venant · The trapezoidal channel · The Lyapunov approach · Stabilization · Feedback control

1 Introduction

Thc Saint-Venant equations [1–3] are extensively employed in engineering applications to simulate shallow water flow [15, 19, 25]. This set of equations constitutes a nonlinear 2×2 hyperbolic system. The stabilization of such systems via boundary control has been studied extensively [20, 21, 28]. Among the methods used for this study Other control strategies include proportional-integral (PI) boundary control [4, 18, 22, 30, 31], backstepping-based full-state feedback [14, 16, 17], and internal stabilization methods [13], the one based on Lyapunov's technique have played a

S. Sore (✉) · Y. Simporé
Laboratoire de Mathématiques, Informatique et Application (LAMIA), Université Norbert ZONGO, Koudougou, Burkina Faso
e-mail: seydousore27@gmail.com

B. M. Ndiaye
Laboratoire de Mathématiques de la Décision et d'Analyse Numérique (LMDAN), Université Cheikh Anta Diop, Dakar-fann, Senegal

Y. Simporé
Université Yembila Abdoulaye TOGUYENI, Fada N'gourma, Burkina Faso

H. Kalita et al. (eds.), *Nonlinear Analysis and Computational Techniques*,
Springer Proceedings in Mathematics & Statistics 501,
https://doi.org/10.1007/978-981-96-6038-4_9

crucial role in proving the exponential stability of nonlinear hyperbolic systems. The quadratic Lyapunov function approach, introduced by Coron et al. [9, 10], ensures stability when boundary conditions satisfy sufficient dissipativity properties. More recently [6, 23, 24], the Lyapunov approach has been successfully applied to inhomogeneous Saint-Venant equations, even for arbitrary friction, various cross-sectional profiles, and different slopes.

Despite these advances, except the most recent work [27] which deals with exponential stability for the L^2-norm, the stability of the Saint-Venant equations in trapezoidal channels is a relatively unexamined field, though it is crucial for accurately modeling real-world flow scenarios. Trapezoidal channels are commonly used in engineering due to their balance between construction simplicity and flow efficiency compared to circular channels. However, their non-rectangular, non-prismatic nature introduces additional complexity, as their width and side slope vary along the channel length. Addressing these variations is essential for developing more realistic and effective control strategies.

In this study, we establish a clear Lyapunov control function designed to secure local exponential stability of the nonlinear Saint-Venant equations in the H^2-norm utilizing a feedback strategy. This strategy facilitates the creation of strong static feedback controllers that assure the exponential stability of steady states. Specifically, we address cases where the slope, bottom width, and lateral slope change with respect to space, a scenario frequently encountered in natural waterways. These variations, even over short distances, significantly impact flow behavior and stability.

The rest of the document is structured into four sections: Sect. 2 provides an overview of the nonlinear Saint-Venant equations in a trapezoidal channel. In Sect. 3, we introduce key definitions and present our main result. Section 4 focuses on the exponential stability for the linearized system, where a quadratic Lyapunov function is constructed. Building on these findings, we demonstrate that an extended Lyapunov function can also establish the exponential stability of the nonlinear system, given an appropriate selection of boundary feedback controls. Finally, Sect. 5 presents numerical simulations that validate the theoretical results.

2 The Saint-Venant Model

In a trapezoidal channel, the broadest formulation of the one-dimensional Saint-Venant equations is delineated as follows (see [27]).

$$\begin{aligned} &\partial_t \mathcal{A} + \partial_x \mathcal{A}\mathcal{V} = 0, && L_1 \\ &\partial_t \mathcal{V} + \mathcal{V}\partial_x \mathcal{V} + \frac{g}{T}\partial_x \mathcal{A} - g(S_b - S_f) = 0, && L_2 \end{aligned} \tag{1}$$

Here, $\mathcal{A}$ represents the cross-sectional area, while $\mathcal{V}$ denotes the flow velocity. g denotes the acceleration due to gravity. The parameters $\mathbf{S}_f$ and $\mathbf{S}_b$ denote the friction slope and the channel slope, respectively. The friction slope is given by $S_f = k\frac{\mathcal{V}^2}{\mathcal{A}}$, where k is a friction coefficient. $T(t,x) = \sqrt{b^2(x) + 4m(x)\mathcal{A}(t,x)}$, with m the inverse slope of the channel walls and b the channel width at the bottom.

The steady states $(\mathcal{A}^*, \mathcal{V}^*)$ of the system (1) satisfy the relation

$$\begin{cases} \partial_x\big(\mathcal{A}^*\mathcal{V}^*\big) = 0, \\ \mathcal{V}^*\partial_x\mathcal{V}^* + \dfrac{g}{T^*}\partial_x\mathcal{A}^* = g(S_b - k\dfrac{\mathcal{V}^{*2}}{\mathcal{A}^*}), \end{cases} \tag{2}$$

For simplicity in future notations, let us set

$$T^*(x) = \sqrt{b^{*2}(x) + 4m^*(x)\mathcal{A}^*(x)}, \quad T_0^* = T^*(0), \quad T_L = T^*(L). \tag{3}$$

In this study, our focus is on physically meaningful stationary states, so we assume that

$$\mathcal{A}^* > 0, \quad \mathcal{V}^* > 0 \tag{4}$$

Additionally, we consider the steep slope regime, as it presents the greatest challenge for stabilization. In this case, $\mathcal{V}^*$ tends to decreases while $\mathcal{A}^*$ increase, causing the system to drift away from the fluvial regime threshold, defined by:

$$\frac{g\mathcal{A}^*}{T^*\mathcal{V}^{*2}} = 1.$$

Under these conditions, we obtain:

$$S_b > k\frac{\mathcal{V}^{*2}}{\mathcal{A}^*} \tag{5}$$

By applying (2), we find that V^2 satisfies

$$\partial_x\mathcal{V}^* = \frac{g\mathcal{V}^*(S_b - k\dfrac{\mathcal{V}^{*2}}{\mathcal{A}^*})}{\mathcal{V}^{*2} - \dfrac{g\mathcal{A}^*}{T^*}} \tag{6}$$

It is essential to recognize that the stationary states exhibit non-uniform characteristics. Furthermore, we operate under the assumption that the flow functions within the subcritical regime, where the Froude number is consistently less than 1. In this case, the system requires a prescribed of two boundary conditions, one upstream and the other downstream, to ensure a well-posed formulation. For any $\big(\mathcal{A}^*(0), \mathcal{V}^*(0)\big)$ such that

$$\frac{g\mathcal{A}^*(0)}{T_0^*\mathcal{V}^{*2}(0)} > 1,$$

the problem (2) admits a unique maximal solution, provided that the following condition holds:

$$\frac{g\mathcal{A}^*}{T^*\mathcal{V}^{*2}} > 1. \tag{7}$$

2.1 Linearisation

Here we introduce v and a defined by:

$$v(t,x) = \mathcal{V}(t,x) - \mathcal{V}^*(x), \quad a(t,x) = \mathcal{A}(t,x) - \mathcal{A}^*(x). \tag{8}$$

as perturbation functions and we remember that the linearized form of a function $f(\mathcal{A}, \mathcal{V})$ around the point $(\mathcal{A}^*, \mathcal{V}^*)$ is expressed as:

$$f(\mathcal{A},\mathcal{V}) \approx f(\mathcal{A}^*,\mathcal{V}^*) + \frac{\partial f}{\partial \mathcal{A}}(\mathcal{A}^*,\mathcal{V}^*)a + \frac{\partial f}{\partial \mathcal{V}}(\mathcal{A}^*,\mathcal{V}^*)v \tag{9}$$

Using (8) and (9) in (1), and taking into account the relations (2), L_1 can be linearized as follows:

$$\begin{aligned}
&\partial_t \mathcal{A} + \partial_x \mathcal{A}\mathcal{V} = \partial_t\left(\mathcal{A}^* + a\right) + \partial_x\left(\mathcal{V}^*\mathcal{A}^* + \mathcal{V}^* a + \mathcal{A}^* v\right)\\
&\qquad \Longrightarrow \partial_t \mathcal{A} + \partial_x \mathcal{A}\mathcal{V} = \partial_t a + \partial_x(\mathcal{V}^*\mathcal{A}^*) + \mathcal{V}^*\partial_x a + (\partial_x \mathcal{V}^*)a + \mathcal{A}^*\partial_x v + (\partial_x \mathcal{A}^*)v\\
&\qquad \Longrightarrow \partial_t \mathcal{A} + \partial_x \mathcal{A}\mathcal{V} = \partial_t a + \mathcal{V}^*\partial_x a + \mathcal{A}^*\partial_x v + (\partial_x \mathcal{V}^*)a + (\partial_x \mathcal{A}^*)v
\end{aligned} \tag{10}$$

Similarly, L_2 can be linearized as follows:

$$\begin{aligned}
&\partial_t \mathcal{V} + \mathcal{V}\partial_x \mathcal{V} + \frac{g}{T}\partial_x \mathcal{A} = g(S_b - S_f)\\
&\Rightarrow \partial_t v + (\partial_x \mathcal{V}^*)v + \mathcal{V}^*\partial_x v + \mathcal{V}^*\partial_x \mathcal{V}^* + \frac{g}{T^*}\partial_x a + \frac{g}{T^*}\partial_x \mathcal{A}^*\\
&\qquad - \left(\frac{2mg}{T^{*3}}\partial_x \mathcal{A}^*\right)a - g\left(\frac{k\mathcal{V}^{*2}}{\mathcal{A}^{*2}}a - \frac{2k\mathcal{V}^*}{\mathcal{A}^*}v\right) - g\left(S_b - \frac{k\mathcal{V}^{*2}}{\mathcal{A}^*}\right) = 0\\
&\Rightarrow \partial_t v + \frac{g}{T^*}\partial_x a + \mathcal{V}^*\partial_x v - \left(\frac{2mg}{T^{*3}}\partial_x \mathcal{A}^*\right)a + (\partial_x \mathcal{V}^*)v - g\left(\frac{k\mathcal{V}^{*2}}{\mathcal{A}^{*2}}a - \frac{2k\mathcal{V}^*}{\mathcal{A}^*}v\right)\\
&\qquad + \mathcal{V}^*\partial_x \mathcal{V}^* + \frac{g}{T^*}\partial_x \mathcal{A}^* - g\left(S_b - \frac{k\mathcal{V}^{*2}}{\mathcal{A}^*}\right) = 0\\
&\Rightarrow \partial_t v + \frac{g}{T^*}\partial_x a + \mathcal{V}^*\partial_x v - \left(\frac{2mg}{T^{*3}}\partial_x \mathcal{A}^*\right)a + (\partial_x \mathcal{V}^*)v - g\left(\frac{k\mathcal{V}^{*2}}{\mathcal{A}^{*2}}a - \frac{2k\mathcal{V}^*}{\mathcal{A}^*}v\right) = 0\\
&\Rightarrow \partial_t v + \frac{g}{T^*}\partial_x a + \mathcal{V}^*\partial_x v - \left(g\frac{k\mathcal{V}^{*2}}{\mathcal{A}^{*2}} + \frac{2mg}{T^{*3}}\partial_x \mathcal{A}^*\right)a + \left(\frac{2gk\mathcal{V}^*}{\mathcal{A}^*} + \partial_x \mathcal{V}^*\right)v = 0
\end{aligned} \tag{11}$$

$$\Rightarrow \partial_t v + \frac{g}{T^*}\partial_x a + \mathcal{V}^*\partial_x v$$
$$- \left(g\frac{k\mathcal{V}^{*2}}{\mathcal{A}^{*2}} + \frac{2mg}{T^{*3}}\left[- \frac{(S_b - k\frac{\mathcal{V}^{*2}}{\mathcal{A}^*})g\mathcal{V}^*T^*}{\mathcal{V}^{*2}T^* - g\mathcal{A}^*}\frac{\mathcal{A}^*}{\mathcal{V}^*}\right]\right) a + (\frac{2gk\mathcal{V}^*}{\mathcal{A}^*} + \partial_x \mathcal{V}^*)v = 0$$
$$\Rightarrow \partial_t v + \frac{g}{T^*}\partial_x a + \mathcal{V}^*\partial_x v + \left(gk\frac{\mathcal{V}^{*2}}{\mathcal{A}^{*2}} + \frac{2(S_b - k\frac{\mathcal{V}^{*2}}{\mathcal{A}^*})gm\mathcal{A}^*}{\left(\mathcal{V}^{*2}T^* - g\mathcal{A}^*\right)T^{*2}}\right) a + (\frac{2gk\mathcal{V}^*}{\mathcal{A}^*} + \partial_x \mathcal{V}^*)v = 0 \tag{12}$$

From the relations (10) and (11) we obtain the linear form of the system (1):

$$\begin{pmatrix} \mathrm{a} \\ \mathrm{v} \end{pmatrix}_t + \begin{pmatrix} \mathcal{V}^* & \mathcal{A}^* \\ \frac{g}{T^*} & \mathcal{V}^* \end{pmatrix} \begin{pmatrix} \mathrm{a} \\ \mathrm{v} \end{pmatrix}_x = \begin{pmatrix} -\mathcal{V}_x^* & -\mathcal{A}_x^* \\ f_w^* & -\mathcal{V}_x^* - 2gk\frac{\mathcal{V}^*}{\mathcal{A}^*} \end{pmatrix} \begin{pmatrix} \mathrm{a} \\ \mathrm{v} \end{pmatrix}, \tag{13}$$

with

$$f_w^* = gk\frac{\mathcal{V}^{*2}}{\mathcal{A}^{*2}} + \frac{2(S_b - k\frac{\mathcal{V}^{*2}}{\mathcal{A}^*})m^*g\mathcal{A}^*\mathcal{V}^{*2}}{(\mathcal{V}^{*2}T^* - g\mathcal{A}^*)T^{*2}}, \tag{14}$$

2.2 Riemann Coordinates

The linearized model (13) can be expressed in matrix form as

$$\partial_t w + M_1\partial_x w + Bw = 0, \tag{15}$$

where

$$w = \begin{pmatrix} \mathrm{a} \\ \mathrm{v} \end{pmatrix}, \quad B = \begin{pmatrix} \mathcal{V}_x^* & \mathcal{A}_x^* \\ f_w^* & \mathcal{V}_x^* + 2gk\frac{\mathcal{V}^*}{\mathcal{A}^*} \end{pmatrix}, \quad M_1 = \begin{pmatrix} \mathcal{V}^* & \mathcal{A}^* \\ \frac{g}{T^*} & \mathcal{V}^* \end{pmatrix},$$

with f_w^* is defined by (14).

Under the condition (7), the matrix M_1 possesses two distinct real eigenvalues λ_1 and $-\lambda_2$ with

$$\lambda_1 = \sqrt{\frac{g\mathcal{A}^*}{T^*}} + \mathcal{V}^* > 0, \quad \lambda_2 = \sqrt{\frac{g\mathcal{A}^*}{T^*}} - \mathcal{V}^* > 0. \tag{16}$$

Therefore, the Riemann coordinates for the (15) are specified by

$$\xi_1 = \mathrm{v} + a\sqrt{\frac{g}{\mathcal{A}^*T^*}} \text{ and } \xi_2 = \mathrm{v} - a\sqrt{\frac{g}{\mathcal{A}^*T^*}}, \tag{17}$$

Using these notions we obtain the following sequences of equalities

$$\begin{aligned}
\mathcal{V}_x^* h + \mathcal{A}_x^* v &= \frac{\xi_1 - \xi_2}{2}\sqrt{\frac{\mathcal{A}^*T^*}{g}}\mathcal{V}_x^* + \mathcal{A}_x^*\frac{\xi_1 + \xi_2}{2} \\
&= \frac{1}{2}\left(\sqrt{\frac{\mathcal{A}^*T^*}{g}}\mathcal{V}_x^* + \mathcal{A}_x^*\right)\xi_1 + \frac{1}{2}\left(\mathcal{A}_x^* - \sqrt{\frac{\mathcal{A}^*T^*}{g}}\mathcal{V}_x^*\right)\xi_2 \\
&= \frac{1}{2}\left(\sqrt{\frac{\mathcal{A}^*T^*}{g}}\mathcal{V}_x^* - \frac{\mathcal{A}^*}{\mathcal{V}^*}\mathcal{V}_x^*\right)\xi_1 + \frac{1}{2}\left(-\frac{\mathcal{A}^*}{\mathcal{V}^*}\mathcal{V}_x^* - \sqrt{\frac{\mathcal{A}^*T^*}{g}}\mathcal{V}_x^*\right)\xi_2 \\
&= \frac{1}{2}\left(\sqrt{\frac{\mathcal{A}^*T^*}{g}} - \frac{\mathcal{A}^*}{\mathcal{V}^*}\right)\mathcal{V}_x^*\xi_1 - \frac{1}{2}\left(\frac{\mathcal{A}^*}{\mathcal{V}^*} + \sqrt{\frac{\mathcal{A}^*T^*}{g}}\right)\mathcal{V}_x^*\xi_2
\end{aligned}$$

$$= \frac{1}{2}\left(\sqrt{\frac{\mathcal{A}^*T^*}{g}} - \frac{\mathcal{A}^*}{\mathcal{V}^*}\right)\frac{(S_0 - k\frac{\mathcal{V}^{*2}}{\mathcal{A}^*})g\mathcal{V}^*}{\mathcal{V}^{*2} - \frac{g\mathcal{A}^*}{T^*}}\xi_1 - \frac{1}{2}\left(\sqrt{\frac{\mathcal{A}^*T^*}{g}} + \frac{\mathcal{A}^*}{\mathcal{V}^*}\right)\frac{(S_0 - k\frac{\mathcal{V}^{*2}}{\mathcal{A}^*})g\mathcal{V}^*}{\mathcal{V}^{*2} - \frac{g\mathcal{A}^*}{T^*}}\xi_2$$

$$= \frac{1}{2}\left(\mathcal{V}^*\sqrt{g\mathcal{A}^*T^*} - g\mathcal{A}^*\right)\frac{(S_0 - k\frac{\mathcal{V}^{*2}}{\mathcal{A}^*})}{\frac{\mathcal{V}^{*2}T^* - g\mathcal{A}^*}{T^*}}\xi_1 - \frac{1}{2}\left(\mathcal{V}^*\sqrt{g\mathcal{A}^*T^*} + g\mathcal{A}^*\right)\frac{(S_0 - k\frac{\mathcal{V}^{*2}}{\mathcal{A}^*})}{\frac{\mathcal{V}^{*2}T^* - g\mathcal{A}^*}{T^*}}\xi_2$$

$$= \frac{1}{2}\left(\mathcal{V}^*\sqrt{g\mathcal{A}^*T^*} - g\mathcal{A}^*\right)\frac{\left(S_0 - k\frac{\mathcal{V}^{*2}}{\mathcal{A}^*}\right)T^*}{\mathcal{V}^{*2}T^* - g\mathcal{A}^*}\xi_1 - \frac{1}{2}\left(\frac{\mathcal{A}^*}{\mathcal{V}^*} + \sqrt{\frac{\mathcal{A}^*T^*}{g}}\right)\mathcal{V}_x^*\xi_2$$

$$\begin{aligned}
&= \frac{\sqrt{g\mathcal{A}^*}}{2}\left(\mathcal{V}^*\sqrt{T^*} - \sqrt{g\mathcal{A}^*}\right)\frac{\left(S_0 - k\frac{\mathcal{V}^{*2}}{\mathcal{A}^*}\right)T^*}{\mathcal{V}^{*2}T^* - g\mathcal{A}^*}\xi_1 - \frac{\sqrt{g\mathcal{A}^*}}{2}\left(\mathcal{V}^*\sqrt{T^*} + \sqrt{g\mathcal{A}^*}\right)\frac{\left(S_0 - k\frac{\mathcal{V}^{*2}}{\mathcal{A}^*}\right)T^*}{\mathcal{V}^{*2}T^* - g\mathcal{A}^*}\xi_2 \\
&= \frac{\sqrt{g\mathcal{A}^*}}{2}\frac{\left(S_0 - k\frac{\mathcal{V}^{*2}}{\mathcal{A}^*}\right)T^*}{\mathcal{V}^*\sqrt{T^*} + \sqrt{g\mathcal{A}^*}}\xi_1 - \frac{\sqrt{g\mathcal{A}^*}}{2}\left(\mathcal{V}^*\sqrt{T^*} + \sqrt{g\mathcal{A}^*}\right)\frac{\left(S_0 - k\frac{\mathcal{V}^{*2}}{\mathcal{A}^*}\right)T^*}{\mathcal{V}^{*2}T^* - g\mathcal{A}^*}\xi_2
\end{aligned}$$

$$= \frac{\left(S_0 - k\frac{\mathcal{V}^{*2}}{\mathcal{A}^*}\right)\sqrt{g\mathcal{A}^*T^*}}{2\left(\mathcal{V}^*\sqrt{T^*} + \sqrt{g\mathcal{A}^*}\right)}\xi_1 - \frac{\left(S_0 - k\frac{\mathcal{V}^{*2}}{\mathcal{A}^*}\right)\sqrt{g\mathcal{A}^*T^*}}{2\left(\mathcal{V}^*\sqrt{T^*} - \sqrt{g\mathcal{A}^*}\right)}\xi_2$$

and

$$
\begin{aligned}
f^*(w)h + \big(\mathcal{V}_x^* + 2gk\frac{\mathcal{V}^*}{\mathcal{A}^*}\big)v &= \frac{\xi_1 - \xi_2}{2}\sqrt{\frac{\mathcal{A}^*T^*}{g}}f^*(w) + \left(\mathcal{V}_x^* + 2gk\frac{\mathcal{V}^*}{\mathcal{A}^*}\right)\frac{\xi_1 + \xi_2}{2} \\
&= \frac{1}{2}\left(\sqrt{\frac{\mathcal{A}^*T^*}{g}}f^*(w) + \mathcal{V}_x^* + 2gk\frac{\mathcal{V}^*}{\mathcal{A}^*}\right)\xi_1 \\
&\quad + \frac{1}{2}\left(\mathcal{V}_x^* + 2gk\frac{\mathcal{V}^*}{\mathcal{A}^*} - \sqrt{\frac{\mathcal{A}^*T^*}{g}}f^*(w)\right)\xi_2.
\end{aligned} \tag{18}
$$

The first part of (18) gives:

$$
\begin{aligned}
&\frac{1}{2}\left(\sqrt{\frac{\mathcal{A}^*T^*}{g}}f^*(w) + \mathcal{V}_x^* + 2gk\frac{\mathcal{V}^*}{\mathcal{A}^*}\right)\xi_1 \\
&= \frac{1}{2}\left(-\sqrt{\frac{\mathcal{A}^*T^*}{g}}\left(gk\frac{\mathcal{V}^{*2}}{\mathcal{A}^{*2}} + \frac{2mg\mathcal{A}^*\mathcal{V}^{*2}(k\frac{\mathcal{V}^{*2}}{\mathcal{A}^*} - S_0)}{\left(\mathcal{V}^{*2}T^* - g\mathcal{A}^*\right)T^{*2}}\right) + \frac{g\mathcal{V}^*\left(S_0 - k\frac{\mathcal{V}^{*2}}{\mathcal{A}^*}\right)T^*}{\mathcal{V}^{*2}T^* - g\mathcal{A}^*} + 2gk\frac{\mathcal{V}^*}{\mathcal{A}^*}\right)\xi_1 \\
&= \frac{1}{2}\left(-\sqrt{g\mathcal{A}^*}T^*\left(k\frac{\mathcal{V}^{*2}}{\mathcal{A}^{*2}} + \frac{2m\mathcal{V}^{*2}\mathcal{A}^*(k\frac{\mathcal{V}^{*2}}{\mathcal{A}^*} - S_0)}{\left(\mathcal{V}^{*2}T^* - g\mathcal{A}^*\right)T^{*2}}\right) + \frac{g\mathcal{V}^*\left(S_0 - k\frac{\mathcal{V}^{*2}}{\mathcal{A}^*}\right)T^*}{\mathcal{V}^{*2}T^* - g\mathcal{A}^*} + 2gk\frac{\mathcal{V}^*}{\mathcal{A}^*}\right)\xi_1 \\
&= \frac{1}{2}\left[k\frac{\mathcal{V}^*}{\mathcal{A}^*}\left(-\frac{\mathcal{V}^*}{\mathcal{A}^*}\sqrt{g\mathcal{A}^*}T^* + 2g\right) + (S_0 - k\frac{\mathcal{V}^{*2}}{\mathcal{A}^*})\left(\frac{2m\mathcal{V}^{*2}\mathcal{A}^*\sqrt{g\mathcal{A}^*}T^*}{\left(\mathcal{V}^{*2}T^* - g\mathcal{A}^*\right)T^{*2}} + \frac{g\mathcal{V}^*T^*}{\mathcal{V}^{*2}T^* - g\mathcal{A}^*}\right)\right]\xi_1 \\
&= \frac{1}{2}\left[k\frac{\mathcal{V}^{*2}}{\mathcal{A}^{*2}}\left(-\sqrt{g\mathcal{A}^*}T^* + \frac{2g\mathcal{A}^*}{\mathcal{V}^*}\right) + (S_0 - k\frac{\mathcal{V}^{*2}}{\mathcal{A}^*})\left(\frac{2m\mathcal{V}^{*2}\mathcal{A}^*\sqrt{g\mathcal{A}^*}T^* + g\mathcal{V}^*T^*T^{*2}}{\left(\mathcal{V}^{*2}T^* - g\mathcal{A}^*\right)T^{*2}}\right]\xi_1 \\
&= \frac{1}{2}\left[k\frac{\mathcal{V}^{*2}}{\mathcal{A}^{*2}}\left(-\sqrt{g\mathcal{A}^*}T^* + \frac{2g\mathcal{A}^*}{\mathcal{V}^*}\right) + (S_0 - k\frac{\mathcal{V}^{*2}}{\mathcal{A}^*})\left(\frac{2m\mathcal{V}^{*2}\mathcal{A}^*\sqrt{g\mathcal{A}^*} + g\mathcal{V}^*T^{*-2/4}T^{*2}}{\left(\mathcal{V}^{*2}T^* - g\mathcal{A}^*\right)T^{*2}T^{*-2/4}}\right]\xi_1 \\
&= \frac{1}{2}\left[\left(\frac{2\mathcal{A}^*g}{\mathcal{V}^*} - \sqrt{g\mathcal{A}^*}T^*\right)k\frac{\mathcal{V}^{*2}}{\mathcal{A}^{*2}} + \frac{g\mathcal{V}^*T^{*3/2} + 2m\mathcal{V}^{*2}\mathcal{A}^*\sqrt{g\mathcal{A}^*}}{\left(\mathcal{V}^{*2}T^* - g\mathcal{A}^*\right)T^{*3/2}}\left(S_0 - k\frac{\mathcal{V}^{*2}}{\mathcal{A}^*}\right)\right]\xi_1
\end{aligned}
$$

and the second part of (18) gives:

$$
\frac{1}{2}\left(-\sqrt{\frac{\mathcal{A}^*T^*}{g}}f^*(w) + \mathcal{V}_x^* + 2gk\frac{\mathcal{V}^*}{\mathcal{A}^*}\right)\xi_2
$$

$$= \frac{1}{2}\left(\sqrt{\frac{\mathcal{A}^*T^*}{g}}\left(gk\frac{\mathcal{V}^{*2}}{\mathcal{A}^{*2}} + \frac{2mg\mathcal{V}^{*2}\mathcal{A}^*(k\frac{\mathcal{V}^{*2}}{\mathcal{A}^*} - S_0)}{\left(\mathcal{V}^{*2}T^* - g\mathcal{A}^*\right)T^{*2}}\right) + \frac{g\mathcal{V}^*\left(S_0 - k\frac{\mathcal{V}^{*2}}{\mathcal{A}^*}\right)T^*}{\mathcal{V}^{*2}T^* - g\mathcal{A}^*} + 2gk\frac{\mathcal{V}^*}{\mathcal{A}^*}\right)\xi_2$$

$$= \frac{1}{2}\left(\sqrt{g\mathcal{A}^*}T^*\left(k\frac{\mathcal{V}^{*2}}{\mathcal{A}^{*2}} + \frac{2m\mathcal{V}^{*2}\mathcal{A}^*(k\frac{\mathcal{V}^{*2}}{\mathcal{A}^*} - S_0)}{\left(\mathcal{V}^{*2}T^* - g\mathcal{A}^*\right)T^{*2}}\right) + \frac{g\mathcal{V}^*\left(S_0 - k\frac{\mathcal{V}^{*2}}{\mathcal{A}^*}\right)T^*}{\mathcal{V}^{*2}T^* - g\mathcal{A}^*} + 2gk\frac{\mathcal{V}^*}{\mathcal{A}^*}\right)\xi_2$$

$$= \frac{1}{2}\left[k\frac{\mathcal{V}^*}{\mathcal{A}^*}\left(\frac{\mathcal{V}^*}{\mathcal{A}^*}\sqrt{g\mathcal{A}^*}T^* + 2g\right) + (k\frac{\mathcal{V}^{*2}}{\mathcal{A}^*} - S_0)\left(\frac{2m\mathcal{V}^{*2}\mathcal{A}^*\sqrt{g\mathcal{A}^*}T^*}{\left(\mathcal{V}^{*2}T^* - g\mathcal{A}^*\right)T^{*2}} + \frac{g\mathcal{V}^*T^*}{\mathcal{V}^{*2}T^* - g\mathcal{A}^*}\right)\right]\xi_2$$

$$= \frac{1}{2}\left[k\frac{\mathcal{V}^{*2}}{\mathcal{A}^{*2}}\left(\sqrt{g\mathcal{A}^*}T^* + \frac{2g\mathcal{A}^*}{\mathcal{V}^*}\right) + (k\frac{\mathcal{V}^{*2}}{\mathcal{A}^*} - S_0)\left(\frac{2m\mathcal{V}^{*2}\mathcal{A}^*\sqrt{g\mathcal{A}^*}T^* + g\mathcal{V}^*T^*T^{*2}}{\left(\mathcal{V}^{*2}T^* - g\mathcal{A}^*\right)T^{*2}}\right]\xi_2$$

$$= \frac{1}{2}\left[k\frac{\mathcal{V}^{*2}}{\mathcal{A}^{*2}}\left(\sqrt{g\mathcal{A}^*}T^* + \frac{2g\mathcal{A}^*}{\mathcal{V}^*}\right) + (k\frac{\mathcal{V}^{*2}}{\mathcal{A}^*} - S_0)\left(\frac{2m\mathcal{V}^{*2}\mathcal{A}^*\sqrt{g\mathcal{A}^*} + g\mathcal{V}^*T^{*-2/4}T^{*2}}{\left(\mathcal{V}^{*2}T^* - g\mathcal{A}^*\right)T^{*2}T^{*-2/4}}\right]\xi_2$$

$$= \frac{1}{2}\left[\left(\frac{2g\mathcal{A}^*}{\mathcal{V}^*} + \sqrt{g\mathcal{A}^*}T^*\right)k\frac{\mathcal{V}^{*2}}{\mathcal{A}^{*2}} - \frac{g\mathcal{V}^*T^{*3/2} + 2m\mathcal{V}^{*2}\mathcal{A}^*\sqrt{g\mathcal{A}^*}}{\left(\mathcal{V}^{*2}T^* - g\mathcal{A}^*\right)T^{*3/2}}(S_0 - k\frac{\mathcal{V}^{*2}}{\mathcal{A}^*})\right]\xi_2.$$

Thus, the system (15) can be written as follows:

$$\begin{pmatrix}\xi_1\\ \xi_2\end{pmatrix}_t + \begin{pmatrix}\lambda_1 & 0\\ 0 & -\lambda_2\end{pmatrix}\begin{pmatrix}\xi_1\\ \xi_2\end{pmatrix}_x = \begin{pmatrix}\gamma_1 & \delta_1\\ \gamma_2 & \delta_2\end{pmatrix}\begin{pmatrix}\xi_1\\ \xi_2\end{pmatrix}, \tag{19}$$

where

$$\gamma_1(x) = -\frac{\left(S_b - k\frac{\mathcal{V}^{*2}}{\mathcal{A}^*}\right)\sqrt{g\mathcal{A}^*}T^*}{2\left(\mathcal{V}^*\sqrt{T^*} + \sqrt{g\mathcal{A}^*}\right)}, \quad \delta_1(x) = \frac{\left(S_b - k\frac{\mathcal{V}^{*2}}{\mathcal{A}^*}\right)\sqrt{g\mathcal{A}^*}T^*}{2\left(\mathcal{V}^*\sqrt{T^*} - \sqrt{g\mathcal{A}^*}\right)},$$

$$\gamma_2(x) = -\frac{1}{2}\left[k\left(\frac{2g\mathcal{A}^*}{\mathcal{V}^*} - \sqrt{g\mathcal{A}^*T^*}\right)\frac{\mathcal{V}^{*2}}{\mathcal{A}^{*2}} + \frac{g\mathcal{V}^*T^{*\frac{3}{2}} - 2m^*\mathcal{A}^*\mathcal{V}^{*2}\sqrt{g\mathcal{A}^*}}{\left(\mathcal{V}^{*2}T^* - g\mathcal{A}^*\right)T^{*\frac{3}{2}}}\left(S_b - k\frac{\mathcal{V}^{*2}}{\mathcal{A}^*}\right)\right], \tag{20}$$

$$\delta_2(x) = -\frac{1}{2}\left[k\left(\frac{2g\mathcal{A}^*}{\mathcal{V}^*} + \sqrt{g\mathcal{A}^*T^*}\right)\frac{\mathcal{V}^{*2}}{\mathcal{A}^{*2}} + \frac{g\mathcal{V}^*T^{*\frac{3}{2}} + 2m^*\mathcal{A}^*\mathcal{V}^{*2}\sqrt{g\mathcal{A}^*}}{\left(\mathcal{V}^{*2}T^* - g\mathcal{A}^*\right)T^{*\frac{3}{2}}}\left(S_b - k\frac{\mathcal{V}^{*2}}{\mathcal{A}^*}\right)\right].$$

In steep-slope fluvial regimes, γ_1 and δ_2 are negative.

3 The Main Result

Our objective is to establish the exponential stability of the steady states of the nonlinear system (1) under specific boundary conditions given by:

$$\mathcal{V}(t,0) = \mathcal{K}\big(\mathcal{A}(t,0)\big), \quad \mathcal{V}(t,L) = \mathcal{K}\big(\mathcal{A}(t,L)\big), \tag{21}$$

and $\mathcal{K} : \mathbb{R} \to \mathbb{R}$ is a function of class C^2. The condition (21) is determined by physical control devices positioned at the ends of the canal, such as movable spillways or adjustable hydraulic gates used in irrigation channels.

Let

$$\mathcal{A}(0,x) = \mathcal{A}_0, \quad \mathcal{V}(0,x) = \mathcal{V}_0, \quad x \in [0,L], \tag{22}$$

be a given initial condition, we assume the compatibility terms apply below

$$\mathcal{V}_0(0) - \mathcal{K}(\mathcal{A}_0(0)) = 0, \quad \mathcal{V}_0(L) - \mathcal{K}\big(\mathcal{A}_0(L)\big) = 0, \tag{23}$$

$$\partial_x\big(\frac{\mathcal{V}_0^2}{2}\big)(0) + \big(\frac{g}{\sqrt{b^2+4m\mathcal{A}_0}}\partial_x\mathcal{A}_0\big)(0) - g\left(S_b - \frac{k\mathcal{V}_0^2}{\mathcal{A}_0}\right)(0) - \mathcal{K}'(\mathcal{A}_0(0))\partial_x(\mathcal{A}_0\mathcal{V}_0)(0) = 0, \tag{24}$$

$$\partial_x\big(\frac{\mathcal{V}_0^2}{2}\big)(L) + \big(\frac{g}{\sqrt{b^2+4m\mathcal{A}_0}}\partial_x\mathcal{A}_0\big)(L) - g\left(S_b - \frac{k\mathcal{V}_0^2}{\mathcal{A}_0}\right)(L) - \mathcal{K}'(\mathcal{A}_0(0))\partial_x(\mathcal{A}_0\mathcal{V}_0)(L) = 0, \tag{25}$$

For sufficiently small initial data, the conditions (23) (24) and (25) ensure that the system (1), (21) and (22) is well posed (see [5], Appendix B).

We have the main result given below.

Theorem 1 *As soon as the boundary conditions* (21) *satisfy*

$$\mathcal{K}'(\mathcal{A}^*(0)) \in \left(\frac{-g}{\mathcal{V}^*(0)T_0^*}, -\frac{\mathcal{V}^*(0)}{\mathcal{A}^*(0)}\right) \text{ and } \mathcal{K}'(\mathcal{A}^*(L)) \in \mathbb{R}\diagdown\left(\frac{-g}{\mathcal{V}^*(L)T_L^*}, -\frac{\mathcal{V}^*(L)}{\mathcal{A}^*(L)}\right), \tag{26}$$

the nonlinear system (1) *is exponentially stable for the norm H^2. In order to prove this theorem, we start by giving preliminaries and the proof will finally be given in Sect. 4.2.*

4 Exponential Stability Concerning the H^2-Norm Within a Trapezoidal Channel

4.1 Exponential Stability for the L^2-Norm of the Linearized System in the L^2-Norm

Let we consider the linear system (13) and the following linearization of boundary conditions (21)

$$\mathrm{v}(t,0) = u_0\mathrm{a}(t,0), \quad \mathrm{v}(t,L) = u_1\mathrm{a}(t,L), \tag{27}$$

where

$$u_0 = \mathcal{K}'(\mathcal{A}^*(0)), \quad u_1 = \mathcal{K}'\big(\mathcal{A}^*(L)\big). \tag{28}$$

and the following initial condition

$$\mathrm{a}(0,x) = \mathrm{a}_0(x) \in L^2((0,L);\mathbb{R}), \quad \mathrm{v}(0,x) = \mathrm{v}_0(x) \in L^2((0,L);\mathbb{R}), \tag{29}$$

According to Appendix A of [5], the problem (13), (27), and (29) is well-posed, thus, the investigation of the exponential stability of the linearized system is reframed as a null stabilization challenge (a, b).

In matrix form, the linearized Saint-Venant equations (13) in its form (19) gives

$$\partial_t \xi + \Lambda \partial_x \xi - M\xi = 0, \tag{30}$$

with

$$\xi = \begin{pmatrix} \xi_1 \\ \xi_2 \end{pmatrix}, \quad \Lambda = \begin{pmatrix} \lambda_1 & 0 \\ 0 & -\lambda_2 \end{pmatrix}, \quad M = \begin{pmatrix} \gamma_1 & \delta_1 \\ \gamma_2 & \delta_2 \end{pmatrix} \quad \text{and} \quad |\Lambda| = \begin{pmatrix} \lambda_1 & 0 \\ 0 & \lambda_2 \end{pmatrix}.$$

According to (17) and (27) we have

$$\begin{aligned}
&\mathrm{v}(t,0) = u_0 \mathrm{h}(t,0) \\
&\Longleftrightarrow \frac{1}{2}\Big(\xi_1(t,0) + \xi_2(t,0)\Big) = \frac{u_0}{2}\Big(\xi_1(t,0) - \xi_2(t,0)\Big)\sqrt{\frac{\mathcal{A}^*(0)T_0^*}{g}} \\
&\Longleftrightarrow \left(1 - u_0\sqrt{\frac{\mathcal{A}^*(0)T_0^*}{g}}\right)\xi_1(t,0) = -\left(1 + u_0\sqrt{\frac{\mathcal{A}^*(0)T_0^*}{g}}\right)\xi_2(t,0)
\end{aligned}$$

$$\Longleftrightarrow \xi_1(t,0) = \frac{u_0\sqrt{\dfrac{\mathcal{A}^*(0)T_0^*}{g}} + 1}{u_0\sqrt{\dfrac{\mathcal{A}^*(0)T_0^*}{g}} - 1}\xi_2(t,0)$$

$$\Longleftrightarrow \xi_1(t,0) = \frac{u_0\sqrt{\mathcal{A}^*(0)T_0^*} + \sqrt{g}}{u_0\sqrt{\mathcal{A}^*(0)T_0^*} - \sqrt{g}}\xi_2(t,0)$$

$$\Longleftrightarrow \xi_1(t,0) = \frac{u_0\mathcal{A}^*(0)T_0^* + \sqrt{g\mathcal{A}^*(0)T_0^*}}{u_0\mathcal{A}^*(0)T_0^* - \sqrt{g\mathcal{A}^*(0)T_0^*}}\xi_2(t,0), \tag{31}$$

and

$$\mathrm{v}(t,L)=u_1\mathrm{h}(t,L)\Longleftrightarrow\frac{\xi_1(t,L)+\xi_2(t,L)}{2}=u_1\frac{\xi_1(t,L)-\xi_2(t,L)}{2}\sqrt{\frac{T_L^*\mathcal{A}^*(L)}{g}}$$

$$\Longleftrightarrow\left(1-u_1\sqrt{\frac{T_L^*\mathcal{A}^*(L)}{g}}\right)\xi_1(t,L)=-\left(1+u_1\sqrt{\frac{T_L^*\mathcal{A}^*(L)}{g}}\right)\xi_2(t,L)$$

$$\Longleftrightarrow\xi_1(t,L)=\frac{u_1\sqrt{\frac{T_L^*\mathcal{A}^*(L)}{g}}+1}{u_1\sqrt{\frac{T_L^*\mathcal{A}^*(L)}{g}}-1}\xi_2(t,L)$$

$$\Longleftrightarrow\xi_1(t,L)=\frac{u_1\sqrt{T_L^*\mathcal{A}^*(L)}+\sqrt{g}}{u_1\sqrt{T_L^*\mathcal{A}^*(L)}-\sqrt{g}}\xi_2(t,L)$$

$$\Longleftrightarrow\xi_2(t,L)=\frac{u_1T_L^*\mathcal{A}^*(L)-\sqrt{gT_L^*\mathcal{A}^*(L)}}{u_1T_L^*\mathcal{A}^*(L)+\sqrt{gT_L^*\mathcal{A}^*(L)}}\xi_1(t,L). \tag{32}$$

Therefore, using (31) and (32), the boundary conditions for system (30) are obtained as follows.

$$\begin{pmatrix}\xi_1(t,0)\\ \xi_2(t,L)\end{pmatrix}=\underbrace{\begin{pmatrix}0 & m_0\\ m_1 & 0\end{pmatrix}}_{K}\begin{pmatrix}\xi_1(t,L)\\ \xi_2(t,0)\end{pmatrix}, \tag{33}$$

where

$$m_0=\frac{u_0\sqrt{\mathcal{A}^*(0)T_0^*}+\sqrt{g}}{u_0\sqrt{\mathcal{A}^*(0)T_0^*}-\sqrt{g}},\quad m_1=\frac{u_1\sqrt{T_L^*\mathcal{A}^*(L)}+\sqrt{g}}{u_1\sqrt{T_L^*\mathcal{A}^*(L)}-\sqrt{g}} \tag{34}$$

and the initial condition

$$\xi(0,x)=\xi^0(x),\quad x\in(0,L). \tag{35}$$

To apply the Lyapunov stability technique under boundary feedback control, let us consider the system written in the form (30), (33), (35) and introduce the following Lyapunov function:

$$\mathcal{L}=\int_0^L\xi^T P(x)\xi dx. \tag{36}$$

$$P(x)=\begin{pmatrix}P_1e^{-\tau x} & 0\\ 0 & P_2e^{\tau x}\end{pmatrix}$$

with $\tau>0$ and $P_1, P_2\in C^1\big([0,L]\to(0,+\infty)\big)$ two real positive functions.

The time derivative of $\mathcal{L}$ with respect to the solutions of (36) is given by

$$\dot{\mathcal{L}} = J_1 + J_2$$

with

$$J_1 = -\left[\xi^T \Lambda P(x)\xi\right]_0^L \text{ and } J_2 = \int_0^L \xi^T \left(-\tau|\Lambda|P(x) + M^T P(x) + P(x)M\right)\xi dx.$$

Let Ω be the set of 2×2 diagonal real matrices with strictly positive diagonal elements and $\mathcal{P}$ denotes the set of matrices $P \in \Omega$ such that the quadratic form defined by $M^T P + PM$ is negative semidefinite.

We define the following norm for the matrix K:

$$\rho(K) \triangleq \inf\left\{\|\Delta K \Delta^{-1}\|,\ \Delta \in \left\{\Delta = \operatorname{diag}\{\sqrt{P_1\lambda_1}, \sqrt{P_2\lambda_2}\},\ \operatorname{diag}\{P_1, P_2\} \in \mathcal{P}\right\}\right\},$$

where $\|\cdot\|$ denotes the standard matrix 2-norm.

According to [12], we have

Theorem 2 *The linear system* (30), (33) *and* (35) *is exponentially stable when* M *is diagonally marginally stable and the dissipative boundary condition* $\rho(K) < 1$ *is verified*

Based on this theorem, we establish the proposition below

Proposition 1 *For the linearized system described by Eqs.* (13), (27) *and* (29), *provided that the boundary conditions satisfy*

$$u_0 \in \left(\frac{-g}{\mathcal{V}^*(0)T_0^*}, -\frac{\mathcal{V}^*(0)}{\mathcal{A}^*(0)}\right), \quad \text{and} \quad u_1 \in \mathbb{R}\setminus\left(\frac{-g}{\mathcal{V}^*(L)T_L^*}, -\frac{\mathcal{V}^*(L)}{\mathcal{A}^*(L)}\right) \tag{37}$$

Consequently, there exist constants $\alpha_1 > 0$, $\alpha > 0$, *and functions* $g_1, g_2 \in C^1\big([0, L] \to (0, +\infty)\big)$, *ensuring that the control Lyapunov function candidate*

$$\mathcal{L}(a, v) = \int_0^L (g_1 + g_2)\left(\frac{g}{\mathcal{A}^*T^*}a^2 + 2\frac{g_1 - g_2}{g_1 + g_2}\sqrt{\frac{g}{\mathcal{A}^*T^*}}va + v^2\right)dx \tag{38}$$

verifies:

$$\mathcal{L}(a, v) \geq \alpha_1\left(\|a\|^2_{L^2(0,L)} + \|v\|^2_{L^2(0,L)}\right) \tag{39}$$

for any element (a, v) *of* $L^2((0, L); \mathbb{R}^2)$, *where* $L^2(0, L)$ *denotes the space* $L^2((0, L); \mathbb{R})$. *Furthermore, if* $(a, v)^T$ *is a solution of the system* (13), (27), *and* (29), *then*

$$\frac{d}{dt}\Big(\mathcal{L}\big(a(t, \cdot), v(t, \cdot)\big)\Big) \leq -\alpha\mathcal{L}\big(a(t, \cdot), v(t, \cdot)\big)$$

In the distributional sense, this inequality implies that the linearized system (13), (27), *and* (29) *is exponentially stable in the* L^2*-norm.*

Proof To apply Theorem 2, it is necessary to:

- Determine $P = \begin{pmatrix} P_1 & 0 \\ 0 & P_2 \end{pmatrix}$ such that $\xi^T\left(M^T P + PM\right)\xi \leq 0$,
- Determine u_0 and u_1 such that $\rho(K) < 1$.

For choice of P, we have :

$$P_1 = \lambda_1 > 0, \qquad P_2 = \lambda_2 > 0. \tag{40}$$

where λ_1 and λ_2 are given by (16). Furthermore by using (5) and (7) we obtain

$$\gamma_1 < 0, \qquad \delta_2 < 0 \qquad \text{and} \qquad \delta_2\gamma_1\lambda_1\lambda_2 > \left(\delta_1\lambda_1 + \gamma_2\lambda_2\right)^2, \tag{41}$$

and

$$\xi^T\left(M^T P + PM\right)\xi = -2|\gamma_1||\lambda_1\left[\xi_1 + \left(\frac{\delta_1\lambda_1 + \gamma_2\lambda_2}{2\gamma_1\lambda_1}\right)\xi_2\right]^2 - \frac{4\delta_2\gamma_1\lambda_1\lambda_2 - \left(\delta_1\lambda_1 + \gamma_2\lambda_2\right)^2}{2|\gamma_1|\lambda_1}\xi_2^2 \;\; \leq 0, \tag{42}$$

therefore $\xi^T\left(M^T P + PM\right)\xi$ is a negative semi definite quadratic form.
Moreover, by using (40), we have

$$\Delta K \Delta^{-1} = \begin{pmatrix} 0 & m_0\dfrac{\lambda_1(0)}{\lambda_2(0)} \\ m_1\dfrac{\lambda_2(L)}{\lambda_1(L)} & 0 \end{pmatrix}, \tag{43}$$

therefore, $\rho(K) < 1$ is equivalent to determine m_0 and m_1 so that we have

$$m_0^2\left(\frac{\lambda_1(0)}{\lambda_2(0)}\right)^2 < 1 \qquad \text{and} \qquad m_1^2\left(\frac{\lambda_2(L)}{\lambda_1(L)}\right)^2 < 1. \tag{44}$$

- $m_0^2\dfrac{\lambda_1^2(0)}{\lambda_2^2(0)} < 1 \Rightarrow \left(m_0\dfrac{\lambda_1(0)}{\lambda_2(0)} - 1\right)\left(m_0\dfrac{\lambda_1(0)}{\lambda_2(0)} + 1\right) < 0$

$$\Rightarrow \left(\frac{u_0\sqrt{A^*(0)T_0^*} + \sqrt{g}}{u_0\sqrt{A^*(0)T_0^*} - \sqrt{g}} \times \frac{\sqrt{\dfrac{gA^*(0)}{T_0^*}} + \mathcal{V}^*(0)}{\sqrt{\dfrac{gA^*(0)}{T_0^*}} - \mathcal{V}^*(0)} - 1\right)$$

$$\left(\frac{u_0\sqrt{\mathcal{A}^*(0)T_0^*}+\sqrt{g}}{u_0\sqrt{\mathcal{A}^*(0)T_0^*}-\sqrt{g}}\times\frac{\sqrt{\dfrac{g\mathcal{A}^*(0)}{T_0^*}}+\mathcal{V}^*(0)}{\sqrt{\dfrac{g\mathcal{A}^*(0)}{T_0^*}}-\mathcal{V}^*(0)}+1\right)<0 \quad (45)$$

$$\Rightarrow\left(\frac{u_0\sqrt{\mathcal{A}^*(0)T_0^*}+\sqrt{g}}{u_0\sqrt{\mathcal{A}^*(0)T_0^*}-\sqrt{g}}\times\frac{\sqrt{g}\mathcal{A}^*(0)+\sqrt{\mathcal{A}^*(0)T_0^*}\mathcal{V}^*(0)}{\sqrt{g}\mathcal{A}^*(0)-\sqrt{\mathcal{A}^*(0)T_0^*}\mathcal{V}^*(0)}-1\right)$$
$$\left(\frac{u_0\sqrt{\mathcal{A}^*(0)T_0^*}+\sqrt{g}}{u_0\sqrt{\mathcal{A}^*(0)T_0^*}-\sqrt{g}}\times\frac{\sqrt{g}\mathcal{A}^*(0)+\sqrt{\mathcal{A}^*(0)T_0^*}\mathcal{V}^*(0)}{\sqrt{g}\mathcal{A}^*(0)-\sqrt{\mathcal{A}^*(0)T_0^*}\mathcal{V}^*(0)}+1\right)<0$$

$$\Rightarrow\frac{2u_0\mathcal{A}^*(0)\mathcal{V}^*(0)T_0^*+2g\mathcal{A}^*(0)}{\left(u_0\sqrt{\mathcal{A}^*(0)T_0^*}-\sqrt{g}\right)\left(\sqrt{g}\mathcal{A}^*(0)-\sqrt{\mathcal{A}^*(0)T_0^*}\mathcal{V}^*(0)\right)}$$
$$\frac{2u_0\sqrt{g}\mathcal{A}^*(0)\sqrt{\mathcal{A}^*(0)T_0^*}+2\sqrt{g}\sqrt{\mathcal{A}^*(0)T_0^*}\mathcal{V}^*(0)}{\left(u_0\sqrt{\mathcal{A}^*(0)T_0^*}-\sqrt{g}\right)\left(\sqrt{g}\mathcal{A}^*(0)-\sqrt{\mathcal{A}^*(0)T_0^*}\mathcal{V}^*(0)\right)}<0$$

$$\Rightarrow 4\mathcal{A}^*(0)\sqrt{g}\mathcal{A}^*(0)\sqrt{\mathcal{A}^*(0)T_0^*}\left(u_0\mathcal{V}^*(0)T_0^*+g\right)\left(u_0\mathcal{A}^*(0)+\mathcal{V}^*(0)\right)<0$$
$$\Rightarrow\left(u_0\mathcal{V}^*(0)T_0^*+g\right)\left(u_0\mathcal{A}^*(0)+\mathcal{V}^*(0)\right)<0$$

$$\Rightarrow u_0\in\left(-\frac{g}{\mathcal{V}^*(0)T_0^*},-\frac{\mathcal{V}^*(0)}{\mathcal{A}^*(0)}\right) \quad (46)$$

we have also

- $m_1^2\frac{\lambda_2^2(L)}{\lambda_1^2(L)}<1\Rightarrow\left(m_1\frac{\lambda_2(L)}{\lambda_1(L)}-1\right)\left(m_1\frac{\lambda_2(L)}{\lambda_1(L)}+1\right)<0$

$$\Rightarrow\left(\frac{u_1\sqrt{T_L^*\mathcal{A}^*(L)}-\sqrt{g}}{u_1\sqrt{T_L^*\mathcal{A}^*(L)}+\sqrt{g}}\times\frac{\sqrt{\dfrac{g\mathcal{A}^*(L)}{T_L^*}}-\mathcal{V}^*(L)}{\sqrt{\dfrac{g\mathcal{A}^*(L)}{T_L^*}}+\mathcal{V}^*(L)}-1\right)$$

$$\left(\frac{u_1\sqrt{T_L^*\mathcal{A}^*(L)}-\sqrt{g}}{u_1\sqrt{T_L^*\mathcal{A}^*(L)}+\sqrt{g}}\times\frac{\sqrt{\frac{g\mathcal{A}^*(L)}{T_L^*}}-\mathcal{V}^*(L)}{\sqrt{\frac{g\mathcal{A}^*(L)}{T_L^*}}+\mathcal{V}^*(L)}+1\right)<0 \tag{47}$$

$$\Rightarrow\left(\frac{u_1\sqrt{T_L^*\mathcal{A}^*(L)}-\sqrt{g}}{u_1\sqrt{T_L^*\mathcal{A}^*(L)}+\sqrt{g}}\times\frac{\sqrt{g}\mathcal{A}^*(L)-\sqrt{T_L^*\mathcal{A}^*(L)}\mathcal{V}^*(L)}{\sqrt{g}\mathcal{A}^*(L)+\sqrt{T_L^*\mathcal{A}^*(L)}\mathcal{V}^*(L)}-1\right)$$
$$\left(\frac{u_1\sqrt{T_L^*\mathcal{A}^*(L)}-\sqrt{g}}{u_1\sqrt{T_L^*\mathcal{A}^*(L)}+\sqrt{g}}\times\frac{\sqrt{g}\mathcal{A}^*(L)-\sqrt{T_L^*\mathcal{A}^*(L)}\mathcal{V}^*(L)}{\sqrt{g}\mathcal{A}^*(L)+\sqrt{T_L^*\mathcal{A}^*(L)}\mathcal{V}^*(L)}+1\right)<0$$

$$\Rightarrow\frac{-2u_1\mathcal{A}^*(L)\mathcal{V}^*(L)T_L^*-2g\mathcal{A}^*(L)}{\left(u_1\sqrt{T_L^*\mathcal{A}^*(L)}+\sqrt{g}\right)\left(\sqrt{g}\mathcal{A}^*(L)+\sqrt{T_L^*\mathcal{A}^*(L)}\mathcal{V}^*(L)\right)}$$
$$\frac{2u_1\sqrt{g}\mathcal{A}^*(L)\sqrt{T_L^*\mathcal{A}^*(L)}-2\sqrt{g}\sqrt{T_L^*\mathcal{A}^*(L)}\mathcal{V}^*(L)}{\left(u_1\sqrt{T_L^*\mathcal{A}^*(L)}+\sqrt{g}\right)\left(\sqrt{g}\mathcal{A}^*(L)+\sqrt{T_L^*\mathcal{A}^*(L)}\mathcal{V}^*(L)\right)}<0$$

$$\Rightarrow -4\sqrt{g}\mathcal{A}^*(L)\mathcal{A}^*(L)\sqrt{T_L^*\mathcal{A}^*(L)}\left(u_1\mathcal{V}^*(L)T_L^*+g\right)\left(u_1\mathcal{A}^*(L)+\mathcal{V}^*(L)\right)<0$$

$$\Rightarrow\left(u_1\mathcal{V}^*(L)T_L^*+g\right)\left(u_1\mathcal{A}^*(L)+\mathcal{V}^*(L)\right)>0$$

$$\Rightarrow u_1\in\mathbb{R}\diagdown\left(\frac{-g}{\mathcal{V}^*(L)T_L^*},-\frac{\mathcal{V}^*(L)}{\mathcal{A}^*(L)}\right). \tag{48}$$

Let us define

$$g_1\triangleq\lambda_1e^{-\tau x},\qquad g_2\triangleq\lambda_2e^{\tau x}, \tag{49}$$

and consider the relations (17) and (49) for any $v\in L^2((0,L);\mathbb{R})$ and $a\in L^2((0,L);\mathbb{R})$, we obtain the transformations below.

$$\mathcal{L}=\int_0^L\xi^TP(x)\xi dx$$

$$= \int_0^L (\xi_1\ \xi_2) \begin{pmatrix} P_1 e^{-\tau x} & 0 \\ 0 & P_2 e^{\tau x} \end{pmatrix} \begin{pmatrix} \xi_1 \\ \xi_2 \end{pmatrix} dx$$

$$= \int_0^L P_1 e^{-\tau x} \xi_1^2 + P_2 e^{\tau x} \xi_2^2 dx$$

$$= \int_0^L \lambda_1 e^{-\tau x} \xi_1^2 + \lambda_2 e^{\tau x} \xi_2^2 dx$$

$$= \int_0^L \lambda_1 e^{-\tau x} \big(v + a\sqrt{\frac{g}{\mathcal{A}^* T^*}}\big)^2 + \lambda_2 e^{\tau x} \big(v - a\sqrt{\frac{g}{\mathcal{A}^* T^*}}\big)^2 dx$$

$$= \int_0^L (g_1 + g_2) \left(\frac{g}{\mathcal{A}^* T^*} a^2 + 2 \frac{g_1 - g_2}{g_1 + g_2} \sqrt{\frac{g}{\mathcal{A}^* T^*}} va + v^2 \right) dx.$$

So, we have (38). Moreover, According to the result (42), there exists $\alpha > 0$ so that we have (39).

4.2 *Exponential Stability of the Steady-State of the Nonlinear System in H^2–Norm*

The proof of Theorem 1 follows easily by using the following theorem:

Theorem 3 *The steady-state* $(\mathcal{A}^*, \mathcal{V}^*)$ *of the system* (1) *and* (21) *is exponentially stable for the* H^2*-norm if*

- *There exists two functions* $P_1, P_2 \in C^1\big([0, L] \to (0, +\infty)\big)$ *such that*

$$4\delta_2 \gamma_1 P_1(x) P_2(x) - \big(\delta_1 P_1(x) + \gamma_2 P_2(x)\big)^2 > 0 \tag{50}$$

  for any $x \in [0, L]$, *where* $\gamma_1(x)$, $\gamma_2(x)$, $\delta_1(x)$ *and* δ_2 *are given by* (20),
- *and*

$$\frac{\lambda_1(0) P_1(0)}{\lambda_2(0) P_2(0)} \left(\frac{u_0 \sqrt{\mathcal{A}^*(0) T_0^*} + \sqrt{g}}{u_0 \sqrt{\mathcal{A}^*(0) T_0^*} - \sqrt{g}} \right)^2 < 1,$$

$$\frac{\lambda_2(L) P_2(L)}{\lambda_1(L) P_1(L)} \left(\frac{u_1 \sqrt{T_L^* \mathcal{A}^*(L)} - \sqrt{g}}{u_1 \sqrt{T_L^* \mathcal{A}^*(L)} + \sqrt{g}} \right)^2 < 1, \tag{51}$$

  where u_0 *and* u_1 *are given by* (28) *respectively.*

Proof Consider the nonlinear system (1) and apply the change of variables (17) as follows:

$$z_1 = v + a\sqrt{\frac{g}{\mathcal{A}^* T^*}}, \quad z_2 = v - a\sqrt{\frac{g}{\mathcal{A}^* T^*}}, \tag{52}$$

where a and v are the perturbations given by (8).

The nonlinear system (1) can first be rewritten as follows :

$$\partial_t Z + A_1(Z, x)\partial_x Z + B_1(Z, x) = 0. \tag{53}$$

Then as:

$$\partial_t Z + A_1(Z, x)\partial_x Z - M_1(Z, x)(Z, x) = 0, \tag{54}$$

$$\begin{pmatrix} z_1(t,0) \\ z_2(t,L) \end{pmatrix} = G\begin{pmatrix} z_1(t,L) \\ z_2(t,0) \end{pmatrix} \tag{55}$$

with

$$Z = \begin{pmatrix} z_1 \\ z_2 \end{pmatrix}, \quad A_1(0,x) = \begin{pmatrix} \lambda_1 & 0 \\ 0 & -\lambda_2 \end{pmatrix}, B_1(0,x) = 0, \quad M_1(0,x) := \frac{\partial B_1}{\partial Z}(0,x) = \begin{pmatrix} \gamma_1 & \delta_1 \\ \gamma_2 & \delta_2 \end{pmatrix},$$

and

$$G(0) = 0, \qquad K := G'(0) = \begin{pmatrix} 0 & m_0 \\ m_1 & 0 \end{pmatrix}, \tag{56}$$

with γ_1, γ_2, δ_1 and δ_2 defined by (20), and m_0 and m_1 by (34). The condition for the steady state ($\mathcal{A}^*$, $\mathcal{V}^*$) of the system (54), (55) to be exponentially stable, which corresponds to finding a Lyapunov function for the H^2 norm of the form (36), is to find

$$P(x) = \begin{pmatrix} P_1(x) & 0 \\ 0 & P_2(x) \end{pmatrix}$$

such that $Z^T\big(M_1^T P(x) + P(x)M_1\big)Z$ is a negative semi-definite quadratic form and such that the boundary conditions (55) are dissipative. But, $Z^T\big(M_1^T P(x) + P(x)M_1\big)Z$ is a negative semi-definite quadratic form as soon as $4\delta_2\gamma_1 P_1(x)P_2(x) - \big(\delta_1 P_1(x) + \gamma_2 P_2(x)\big)^2 > 0, \forall x \in [0, L]$ since

$$Z^T\big(M_1^T P(x) + P(x)M_1\big)Z = -2|\gamma_1|P_1(x)\Big[z_1 + \Big(\frac{\delta_1 P_1(x) + \gamma_2 P_2(x)}{2\gamma_1 P_1(x)}\Big)z_2\Big]^2 \tag{57}$$

$$-\Big[\frac{4\delta_2\gamma_1 P_1(x)P_2(x) - \big(\delta_1 P_1(x) + \gamma_2 P_2(x)\big)^2}{2\gamma_1 P_1(x)}\Big]z_2^2, \tag{58}$$

and under the hypothesis (51), we have:

$$\frac{\lambda_1(0)P_1(0)}{\lambda_2(0)P_2(0)}\left(\frac{k_0\sqrt{\mathcal{A}^*(0)T_0^*}+\sqrt{g}}{k_0\sqrt{\mathcal{A}^*(0)T_0^*}-\sqrt{g}}\right)^2<1\Longrightarrow\left(\frac{k_0\sqrt{\mathcal{A}^*(0)T_0^*}+\sqrt{g}}{k_0\sqrt{\mathcal{A}^*(0)T_0^*}-\sqrt{g}}\sqrt{\frac{\lambda_1(0)P_1(0)}{\lambda_2(0)P_2(0)}}\right)^2<1$$

$$\Longrightarrow\left(m_0\sqrt{\frac{\lambda_1(0)P_1(0)}{\lambda_2(0)P_2(0)}}\right)^2<1 \tag{59}$$

and

$$\frac{\lambda_2(L)P_2(L)}{\lambda_1(L)P_1(L)}\left(\frac{k_1\sqrt{T_L^*\mathcal{A}^*(L)}-\sqrt{g}}{k_1\sqrt{T_L^*\mathcal{A}^*(L)}+\sqrt{g}}\right)^2<1\Longrightarrow\left(\frac{k_1\sqrt{T_L^*\mathcal{A}^*(L)}-\sqrt{g}}{k_1\sqrt{T_L^*\mathcal{A}^*(L)}+\sqrt{g}}\sqrt{\frac{\lambda_2(L)P_2(L)}{\lambda_1(L)P_1(L)}}\right)^2<1,$$

$$\Longrightarrow\left(m_1\sqrt{\frac{\lambda_2(L)P_2(L)}{\lambda_1(L)P_1(L)}}\right)^2<1. \tag{60}$$

Inequalities (59) and (60) ensure that the boundary conditions (55) are dissipative ($\rho(K)<1$). Indeed,

$$\Delta K\Delta^{-1}=\begin{pmatrix}0 & m_0\sqrt{\dfrac{\lambda_1(0)P_1(0)}{\lambda_2(0)P_2(0)}}\\ m_1\sqrt{\dfrac{\lambda_2(L)P_2(L)}{\lambda_1(L)P_1(L)}} & 0\end{pmatrix}. \tag{61}$$

□

Proof (of Theorem 1). Under the hypothesis (26) of the Theorem 1, we have:

$$\mathcal{K}'(\mathcal{A}^*(0))\in\left(\frac{-g}{\mathcal{V}^*(0)T_0^*},-\frac{\mathcal{V}^*(0)}{\mathcal{A}^*(0)}\right)\Rightarrow k_0\in\left(-\frac{g}{\mathcal{V}^*(0)T_0^*},-\frac{\mathcal{V}^*(0)}{\mathcal{A}^*(0)}\right)$$

$$\Rightarrow 4\mathcal{A}^*(0)\sqrt{g}\mathcal{A}^*(0)\sqrt{\mathcal{A}^*(0)T_0^*}\left(k_0\mathcal{V}^*(0)T_0^*+g\right)\left(k_0\mathcal{A}^*(0)+\mathcal{V}^*(0)\right)<0$$

$$\Rightarrow\left(k_0\mathcal{V}^*(0)T_0^*+g\right)\left(k_0\mathcal{A}^*(0)+\mathcal{V}^*(0)\right)<0$$

$$\Rightarrow\frac{2k_0\mathcal{A}^*(0)\mathcal{V}^*(0)T_0^*+2g\mathcal{A}^*(0)}{\left(k_0\sqrt{\mathcal{A}^*(0)T_0^*}-\sqrt{g}\right)\left(\sqrt{g}\mathcal{A}^*(0)-\sqrt{\mathcal{A}^*(0)T_0^*}\mathcal{V}^*(0)\right)}$$

$$\frac{2k_0\sqrt{g}\mathcal{A}^*(0)\sqrt{\mathcal{A}^*(0)T_0^*}+2\sqrt{g}\sqrt{\mathcal{A}^*(0)T_0^*}\mathcal{V}^*(0)}{\left(k_0\sqrt{\mathcal{A}^*(0)T_0^*}-\sqrt{g}\right)\left(\sqrt{g}\mathcal{A}^*(0)-\sqrt{\mathcal{A}^*(0)T_0^*}\mathcal{V}^*(0)\right)}<0 \tag{62}$$

$$\Rightarrow$$

$$\left(\frac{k_0\sqrt{\mathcal{A}^*(0)T_0^*}+\sqrt{g}}{k_0\sqrt{\mathcal{A}^*(0)T_0^*}-\sqrt{g}}\times\frac{\sqrt{g}\mathcal{A}^*(0)+\sqrt{\mathcal{A}^*(0)T_0^*}\mathcal{V}^*(0)}{\sqrt{g}\mathcal{A}^*(0)-\sqrt{\mathcal{A}^*(0)T_0^*}\mathcal{V}^*(0)}-1\right)$$

$$\left(\frac{k_0\sqrt{\mathcal{A}^*(0)T_0^*}+\sqrt{g}}{k_0\sqrt{\mathcal{A}^*(0)T_0^*}-\sqrt{g}}\times\frac{\sqrt{g}\mathcal{A}^*(0)+\sqrt{\mathcal{A}^*(0)T_0^*}\mathcal{V}^*(0)}{\sqrt{g}\mathcal{A}^*(0)-\sqrt{\mathcal{A}^*(0)T_0^*}\mathcal{V}^*(0)}+1\right)<0$$

$$\Rightarrow\left(\frac{k_0\sqrt{\mathcal{A}^*(0)T_0^*}+\sqrt{g}}{k_0\sqrt{\mathcal{A}^*(0)T_0^*}-\sqrt{g}}\frac{\sqrt{\frac{g\mathcal{A}^*(0)}{T_0^*}}+\mathcal{V}^*(0)}{\sqrt{\frac{g\mathcal{A}^*(0)}{T_0^*}}-\mathcal{V}^*(0)}-1\right)$$

$$\left(\frac{k_0\sqrt{\mathcal{A}^*(0)T_0^*}+\sqrt{g}}{k_0\sqrt{\mathcal{A}^*(0)T_0^*}-\sqrt{g}}\times\frac{\sqrt{\frac{g\mathcal{A}^*(0)}{T_0^*}}+\mathcal{V}^*(0)}{\sqrt{\frac{g\mathcal{A}^*(0)}{T_0^*}}-\mathcal{V}^*(0)}+1\right)<0 \tag{63}$$

From (62) and (16), we obtain

$$\left(\frac{k_0\sqrt{\mathcal{A}^*(0)T_0^*}+\sqrt{g}}{k_0\sqrt{\mathcal{A}^*(0)T_0^*}-\sqrt{g}}\times\frac{\lambda_1(0)}{\lambda_2(0)}-1\right)\left(\frac{k_0\sqrt{\mathcal{A}^*(0)T_0^*}+\sqrt{g}}{k_0\sqrt{\mathcal{A}^*(0)T_0^*}-\sqrt{g}}\times\frac{\lambda_1(0)}{\lambda_2(0)}+1\right)<0$$

$$\Rightarrow\left(\frac{k_0\sqrt{\mathcal{A}^*(0)T_0^*}+\sqrt{g}}{k_0\sqrt{\mathcal{A}^*(0)T_0^*}-\sqrt{g}}\right)^2\times\left(\frac{\lambda_1(0)}{\lambda_2(0)}\right)^2-1<0$$

$$\Rightarrow\left(\frac{k_0\sqrt{\mathcal{A}^*(0)T_0^*}+\sqrt{g}}{k_0\sqrt{\mathcal{A}^*(0)T_0^*}-\sqrt{g}}\right)^2\times\left(\frac{\lambda_1(0)}{\lambda_2(0)}\right)^2<1$$

$$\Rightarrow\left(\frac{k_0\sqrt{\mathcal{A}^*(0)T_0^*}+\sqrt{g}}{k_0\sqrt{\mathcal{A}^*(0)T_0^*}-\sqrt{g}}\right)^2\times\frac{\lambda_1(0)P_1(0)}{\lambda_2(0)P_2(0)}<1, \tag{64}$$

$$\text{avec } P_1(x)=\lambda_1(x) \text{ et } P_2(x)=\lambda_2(x) \tag{65}$$

We also have

$$\mathcal{K}'(\mathcal{A}^*(L))\in\mathbb{R}\diagdown\left[\frac{-g}{\mathcal{V}^*(L)T_L^*},-\frac{\mathcal{V}^*(L)}{\mathcal{A}^*(L)}\right]\Rightarrow k_1\in\mathbb{R}\diagdown\left[\frac{-g}{\mathcal{V}^*(L)T_L^*},-\frac{\mathcal{V}^*(L)}{\mathcal{A}^*(L)}\right]$$

$$\Rightarrow\big(k_1\mathcal{V}^*(L)T_L^*+g\big)\big(k_1\mathcal{A}^*(L)+\mathcal{V}^*(L)\big)\geq 0$$

$$\Rightarrow -4\sqrt{g}\mathcal{A}^*(L)\mathcal{A}^*(L)\sqrt{T_L^*\mathcal{A}^*(L)}\Big(k_1\mathcal{V}^*(L)T_L^*+g\Big)$$

$$\Big(k_1\mathcal{A}^*(L)+\mathcal{V}^*(L)\Big)<0$$

$$\Rightarrow$$

$$\frac{-2k_1\mathcal{A}^*(L)\mathcal{V}^*(L)T_L^* - 2g\mathcal{A}^*(L)}{\left(k_1\sqrt{T_L^*\mathcal{A}^*(L)} + \sqrt{g}\right)\left(\sqrt{g}\mathcal{A}^*(L) + \sqrt{T_L^*\mathcal{A}^*(L)}\mathcal{V}^*(L)\right)}$$
$$\frac{2k_1\sqrt{g}\mathcal{A}^*(L)\sqrt{T_L^*\mathcal{A}^*(L)} - 2\sqrt{g}\sqrt{T_L^*\mathcal{A}^*(L)}\mathcal{V}^*(L)}{\left(k_1\sqrt{T_L^*\mathcal{A}^*(L)} + \sqrt{g}\right)\left(\sqrt{g}\mathcal{A}^*(L) + \sqrt{T_L^*\mathcal{A}^*(L)}\mathcal{V}^*(L)\right)} < 0$$
$$\Rightarrow$$
$$\left(\frac{k_1\sqrt{T_L^*\mathcal{A}^*(L)} - \sqrt{g}}{k_1\sqrt{T_L^*\mathcal{A}^*(L)} + \sqrt{g}}\,\frac{\sqrt{g}\mathcal{A}^*(L) - \sqrt{T_L^*\mathcal{A}^*(L)}\mathcal{V}^*(L)}{\sqrt{g}\mathcal{A}^*(L) + \sqrt{T_L^*\mathcal{A}^*(L)}\mathcal{V}^*(L)} - 1\right)$$
$$\left(\frac{k_1\sqrt{T_L^*\mathcal{A}^*(L)} - \sqrt{g}}{k_1\sqrt{T_L^*\mathcal{A}^*(L)} + \sqrt{g}}\,\frac{\sqrt{g}\mathcal{A}^*(L) - \sqrt{T_L^*\mathcal{A}^*(L)}\mathcal{V}^*(L)}{\sqrt{g}\mathcal{A}^*(L) + \sqrt{T_L^*\mathcal{A}^*(L)}\mathcal{V}^*(L)} + 1\right) < 0$$
$$\Rightarrow \left(\frac{k_1\sqrt{T_L^*\mathcal{A}^*(L)} - \sqrt{g}}{k_1\sqrt{T_L^*\mathcal{A}^*(L)} + \sqrt{g}} \times \frac{\sqrt{\frac{g\mathcal{A}^*(L)}{T_L^*}} - \mathcal{V}^*(L)}{\sqrt{\frac{g\mathcal{A}^*(L)}{T_L^*}} + \mathcal{V}^*(L)} - 1\right)$$
$$\left(\frac{k_1\sqrt{T_L^*\mathcal{A}^*(L)} - \sqrt{g}}{k_1\sqrt{T_L^*\mathcal{A}^*(L)} + \sqrt{g}} \times \frac{\sqrt{\frac{g\mathcal{A}^*(L)}{T_L^*}} - \mathcal{V}^*(L)}{\sqrt{\frac{g\mathcal{A}^*(L)}{T_L^*}} + \mathcal{V}^*(L)} + 1\right) < 0. \tag{66}$$

From (66) and (16), we obtain

$$\left(\frac{k_1\sqrt{T_L^*\mathcal{A}^*(L)} - \sqrt{g}}{k_1\sqrt{T_L^*\mathcal{A}^*(L)} + \sqrt{g}} \times \frac{\lambda_2(L)}{\lambda_1(L)} - 1\right)$$
$$\left(\frac{k_1\sqrt{T_L^*\mathcal{A}^*(L)} - \sqrt{g}}{k_1\sqrt{T_L^*\mathcal{A}^*(L)} + \sqrt{g}} \times \frac{\lambda_2(L)}{\lambda_1(L)} + 1\right) < 0$$

$$\Rightarrow \left(\frac{k_1\sqrt{T_L^*\mathcal{A}^*(L)} - \sqrt{g}}{k_1\sqrt{T_L^*\mathcal{A}^*(L)} + \sqrt{g}}\right)^2 \times \left(\frac{\lambda_2(L)}{\lambda_1(L)}\right)^2 - 1 < 0$$

$$\Rightarrow \left(\frac{k_1\sqrt{T_L^*\mathcal{A}^*(L)} - \sqrt{g}}{k_1\sqrt{T_L^*\mathcal{A}^*(L)} + \sqrt{g}}\right)^2 \times \left(\frac{\lambda_2(L)}{\lambda_1(L)}\right)^2 < 1$$

$$\Rightarrow \left(\frac{k_1\sqrt{T_L^*\mathcal{A}^*(L)} - \sqrt{g}}{k_1\sqrt{T_L^*\mathcal{A}^*(L)} + \sqrt{g}}\right)^2 \times \frac{\lambda_2(L)P_2(L)}{\lambda_1(L)P_1(L)} < 1. \tag{67}$$

Furthermore, we have:

$$\delta_2\gamma_1\lambda_1\lambda_2 > \left(\delta_1\lambda_1 + \gamma_2\lambda_2\right)^2 \Longrightarrow \delta_2\gamma_1\lambda_1\lambda_2 - \left(\delta_1\lambda_1 + \gamma_2\lambda_2\right)^2 > 0$$
$$\Longrightarrow 4\delta_2\gamma_1 P_1(x)P_2(x) - \left(\delta_1 P_1(x) + \gamma_2 P_2(x >)\right)^2 > 0 \quad (68)$$

where P_1 and P_2 are defined by (65).

Thus, from (64), (67) and (68) we deduce from Theorem 3 that the steady state $(\mathcal{A}^*, \mathcal{V}^*)$ of (1) and (21) is exponentially stable for the norm H^2. □

5 Numerical Simulations

To illustrate the performance of the suggested selection, the boundary controls (26), for stabilizing the system (1) for the H^2 norm, we consider a steady-state flow in a trapezoidal channel of length $L = 2000\,\text{m}$ with $m = 0.5$, $b = 4$, $S_b = 2k\mathcal{V}^{*2}(0)/\mathcal{A}^*(0)$, and $k = 0.002$. Numerical simulations in the H^2 norm of the solutions of system (1) subjected to the boundary conditions (21) using the selection defined in (26) are performed.

The initial condition is given by $\mathcal{A}^*(0) = 10\,\text{m}^2$ and $\mathcal{V}^*(0) = 1\,\text{m/s}$. The initial perturbations are 0.4 m in the cross-sectional area of the water and 0.1 m/s in velocity. With $\mathcal{K}'\left(\mathcal{A}(t, 0)\right) = -1.1\mathcal{V}^*(0)/\mathcal{A}^*(0)$ and $\mathcal{K}'\left(\mathcal{A}(t, L)\right) = -0.9\mathcal{V}^*(L)/\mathcal{A}^*(L)$. It is worth noting that these values are chosen to satisfy the conditions defined in (26).

The result is shown by Fig. 1, where the blue curve represents the variation of the H^2 norm of the perturbations for the cross-sectional area of the water in the channel as a function of time, and the red curve represents the variation of the H^2

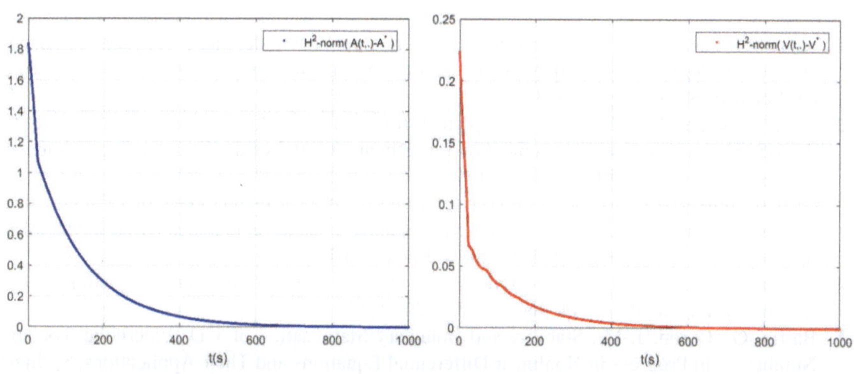

Fig. 1 Convergence of the cross-sectional area of the water and velocity

norm of the perturbations for the velocity as a function of time. It is observed that the perturbations in velocity and water cross-sectional area tend to zero as time increases. This implies the stability of the nonlinear system.

6 Conclusion

To summarize, we have effectively resolved the challenge of achieving exponential stability in the H^2 norm for the Saint-Venant equations in a trapezoidal channel, which is equipped at both ends with physical devices that facilitate feedback controls. We have determined a specific boundary condition that guarantees this exponential stability. Initially, we established a result of exponential stability in the L^2 norm for the associated linearized system by constructing a Lyapunov function. By further developing this function, we were able to derive the necessary conditions for the nonlinear system in the H^2 norm through appropriate boundary conditions. Additionally, we provide numerical simulations to validate the effectiveness of the control strategies implemented for system stabilization.

Data Availability and Conflicts of Interest

No data were utilized in this study, and the authors declare that there are no conflicts of interest related to the publication of this paper.

References

1. de Saint-Venant, A.B.: Théorie du mouvement non permanent des eaux, avec application aux crues des rivières et à l'introduction des marés dans leur lit. Comptes Rendus de l'Académie des Sciences **53**, 147–154 (1871)
2. de Saint-Venant, A.B.: Mémoire sur la perte de force vive d'un fluide aux endroits où sa section d'écoulement augmente brusquement ou rapidement. Mémoires de l'Académie des Sciences de l'Institut de France **44**, 193–243 (1888)
3. de Saint-Venant, A.B.: Mémoire sur la prise en considération de la force centrifuge dans le calcul du mouvement des eaux courantes et sur la distinction des torrents et des rivières. Mémoires de l'Académie des Sciences de l'Institut de France **44**, 245–273 (1888)
4. Bastin, G., Coron, J.-M.: Exponential stability of PI control for Saint-Venant equations with a friction term. IFAC Proc. Vol. **46**(26), 221–226 (2013)
5. Bastin, G., Coron, J.-M.: Stability and Boundary Stabilisation of 1-D Hyperbolic Systems. Number 88 in Progress in Nonlinear Differential Equations and Their Applications. Springer (2016)
6. Bastin, G., Coron, J.-M.: A quadratic Lyapunov function for hyperbolic density–velocity systems with nonuniform steady states. Syst. Control Lett. **104**, 66–71 (2017)
7. Bastin, G., Coron, J.-M., d'Andréa-Novel, B.: On Lyapunov stability of linearised Saint-Venant equations for a sloping channel. Netw. Heterogen. Media **4**(2), 177–187 (2009)

8. Coron, J.-M., Bastin, G., d'Andréa-Novel, B.: Dissipative boundary conditions for one-dimensional nonlinear hyperbolic systems. SIAM J. Control Optim. **47**(3), 1460–1498 (2008)
9. Coron, J.-M., d'Andréa-Novel, B., Bastin, G.: A Lyapunov approach to control irrigation canals modeled by Saint-Venant equations. In: CD-Rom Proceedings, Paper F1008-5, ECC99, Karlsruhe, Germany, pp. 3178–3183 (1999)
10. Coron, J.-M., d'Andréa-Novel, B., Bastin, G.: A strict Lyapunov function for boundary control of hyperbolic systems of conservation laws. IEEE Trans. Autom. Control **52**(1), 2–11 (2007)
11. Coron, J.-M., Nguyen, H.-M.: Dissipative boundary conditions for nonlinear 1-D hyperbolic systems: sharp conditions through an approach via time-delay systems. SIAM J. Math. Anal. **47**(3), 2220–2240 (2015)
12. Diagne, A., Bastin, G., Coron, J.-M.: Lyapunov exponential stability of 1-D linear hyperbolic systems of balance laws. Automatica **48**(1), 109–114 (2012)
13. Coron, J.M., Hayat, A., Xiang, S. et coll. Stabilisation du système de réservoir d'eau linéarisée. Arch Rational Mech Anal **244**, 1019–1097 (2022)
14. Coron, J.-M., Vazquez, R., Krstic, M., Bastin, G.: Local exponential H^2 stabilization of a 2×2 quasilinear hyperbolic system using backstepping. SIAM J. Control Optim. **51**(3), 2005–2035 (2013)
15. de Halleux, J., Prieur, C., Coron, J.-M., d'Andréa-Novel, B., Bastin, G.: Boundary feedback control in networks of open channels. Automatica J. IFAC **39**(8), 1365–1376 (2003)
16. Diagne, A., Diagne, M., Tang, S., Krstic, M.: Backstepping stabilization of the linearized Saint-Venant-Exner model. Automatica J. IFAC **76**, 345–354 (2017)
17. Diagne, A., Tang, S., Diagne, M., Krstic, M.: State feedback stabilization of the linearized bilayer saint-venant model. IFAC-PapersOnLine **49**(8), 130–135 (2016)
18. Dos Santos, V., Bastin, G., Coron, J.-M., d'Andréa-Novel, B.: Boundary control with integral action for hyperbolic systems of conservation laws: stability and experiments. Automatica J. IFAC **44**(5), 1310–1318 (2008)
19. Dos Santos, V., Prieur, C.: Boundary control of open channels with numerical and experimental validations. IEEE Trans. Control Syst. Technol. **16**(6), 1252–1264 (2008)
20. Greenberg, J.M., Tsien, L.T.: The effect of boundary damping for the quasilinear wave equation. J. Differ. Equ. **52**(1), 66–75 (1984)
21. Hayat, A.: Boundary stability of 1-D nonlinear inhomogeneous hyperbolic systems for the C^1 norm. SIAM J. Control Optim. **57**(6), 3603–3638 (2019)
22. Hayat, A.: PI controllers for the general Saint-Venant equations (2019). Preprint, hal-01827988
23. Hayat, A., Shang, P.: Exponential stability of density-velocity systems with boundary conditions and source term for the H^2-norm (2019). Preprint, hal-02190778
24. Hayat, A., Shang, P.: A quadratic Lyapunov function for Saint-Venant equations with arbitrary friction and space-varying slope. Automatica J. IFAC **100**, 52–60 (2019)
25. Leugering, G., Schmidt, J.P.G.: On the modeling and stabilization of flows in open channel networks. SIAM J. Control Optim. **41**(1), 164–180 (2002)
26. Gugat, M., Herty, M.: Existence of classical solutions and feedback stabilization for the flow in gas networks. ESAIM: COCV **17**(1), 28–51 (2011)
27. Sore, S., Ndiaye, B., Simpore, Y.: Exponential stability of the shallow water equations in trapezoidal channel. Pan-Am. J. Math. (2024). 3. 9. 10.28919/cpr-pajm/3-9
28. Li, T., Rao, B., Wang, Z.: Exact boundary controllability and observability for first order quasilinear hyperbolic systems with a kind of nonlocal boundary conditions. Discrete Contin. Dyn. Syst. **28**(1), 243–257 (2010)
29. Prieur, C., Winkin, J., Bastin, G.: Robust boundary control of systems of conservation laws. Math. Control Signals Syst. **20**(2), 173–197 (2008)
30. Trinh, N.-T., Andrieu, V., Xu, C.-Z.: Output regulation for a cascaded network of 2×2 hyperbolic systems with PI controller. Automatica J. IFAC **91**, 270–278 (2018)
31. Xu, C.-Z., Sallet, G.: Multivariable boundary PI control and regulation of a fluid flow system. Math. Control Relat. Fields **4**(4), 501–520 (2014)

On Some Fixed Point Results of Banach Type Contraction in Controlled G-Metric Spaces

John Pamba and Talat Nazir

Abstract In this paper, we introduce a new type of metric spaces called controlled G-metric spaces, denoted, G_ζ, which are a generalisation of G-metric spaces. This is an extension of the work by Aghajani et al. in the article "*Common fixed point of generalised weak contractive mappings in partially ordered G_b-metric spaces. Filomat 2014, 28, 1087–1101*". We do this by employing a control function $\zeta(x, y, z)$ to the right-hand side of the G_b-triangle inequality so that the new triangle inequality becomes $G_\zeta(x, y, z) \leq \zeta(x, a, a)G_\zeta(x, a, a) + \zeta(a, y, z)G_\zeta(a, y, z)$ for all $a, x, y, z \in X$. Examples of controlled $G-$metric spaces which are not G_b-metric spaces in the sense of Aghajani et al. are given to show that our extension of $G-$metric spaces is different. We give some convergence properties on this new space. In order to further illustrate the usefulness of this new structure, a generalised Banach contraction principle is introduced and new fixed point results are developed and proved. Some examples are presented to support main result proved therein. These results improve, unify and generalize already well-known results on $G-$metric spaces and G_b-metric spaces.

Keywords Metric space · Controlled $G-$metric space · $G-$metric space · G_b–metric space · Banach contraction principle · Fixed point

1 Introduction

Fixed point theory has been one of the most significant research fields in nonlinear functional analysis. It has wide applications in many disciplines such as studying the existence of solutions for differential and integral equations, convergence of

J. Pamba
Department of Mathematics and Statistics, Mukuba University, Kitwe, Zambia
e-mail: john.pamba@mukuba.edu.zm

T. Nazir (✉)
Department of Mathematical Sciences, University of South Africa, Florida, South Africa
e-mail: talatn@unisa.ac.za

H. Kalita et al. (eds.), *Nonlinear Analysis and Computational Techniques*,
Springer Proceedings in Mathematics & Statistics 501,
https://doi.org/10.1007/978-981-96-6038-4_10

many computational methods, in economics, etc. In the recent decades, a lot of researchers have done some significant works in this direction. In order to generalise metric spaces, in 1989, Bakhtin [1] and, in 1993, Czerwik [2] introduced a new distance on a non-empty set, which is called a $b-$metric. In the $b-$metric space, the triangle inequality was generalised from $d(x, y) \leq d(x, z) + d(y, z)$ to $d(x, y) \leq s[d(x, z) + d(y, z)]$ for all $x, y, z \in X$ and $s \geq 1$. Mustafa and Sims [3] introduced a new structure of generalised metric spaces called $G-$metric spaces in 2007. Follow [4–11] for further work on the setting of $G-$matric spaces. Aghajani et al. [12] introduced the concept of a G_b-metric space by generalising the $G-$metric space after modifying the $G-$triangle inequality from $G(x, y, z) \leq G(x, a, a) + G(a, y, z)$ to $G_b(x, y, z) \leq s[G_b(x, a, a) + G_b(a, y, z)]$ for $x, y, z, a \in X$ and $s \geq 1$. A number of properties have been proved on these spaces, see for example in [13–23]. Gupta et al. [24] demonstrated the formation of a generalisation of G_b-metric spaces to extended G_b-metric spaces by means of the results of Kamran et al. [25]. A number of properties under the framework of extended G_b-metric spaces were given in [25].

In 1922, Banach [26] proved that each contraction map in a complete metric space has a unique fixed point and this has been generalised greatly over the past years. There are other forms of generalised metrics such as cone metric [27], F-metric [28], ϕ-metric [29], parametric metric [30] and partial metric [31].

In this chapter, we introduce controlled $G-$metric space as a generalisation of controlled metric spaces, G_b-metric spaces and $G-$metric spaces. The main results are in Sect. 3 where we have defined a controlled $G-$metric space as a new generalised structure of controlled type. Some convergence properties are proved on this new structure. We develop a theorem for the existence of the fixed points of Banach contraction type on controlled $G-$metric spaces. We also present some examples for clear illustration. These results improve, unify and generalise some known results in the existing literature.

2 Preliminaries

In this section, we present some important definitions and concepts from the literature which will be needed in the sequel. We begin with the concept of G-metric spaces according to [3].

Definition 1 ([3]) Let X be a non-empty set and let $G : X \times X \times X \to [0, \infty)$ be a function satisfying the following:

G_1) $G(x, y, z) = 0$ if $x = y = z$;
G_2) $G(x, x, y) > 0$ for all $x, y \in X$ with $x \neq y$;
G_3) $G(x, x, y) \leq G(x, y, z)$, for all $x, y, z \in X$ with $z \neq y$;
G_4) $G(x, y, z) = G(x, z, y) = G(y, z, x) = \cdots$, (symmetry in all three variables);
G_5) $G(x, y, z) \leq G(x, a, a) + G(a, y, z)$, for all $x, y, z, a \in X$ (rectangle inequality).

Then the function G is called a generalised metric or a G-metric on X, and the pair (X, G) is called a $G-$metric space.

Example 1 ([3]) Let $X = \mathbb{R}$. Then $G(x, y, z) = \frac{1}{3}(| x - y | + | y - z | + | x - z |)$, for all $x, y, z \in X$ is a $G-$metric.

Proposition 1 ([4]) *Every G-metric space (X, G) will define a metric space (X, d_G) by $d_G(x, y) = G(x, x, y) + G(y, x, x)$, for all $x, y \in X$.*

Definition 2 ([3]) Let (X, G) be a $G-$metric space, and let $\{a_n\}$ be a sequence of points of X. Then we say that $\{a_n\}$ is G-convergent to $a \in X$ if

$$\lim_{n,m\to\infty} G(a, a_n, a_m) = 0,$$

that is, for any $\epsilon > 0$, there exists $N \in \mathbb{N}$ such that $G(a, a_n, a_m) < \epsilon$ for all $n, m \geq N$. We call a the limit of the sequence and write $a_n \to a$ as $n \to \infty$ or

$$\lim_{n\to\infty} a_n = a.$$

Definition 3 ([3]) Let (X, G) be a $G-$metric space. A sequence $\{a_n\}$ is called a G-Cauchy sequence if for any $\epsilon > 0$, there is $N \in \mathbb{N}$ such that $G(a_n, a_m, a_l) < \epsilon$ for all $n, m, l \geq N$, that is, $G(a_n, a_m, a_l) \to 0$ as $n, m, l \to \infty$.

Definition 4 ([3]) A G-metric space (X, G) is called G-complete if every G-Cauchy sequence is G-convergent and converges to some element in X.

Proposition 2 ([3]) *Let (X, G) be a $G-$metric space. Then (X, G) is a complete space if and only if (X, d_G) is a complete metric space.*

A new distance on a non-empty set which is called a $b-$metric space was introduced by Bakhtin [1] in 1989, and Czerwik [2] in 1993 as a generalisation of a metric space.

Definition 5 ([2]) Let X be a non-empty set and $s \geq 1$ a given real number. A function $d : X \times X \to [0, \infty)$ is called a $b-$metric on X if for all $x, y, z \in X$, the following conditions hold:

b_1) $d(x, y) = 0$ if and only if $x = y$;
b_2) $d(x, y) = d(y, x)$; and
b_3) $d(x, z) \leq s[d(x, y) + d(y, z)]$.

The pair (X, d) is called a $b-$metric space and the real number $s \geq 1$ is called the coefficient of (X, d). Every metric space is a $b-$metric space with coefficient $s = 1$. The converse is generally not true.

Example 2 Let (X, d) be a metric space. Consider $d_b(x, y) = [d(x, y)]^p$, where $p > 1$ is a natural number. Then d_b is a $b-$metric with $s = 2^{p-1}$ but not a metric on X.

Aghajani et al. [12] introduced the concept of a G_b-metric space by combining the concepts of a $G-$metric space and the $b-$metric space in the following way:

Definition 6 ([12]) Let X be a non-empty set and $s \geq 1$ be a given real number. Suppose that $G_b : X \times X \times X \to [0, \infty)$ satisfy the following:

G_{b1}) $G_b(x, y, z) = 0$, if $x = y = z$;
G_{b2}) $0 < G_b(x, x, y)$ for all $x, y \in X$ with $x \neq y$;
G_{b3}) $G_b(x, x, y) \leq G_b(x, y, z)$ for all $x, y, z \in X$ with $y \neq z$;
G_{b4}) $G_b(x, y, z) = G_b(x, z, y) = G_b(y, z, x) = \cdots$, (symmetric in all three variables); and
G_{b5}) $G_b(x, y, z) \leq s[G_b(x, a, a) + G_b(a, y, z)]$ for all $x, y, a \in X$ (rectangular inequality).

Then G_b is called a generalised $b-$metric and the pair (X, G_b) is called a generalised $b-$metric space or G_b-metric space. See [12] for examples.

Example 3 ([12]) Let (X, G) be a $G-$metric space. Consider $G_b(x, y, z) = [G(x, y, z)]^p$, where $p > 1$ is a real number. Then (X, G_b) is a G_b-metric space with $s = 2^{p-1}$.

It was proved by Aghajani et al. [12] that every G_b-metric space is topologically equivalent to a $b-$metric space. As such, we can without challenges extend concepts and results for $b-$metric spaces into G_b-metric spaces. The example below shows that a G_b-metric on X does not need to be a $G-$metric on X.

Example 4 ([12]) Let $X = \mathbb{R}$. The function defined by

$$G_b(x, y, z) = \frac{1}{9}(\mid x - y \mid + \mid y - z \mid + \mid x - z \mid)^2$$

is a G_b-metric on X with $s = 2$, but it is not a $G-$metric on the whole of X. This can be confirmed by taking for $x = 3$, $y = 5$, $z = 7$ and $a = \frac{7}{2}$ so that $G_b(3, 5, 7) = \frac{64}{9}$, $G_b(3, \frac{7}{2}, \frac{7}{2}) = \frac{1}{9}$ and $G_b(\frac{7}{2}, 5, 7) = \frac{49}{9}$, so

$$G_b(3, 5, 7) = \frac{64}{9} > \frac{50}{9} = G_b(3, \frac{7}{2}, \frac{7}{2}) + G_b(\frac{7}{2}, 5, 7).$$

Definition 7 ([12]) Let X be a G_b-metric space. A sequence $\{x_n\}$ in X is said to be

1. G_b-Cauchy sequence if for each $\epsilon > 0$, there exists a positive integer n_0 such that for all $m, n, l \geq n_0$, $G_b(x_n, x_m, x_l) < \epsilon$, and
2. G_b-convergent to a point $x \in X$, if for each $\epsilon > 0$, there exists a positive integer n_0 such that, for all $m, n \geq n_0$, $G_b(x_n, x_m, x) < \epsilon$.

Lemma 1 ([16]) *Let $\{x_n\}$ be a sequence in (X, G_b) such that*

$$G_b(x_{n+1}, x_{n+2}, x_{n+2}) \leq \lambda G_b(x_n, x_{n+1}, x_{n+1}), \tag{1}$$

where $\lambda \in [0, 1)$ and $n = 1, 2, 3, \ldots$. Then, $\{x_n\}$ is a G_b-Cauchy sequence.

In 2018, Mlaiki et al. [32] initiated the concept of controlled metric type spaces by employing a control function $\alpha(x, y)$ of the right-hand side of the triangle inequality metric space.

Definition 8 [32] Let X be a non-empty set and $\alpha : X \times X \to [0, \infty)$. Then a mapping $d_\alpha : X \times X \to [0, \infty)$ is called a controlled metric type, if for all $x, y, z \in X$, it satisfies the following axioms:

(i) $d_\alpha(x, y) = 0$ if and only if $x = y$;
(ii) $d_\alpha(x, y) = d_\alpha(y, x)$;
(iii) $d_\alpha(x, z) \leq \alpha(x, y)d_\alpha(x, y) + \alpha(y, z)d_\alpha(y, z)$.

The pair (X, d_α) is called a controlled metric type space.

The convergence, Cauchy and completeness in controlled metric type spaces are defined as follows:

Definition 9 [32] Let (X, d_α) be a controlled metric type space and $\{x_n\}$ be a sequence in X. Then

1. the sequence $\{x_n\}$ converges to some x in X if for every $\epsilon > 0$, there exists $N \in \mathbb{N}$ such that $d_\alpha(x_n, x) < \epsilon$ for all $n \geq N$. In this case, we write $\lim_{n\to\infty} x_n = x$,
2. the sequence $\{x_n\}$ is Cauchy if for every $\epsilon > 0$, there exists $N \in \mathbb{N}$ such that $d_\alpha(x_m, x_n) < \epsilon$ for all $m, n \geq N$,
3. the controlled metric space (X, d_α) is called complete if every Cauchy sequence is convergent and converges to some element in X.

Definition 10 ([33]) Let (X, d) be a metric space and $T : X \to X$ be a given function. Given any $x_0 \in X$ the orbit of x_0 is defined by the sequence

$$\{x_0,\ x_1 = T(x_0),\ x_2 = T^2(x_0),\ x_3 = T^3(x_0), \ldots\},$$

that is, the sequence $\{x_n\}$ such that for all $n \geq 1$, $x_n = T^n(x_0) = T(x_{n-1})$.

Definition 11 ([33]) Let (X, d) be a metric space and $T : X \to X$ be a given function. A given $x \in X$ is an orbital point of T if $T^n(x) = x$ for some positive integer n.

Theorem 1 ([6]) *Let (X, G) be a $G-$complete $G-$metric space and let $T : X \to X$ be a mapping such that there exists $\lambda \in [0, 1)$ satisfying*

$$G(Tx, Ty, Tz) \leq \lambda G(x, y, z),$$

where $x, y, z \in X$. Then T has a unique fixed point.

One can follow [4, 6] for more results of fixed points in this direction.

3 Controlled $G-$Metric Spaces

By modifying definitions of a $G-$metric space, a G_b-metric space and that of a controlled metric type space, we give the definition of a controlled $G-$metric space as follows:

Definition 12 Let X be a non-empty set and $\zeta : X \times X \times X \to [1, \infty)$ be a function. Then a mapping $G_\zeta : X \times X \times X \to [0, \infty)$ is called a controlled $G-$metric if for all $a, x, y, z \in X$, the following axioms are satisfied:

(i) $G_\zeta(x, y, z) = 0$ if $x = y = z$;
(ii) $G_\zeta(x, x, y) > 0$ if $x \neq y$;
(iii) $G_\zeta(x, x, y) \leq G_\zeta(x, y, z)$, if $z \neq y$;
(iv) $G_\zeta(x, y, z) = G_\zeta(x, z, y) = G_\zeta(y, z, x) = \cdots$, (symmetry in all three variables);
(v) $G_\zeta(x, y, z) \leq \zeta(x, a, a)G_\zeta(x, a, a) + \zeta(a, y, z)G_\zeta(a, y, z)$.(rectangle inequality).

The pair (X, G_ζ) is called a controlled $G-$metric space.

Note that any G_b-metric space is a controlled $G-$metric space with $\zeta = s \geq 1$, but the converse is not generally true.

Example 5 Let $X = \mathbb{R}$. A function $G_\zeta : X \times X \times X \to [0, \infty)$ defined by

$$G_\zeta(x, y, z) = \frac{2}{3} \max\{|x - y|^2, |y - z|^2, |z - x|^2\}$$

for all $x, y, z \in X$ and a function $\zeta : X \times X \times X \to [1, \infty)$ defined as

$$\zeta(x, y, z) = \begin{cases} 2, & if\ y = z \text{ for all } x \in \mathbb{R}; \\ \infty, & if\ y \neq z \text{ for all } x \in \mathbb{R}. \end{cases}$$

Then clearly, $G_\zeta(x, y, z) = 0$ if $x = y = z$. And $G_\zeta(x, x, y) > 0$ if $x \neq y$, by its definition.

Also, $G_\zeta(x, x, y) = \frac{2}{3}|x - y|^2 \leq \frac{2}{3}\max\{|x - y|^2, |y - z|^2, |z - x|^2\} = G_\zeta(x, y, z)$ if $y \neq z$.

$G_\zeta(x, y, z) = \frac{2}{3}\max\{|x - y|^2, |y - z|^2, |z - x|^2\} = G_\zeta(x, z, y)$ and so on, symmetry in all three variables up to all possible combinations.

Lastly, taking the case of $y = z$, we have

$$\begin{aligned} G_\zeta(x, y, z) &= \frac{2}{3}|x - y|^2 \\ &\leq \frac{4}{3}|x - a|^2 + \frac{4}{3}|y - a|^2 \\ &= \zeta(x, a, a)G_\zeta(x, a, a) + \zeta(a, y, z)G_\zeta(a, y, z). \end{aligned}$$

Note that it follows by definition of $\zeta(x, y, z)$ that if $y \neq z$, then (5) in Definition 12 holds (since $\zeta(x, y, z) = \infty$ if $y \neq z$ and so the right-hand side of (5) in Definition 12 goes to ∞). Thus (X, G_ζ) is a controlled $G-$metric space.

Example 6 Let $X = \{1, 2, 3\}$. Define $G_\zeta : X \times X \times X \to [0, \infty)$ by

$$G_\zeta(x, y, z) = \frac{1}{2}\left(|x-y| + |y-z| + |x-z|\right)$$

for all $x, y, z \in X$ and $\zeta : X \times X \times X \to [1, \infty)$ as $\zeta(x, y, z) = xyz$ for all $x, y, z \in X$.

Clearly, $G_\zeta(x, y, z) = 0$ if $x = y = z$, for all $x, y, z \in X$, since $G_\zeta(1, 1, 1) = G_\zeta(2, 2, 2) = G_\zeta(3, 3, 3) = G_\zeta(4, 4, 4) = 0$. Now $G_\zeta(1, 1, 2) = 1 > 0$, $G_\zeta(1, 1, 3) = 2 > 0$, $G_\zeta(2, 2, 1) = 1 > 0$, $G_\zeta(2, 2, 3) = 1 > 0$, $G_\zeta(3, 3, 1) = 2 > 0$, $G_\zeta(3, 3, 2) = 1 > 0$.
Thus, $G_\zeta(x, x, y) > 0$ if $x \neq y$.

Also,
$G_\zeta(1, 1, 2) = 1 < 2 < G_\zeta(1, 2, 3)$, $G_\zeta(1, 1, 3) = 2 = G_\zeta(1, 2, 3)$,
$G_\zeta(2, 2, 1) = 1 < 2 < G_\zeta(1, 2, 3)$, $G_\zeta(2, 2, 3) = 1 < 2 = G_\zeta(1, 2, 3)$,
$G_\zeta(3, 3, 1) = 2 = G_\zeta(1, 2, 3)$, $G_\zeta(3, 3, 2) = 1 < 2 = G_\zeta(1, 2, 3)$.
Hence, $G_\zeta(x, x, y) \leq G_\zeta(x, y, z)$, if $z \neq y$.

Since, $G_\zeta(1, 2, 3) = G_\zeta(1, 3, 2) = G_\zeta(2, 1, 3) = G_\zeta(2, 3, 1) = G_\zeta(3, 1, 2) = G_\zeta(3, 2, 1) = 2$, that is,
$G_\zeta(x, y, z) = G_\zeta(x, z, x) = G_\zeta(y, x, z) = G_\zeta(y, z, x) = G_\zeta(z, y, x) = G_\zeta(z, x, y)$, for all $x, y, z \in X$.

Now

(x, y, z, a)	$G_\zeta(x, y, z)$	$\leq$	$\zeta(x, a, a)G_\zeta(x, a, a)$	+	$\zeta(a, y, z)G_\zeta(a, y, z)$
(1, 1, 2, 3)	1	<	(9) (2)	+	(6) (2)
(1, 2, 2, 3)	1	<	(9) (2)	+	(12) (1)
(1, 3, 3, 2)	2	<	(4) (1)	+	(18) (1)
(1, 2, 3, 1)	2	<	(1) (0)	+	(6) (2)
(1, 2, 3, 2)	2	<	(4) (1)	+	(12) (1)
(1, 2, 3, 3)	2	<	(9) (2)	+	(18) (1)

Hence, $G_\zeta(x, y, z) \leq \zeta(x, a, a)G_\zeta(x, a, a) + \zeta(a, y, z)G_\zeta(a, y, z)$ holds for all $x, y, z, a \in X$. As all the axoims (1)-(5) of Definition 12 are satisfied. Hence (X, G_ζ) is a controlled $G-$metric space.

Example 7 Let (X, d) be a metric space. Define $G_\zeta : X \times X \times X \to [0, \infty)$ by

$$G_\zeta(x, y, z) = \max\{d(x, y), d(y, z), d(x, z)\} \text{ for all } x, y, z \in X$$

and $\zeta : X \times X \times X \to [1, \infty)$ as $\zeta(x, y, z) = \max\{2, G_\zeta(x, y, z)\}$ for all $x, y, z \in X$.

Then (X, G_ζ) is a controlled G-metric space.
Clearly, $G_\zeta(x, y, z) = \max\{d(x, y), d(y, z), d(x, z)\} = 0,\ \text{if } x = y = z.$

If $x \neq y$, then $G_\zeta(x, x, y) = \max\{d(x, x), d(x, y), d(x, x)\} = d(x, y) > 0.$

If $z \neq y$, then
$$\begin{aligned} G_\zeta(x, y, z) &= \max\{d(x, y), d(y, z), d(x, z)\} \\ &\geq \max\{d(x, y)\} \\ &= \max\{d(x, x), d(y, x), d(x, x)\} \\ &= G_\zeta(x, x, y), \end{aligned}$$
that is, if $z \neq y$, then $G_\zeta(x, x, y) \leq G_\zeta(x, y, z)$. $G_\zeta(x, y, z) = \max\{d(x, y),$
$$\begin{aligned} d(y, z), d(x, z)\} &= \max\{d(x, z), d(z, y), d(x, y)\} \\ &= G_\zeta(x, z, y) \\ &= \max\{d(y, z), d(z, x), d(y, x)\} \\ &= G_\zeta(y, z, x), \end{aligned}$$
that is, the symmetry in all three variables holds.

We now prove the rectangle inequality. For $x, y, z, a \in X$,

$$\begin{aligned} &G_\zeta(x, y, z) \\ &= \max\{d(x, y), d(y, z), d(x, z)\} \\ &\leq \max\{d(x, a) + d(a, y), d(y, z), d(x, a) + d(a, z)\} \\ &\leq 2d(x, a) + 2\max\{d(a, y), d(y, z), d(z, a)\} \\ &\leq \max\{2, G_\zeta(x, a, a)\}d(x, a) + \max\{2, G_\zeta(a, y, z)\}\max\{d(a, y), d(y, z), d(z, a)\} \\ &= \zeta(x, a, a)G_\zeta(x, a, a) + \zeta(a, y, z)G_\zeta(a, y, z). \end{aligned}$$

Remark 1 The class of controlled $G-$metric spaces is larger than that of $G-$metric spaces and G_b-metric spaces.

Definition 13 A controlled $G-$metric space is said to be symmetric if $G_\zeta(x, y, y) = G_\zeta(y, x, x)$ for all $x, y \in X$.

Definition 14 Let (X, G_ζ) be a controlled $G-$metric space and $\{x_n\}$ be a sequence in X. Then

(1) the sequence $\{x_n\}$ $G_\zeta-$converges to some x in X if for every $\epsilon > 0$, there exists $N \in \mathbb{N}$ such that $G_\zeta(x_n, x_n, x) < \epsilon$ for all $n \geq N$;
(2) the sequence $\{x_n\}$ is $G_\zeta-$Cauchy if for every $\epsilon > 0$, there exists $N \in \mathbb{N}$ such that $G_\zeta(x_n, x_n, x_m) < \epsilon$ for all $m, n \geq N$;
(3) the controlled $G-$metric space (X, G_ζ) is called complete if every $G_\zeta-$Cauchy sequence is $G_\zeta-$convergent and $G_\zeta-$converges to some element in X.

The following proposition follows as an extension of Proposition 6 in [3].

Proposition 3 *Let (X, G_ζ) be a controlled $G-$metric space and let $\{x_n\}$ be a sequence in X. Then the following statements are equivalent:*

(1) $\{x_n\}$ is $G_\zeta-$convergent to x.
(2) $G_\zeta(x_n, x_n, x) \to 0$ as $n \to \infty$.
(3) $G_\zeta(x_n, x, x) \to 0$ as $n \to \infty$.

(4) $G_\zeta(x_n, x_m, x) \to 0$ *as* $m, n \to \infty$.

Proof The Proof is similar to the Proposition 6 of [3].

Proposition 4 *Let* (X, G_ζ) *be a controlled* $G-$*metric space and let* $\{x_n\}$ *be a sequence in* X. *Then the following statements are equivalent:*

1. $\{x_n\}$ *is* $G_\zeta-$*Cauchy sequence.*
2. $G_\zeta(x_n, x_m, x_m) \to 0$ *as* $m, n \to \infty$.

Proof The Proof is Proposition that follows directly from the Definition 14 (2).

Theorem 2 *Let* (X, G_ζ) *be a complete controlled* $G-$*metric space. Suppose that a mapping* $f : X \times X$ *is satisfying*

$$G_\zeta(fx, fy, fz) \leq kG_\zeta(x, y, z) \tag{2}$$

for every $x, y, z \in X$, *where* $k \in (0, 1)$. *If for* $r_0 \in X$, $r_n = f^n(r_0)$, *we have*

$$\sup_{m\geq 1}\lim_{n\to\infty}\frac{\zeta(r_{n+1}, r_{n+2}, r_{n+3})}{\zeta(r_n, r_{n+1}, r_{n+2})}\zeta(r_{n+1}, r_m, r_m) \leq \frac{1}{k}.$$

Also, if for every $r \in X$, *we have* $\lim\limits_{n\to\infty}\zeta(r, r_n, r_n)$ *and* $\lim\limits_{n\to\infty}\zeta(r_n, r, r)$ *exist and are finite. Then there exists a unique fixed point of mapping* f.

Proof For a sequence $\{r_n\}$ and for an arbitrary point $r_0 \in X$, define $r_1 = fr_0$, $r_2 = fr_1, \ldots, r_n = f^n r_0, \ldots$. From (2), we have

$$\begin{aligned}
G_\zeta(r_n, r_{n+1}, r_{n+1}) &\leq G_\zeta(fr_{n-1}, fr_n, fr_n)\\
&\leq kG_\zeta(r_{n-1}, r_n, r_n)\\
&\leq k^2G_\zeta(r_{n-2}, r_{n-1}, r_{n-1})\\
&\leq \cdots\\
&\leq k^nG_\zeta(r_1, r_0, r_0).
\end{aligned}$$

By condition (5) in Definition 12, we have

$$\begin{aligned}
&G_\zeta(r_n, r_m, r_m)\\
&\leq \zeta(r_n, r_{n+1}, r_{n+1})G_\zeta(r_n, r_{n+1}, r_{n+1}) + \zeta(r_{n+1}, r_m, r_m)G_\zeta(r_{n+1}, r_m, r_m)\\
&\leq \zeta(r_n, r_{n+1}, r_{n+1})G_\zeta(r_n, r_{n+1}, r_{n+1}) + \zeta(r_{n+1}, r_m, r_m)\zeta(r_{n+1}, r_{n+2}, r_{n+2})G_\zeta(r_{n+1}, r_{n+2}, r_{n+2})\\
&\quad+\zeta(r_{n+1}, r_m, r_m)\zeta(r_{n+2}, r_m, r_m)G_\zeta(r_{n+2}, r_m, r_m)\\
&\leq \zeta(r_n, r_{n+1}, r_{n+1})G_\zeta(r_n, r_{n+1}, r_{n+1}) + \zeta(r_{n+1}, r_m, r_m)\zeta(r_{n+1}, r_{n+2}, r_{n+2})G_\zeta(r_{n+1}, r_{n+2}, r_{n+2})\\
&\quad+\zeta(r_{n+1}, r_m, r_m)\zeta(r_{n+2}, r_m, r_m)\zeta(r_{n+2}, r_{n+3}, r_{n+3})G_\zeta(r_{n+2}, r_{n+3}, r_{n+3})\\
&\quad+\zeta(r_{n+1}, r_m, r_m)\zeta(r_{n+2}, r_m, r_m)\zeta(r_{n+3}, r_m, r_m)G_\zeta(r_{n+3}, r_m, r_m)\\
&\leq \zeta(r_n, r_{n+1}, r_{n+1})G_\zeta(r_n, r_{n+1}, r_{n+1}) + \zeta(r_{n+1}, r_m, r_m)\zeta(r_{n+1}, r_{n+2}, r_{n+2})G_\zeta(r_{n+1}, r_{n+2}, r_{n+2})\\
&\quad+\zeta(r_{n+1}, r_m, r_m)\zeta(r_{n+2}, r_m, r_m)\zeta(r_{n+2}, r_{n+3}, r_{n+3})G_\zeta(r_{n+2}, r_{n+3}, r_{n+3})\\
&\quad+\zeta(r_{n+1}, r_m, r_m)\zeta(r_{n+2}, r_m, r_m)\zeta(r_{n+3}, r_m, r_m)\zeta(r_{n+3}, r_{n+4}, r_{n+4})G_\zeta(r_{n+3}, r_{n+4}, r_{n+4})
\end{aligned}$$

$$+\zeta(r_{n+1},r_m,r_m)\zeta(r_{n+2},r_m,r_m)\zeta(r_{n+3},r_m,r_m)\zeta(r_{n+4},r_m,r_m)G_\zeta(r_{n+4},r_m,r_m),$$

that is,

$$\begin{aligned}
&G_\zeta(r_n,r_m,r_m)\\
&\leq \zeta(r_n,r_{n+1},r_{n+1})G_\zeta(r_n,r_{n+1},r_{n+1})+\zeta(r_{n+1},r_m,r_m)\zeta(r_{n+1},r_{n+2},r_{n+2})G_\zeta(r_{n+1},r_{n+2},r_{n+2})\\
&\quad+\zeta(r_{n+1},r_m,r_m)\zeta(r_{n+2},r_m,r_m)\zeta(r_{n+2},r_{n+3},r_{n+3})G_\zeta(r_{n+2},r_{n+3},r_{n+3})\\
&\quad+\zeta(r_{n+1},r_m,r_m)\zeta(r_{n+2},r_m,r_m)\zeta(r_{n+3},r_m,r_m)\zeta(r_{n+3},r_{n+4},r_{n+4})G_\zeta(r_{n+3},r_{n+4},r_{n+4})\\
&\quad+\zeta(r_{n+1},r_m,r_m)\zeta(r_{n+2},r_m,r_m)\zeta(r_{n+3},r_m,r_m)\zeta(r_{n+4},r_m,r_m)G_\zeta(r_{n+4},r_m,r_m)\\
&\leq \cdots\\
&\leq \zeta(r_n,r_{n+1},r_{n+1})G_\zeta(r_n,r_{n+1},r_{n+1})+\sum_{i=n+1}^{m-2}\left(\prod_{j=n+1}^{i}\zeta(r_j,r_m,r_m)\right)\zeta(r_i,r_{i+1},r_{i+1})G_\zeta(r_i,r_{i+1},r_{i+1})\\
&\quad+\prod_{j=n+1}^{m-1}\zeta(r_j,r_m,r_m)G_\zeta(r_{m-1},r_m,r_m)\\
&\leq \zeta(r_n,r_{n+1},r_{n+1})k^nG_\zeta(r_0,r_1,r_1)+\sum_{i=n+1}^{m-2}\left(\prod_{j=n+1}^{i}\zeta(r_j,r_m,r_m)\right)\zeta(r_i,r_{i+1},r_{i+1})k^iG_\zeta(r_0,r_1,r_1)\\
&\quad+\prod_{j=n+1}^{m-1}\zeta(r_j,r_m,r_m)k^{m-1}G_\zeta(r_0,r_1,r_1)\\
&\leq \zeta(r_n,r_{n+1},r_{n+1})k^nG_\zeta(r_0,r_1,r_1)+\sum_{i=n+1}^{m-1}\left(\prod_{j=0}^{i}\zeta(r_j,r_m,r_m)\right)\zeta(r_i,r_{i+1},r_{i+1})k^iG_\zeta(r_0,r_1,r_1).
\end{aligned}$$

Let us take

$$P_I=\sum_{i=0}^{I}\left(\prod_{j=0}^{i}\zeta(r_j,r_m,r_m)\right)\zeta(r_n,r_{n+1},r_{n+1})k^i.$$

Then this implies that

$$G_\zeta(r_n,r_m,r_m)\leq G_\zeta(r_0,r_1,r_1)[k^n\zeta(r_n,r_{n+1},r_{n+1})+(P_{m-1}-P_n)].$$

As $\zeta(x,y,z)\geq 1$ for all $x,y,z\in X$ and by using the ratio test, $\lim_{n\to\infty}P_n$ exists. Hence $\{P_n\}$ is a Cauchy sequence being a sequence of real numbers. It follows that

$$\lim_{n,m\to\infty}G_\zeta(r_n,r_m,r_m)=0,$$

that is, $\{r_n\}$ is a $G_\zeta-$Cauchy sequence in complete (X,G_ζ). Therefore, a point $r\in X$ exists such that $\{r_n\}$ $G_\zeta-$converges to r. Now by condition (5) in Definition 12, we have

$$\begin{aligned}
G_\zeta(r,fr,fr)&\leq \zeta(r,r_n,r_n)G_\zeta(r,r_{n+1},r_{n+1})+\zeta(r_{n+1},fr,fr)G_\zeta(r_{n+1},fr,fr)\\
&=\zeta(r,r_n,r_n)G_\zeta(r,r_{n+1},r_{n+1})+\zeta(r_{n+1},fr,fr)G_\zeta(fr_n,fr,fr)\\
&\leq \zeta(r,r_n,r_n)G_\zeta(r,r_{n+1},r_{n+1})+\zeta(r_n,r,r)kG_\zeta(r_n,r,r).
\end{aligned}$$

Taking the limit as $n \to \infty$ yields $G_\zeta(r, fr, fr) = 0$. This implies that $r = fr$, that is, r is the fixed point of f.

Now, to show the uniqueness of fixed point of f, assume that there is another $r^* \in X$ exists such that $r^* = fr^*$. Then

$$\begin{aligned} G_\zeta(r, r^*, r^*) &= G_\zeta(fr, fr^*, fr^*) \\ &\leq kG_\zeta(r, r^*, r^*), \end{aligned}$$

which is not possible since $k \in (0, 1)$, unless $r = r^*$. Hence, r is the unique fixed point of f. This completes the proof.

The following Example shows the validity of Theorem 2.

Example 8 Let $X = [0, \frac{1}{2}]$, $G_\zeta : X \times X \times X \to [0, \infty)$ be defined as

$$G_\zeta(x, y, z) = |x - y| + |y - z| + |z - x| \text{ for } x, y, z \in X$$

and $\zeta : X \times X \times X \to [1, \infty)$ be defined by $\zeta(x, y, z) = \max\{x, y, z\} + 1$ for $x, y, z \in X$. Then G_ζ is a complete controlled $G-$metric space on X. Since, it is easy to verify conditions (1) to (4) of controlled $G-$metric space on X. For condition (5), we have for $x, y, z \in X$,

$$\begin{aligned} G_\zeta(x, y, z) &= |x - y| + |y - z| + |z - x| \\ &\leq |x - a| + |a - z| + |y - z| + |z - a| + |a - x| \\ &= 2|x - a| + (|a - z| + |y - z| + |z - a|) \\ &\leq (\max\{x, a\} + 1)(2|x - a|) + (\max\{a, y, z\} + 1)(|a - y| + |y - z| + |z - a|) \\ &= \zeta(x, a, a)G_\zeta(x, a, a) + \zeta(a, y, z)G_\zeta(a, y, z). \end{aligned}$$

Define $f : X \to X$ by $ft = \frac{1}{2}t$ for all $t \in X$. Then for $x, y, z \in X$, we have

$$G_\zeta(fx, fy, fz) = G_\zeta(\frac{x}{2}, \frac{y}{2}, \frac{z}{2}) = \frac{1}{2}|x - y| + |y - z| + |z - x| = \frac{1}{2}G_\zeta(x, y, z) = kG_\zeta(x, y, z),$$

where $k = \frac{1}{2}$ is a fixed number in $(0, 1)$.

Now, for any $r \in X$, $r_n = f^n(r)$, we have

$$\begin{aligned} &\sup_{m\geq 1} \lim_{n\to\infty} \frac{\zeta(r_{n+1}, r_{n+2}, r_{n+3})}{\zeta(r_n, r_{n+1}, r_{n+2})} \zeta(r_{n+1}, r_m, r_m) \\ &= \sup_{m\geq 1} \lim_{n\to\infty} \frac{\max\{\frac{r^{n+1}}{2^{n+1}}, \frac{r^{n+2}}{2^{n+2}}, \frac{r^{n+3}}{2^{n+3}}\} + 1}{\max\{\frac{r^n}{2^n}, \frac{r^{n+1}}{2^{n+1}}, \frac{r^{n+2}}{2^{n+2}}\} + 1} \left(\max\{\frac{r^{n+1}}{2^{n+1}}, \frac{r^m}{2^m}, \frac{r^m}{2^m}\} + 1\right) \\ &\leq 2 = \frac{1}{k}, \end{aligned}$$

where $k = \frac{1}{2}$. Clearly, $k \in (0, 1)$. Thus, all the conditions of Theorem 2 are satisfied. Hence, the mapping f has a unique fixed point.

The following example shows that the Theorem 2 is sufficient, since the self-map has a unique fixed point but not satisfies the condition of Theorem 2.

Example 9 Let $X = [\frac{1}{2}, \infty)$. Define $G_\zeta : X \times X \times X \to [0, \infty)$ by

$$G_\zeta(x, y, z) = \frac{1}{3}|x - y| + \frac{1}{3}|y - z| + \frac{1}{3}|x - z|$$

for all $x, y, z \in X$ and define $\zeta : X \times X \times X \to [1, \infty)$ by $\zeta(x, y, z) = x + y + z + 1$ for all $x, y, z \in X$. Then G_ζ is a complete controlled $G-$metric space on X. Since, it is easy to verify conditions (1) to (4) of controlled $G-$metric space on X. For condition (5), we have for $x, y, z \in X$,

$$\begin{aligned} G_\zeta(x, y, z) &= \frac{1}{3}|x - y| + \frac{1}{3}|y - z| + \frac{1}{3}|z - x| \\ &\leq \frac{1}{3}(|x - a| + |a - z| + |y - z| + |z - a| + |a - x|) \\ &= \frac{2}{3}|x - a| + \left(\frac{1}{3}|a - z| + \frac{1}{3}|y - z| + \frac{1}{3}|z - a|\right) \\ &\leq (x + 2a + 1)\left(\frac{2}{3}|x - a|\right) + (a + y + z + 1)\left(\frac{1}{3}|a - y| + \frac{1}{3}|y - z| + \frac{1}{3}|z - a|\right) \\ &= \zeta(x, a, a)G_\zeta(x, a, a) + \zeta(a, y, z)G_\zeta(a, y, z). \end{aligned}$$

Define $f : X \to X$ by $ft = 2t - \frac{1}{2}$ for all $t \in X$. Then for $x, y, z \in X$, we have

$$\begin{aligned} G_\zeta(fx, fy, fz) &= G_\zeta(2x - \frac{1}{2}, 2y - \frac{1}{2}, 2z - \frac{1}{2}) \\ &= 2\left(\frac{1}{3}|x - y| + \frac{1}{3}|y - z| + \frac{1}{3}|z - z|\right) \\ &\nleq kG_\zeta(x, y, z), \end{aligned}$$

for any $k \in (0, 1)$. The map f has a unique fixed point $t = \frac{1}{2}$ but it does not satisfy the conditions of Theorem 2.

4 Conclusion

In this work, we have mainly extended the works of Aghajani et al. [12] by introducing the concept of G_b-metric spaces. On this new structure, a generalised fixed point theory is obtained using the generalised Banach contraction adopted from Gupta et al. [24]. Going forward, works regarding some more applications on this space may be useful.

References

1. Bakhtin, I.A.: The contraction mapping principle in almost metric space. Funct. Anal. **30**, 26–37 (1989)
2. Czerwik, C.: Contraction mappings in b-metric spaces. Acta Math. Inform. Univ. Ostrav. **1**, 5–11 (1993)
3. Mustafa, Z., Sims, B.: A new approach to generalized metric space. J. Nonlinear Convex Anal. **7**(2), 289–297 (2006)
4. Mustafa, Z., Shatanawi, W., Bataineh, M.: Existence of fixed point results in G-metric spaces. Int. J. Math. Math. Sci. **2009**, Article ID 283028, 1–10 (2009)
5. Mustafa, Z.: A new structure for generalized metric spaces—with applications to fixed point theory. Ph.D. Thesis, The University of Newcastle, Australia (2005)
6. Mustafa, Z., Obiedat, H., Awawdeh, F.: Some fixed point theorem for mappings on complete G-metric spaces. Fixed Point Theory Appl. **2008**, Article ID 189870, 1–12 (2008). https://doi.org/10.1155/2008/189870
7. Abbas, M., Nazir, T., Shatanawi, W., Mustafa, Z.: Fixed and related fixed point theorems for three maps in G-metric spaces. Hacet. J. Math. Stat. **41**(2), 291–306 (2012). https://doi.org/10.1186/1687-1812-2012-1
8. Abbas, M., Nazir, T., Vetro, P.: Common fixed point results for three maps in G-metric spaces. FILOMAT **25**(4), 1–17 (2011). https://doi.org/10.2298/FIL1104001A
9. Aydi, H.: A fixed point result involving a generalized weakly contractive condition in $G-$metric spaces. Bull. Math. Anal. Appl. **3**(4), 180–188 (2011)
10. Aydi, H., Shatanawi, W., Vetro, C.: On generalized weakly G-contraction mapping in $G-$metric spaces. Comput. Math. Appl. **62**(8), 4222–4229 (2011)
11. Gaba, Y.U., Agyingi, C.A., Choudhury, B.S., Maity, P.: Generalised Banach contraction mapping principle in generalised metric spaces with a ternary relation. Surv. Math. Appl. **14**, 159–171 (2019)
12. Aghajani, A., Abbas, M., Roshani, J.R.: Common fixed point of generalised weak contractive mappings in partially ordered G_b-metric spaces. FILOMAT **28**(6), 1087–1101 (2014). https://doi.org/10.2298/FIL1406087A
13. Mustafa, Z., Jaradat, M.M.M., Aydi, H., Alrhayyel, A.: Some common fixed points of six mappings on G_b-metric spaces using (E.A) property. Eur. J. Pure Appl. Math. **11**(1), 90–109 (2018)
14. Mustafa, Z., Roshan, J.R., Parvaneh, V.: Existence of a tripled coincidence point in ordered G_b-metric spaces and applications to a system of integral equations. J. Inequal. Appl. **2013**(453), 1–27 (2013)
15. Mustafa, Z., Roshan, J.R., Parvaneh, V.: Coupled coincidence point results for $(\psi, \phi)-$weakly contractive mappings in partially ordered G_b-metric spaces. Fixed Point Theory Appl. **2013**(206), 1–21 (2013)
16. Aydi, H., Rakić, D., Aghajani, A., Dośenović, T., Noorani, M.S.M., Qawaqneh, H.: On fixed point results in G_b-metric spaces. Mathematics **7**(617), 1–19 (2019). https://doi.org/10.3390/math70700617
17. Ege, O.: Complex valued G_b-metric spaces. J. Comput. Anal. Appl. **21**(2), 363–368 (2016)
18. Ege, O.: Some fixed point theorems in complex valued G_b-metric spaces. J. Nonlinear Convex Anal. **18**(11), 1997–2005 (2017)
19. Ege, O., Park, C., Ansari, A.H.: A different approach to complex valued G_b-metric spaces. Adv. Differ. Equ. **2020**(152), 1–13 (2020)
20. Gupta, V., Ege, O., Saini, R., de la Sen, M.: Various fixed point results in complete G_b-metric spaces. Dynam. Syst. Appl. **30**(2), 277–293 (2021)
21. Liang, M., Zhu, C., Chen, C., Wu, Z.: Some new theorems for cyclic contraction in G_b-metric spaces and some applications. Appl. Math. Comput. **346**, 545–558 (2019)
22. Meitel, M.K., Rohen, Y., Verma, R.S.: Some common fixed point theorems for two pairs of weak compatible mappings of type (A) in G_b-metric space. Amer. J. Appl. Math. Stat. **6**(4), 135–140 (2018)

23. Pakhira, S., Hossein, S.K.M.: A new fixed point theorem in G_b-metric space and its application to solve a class of nonlinear matrix equations. J. Comput. Appl. Math. **437** Article ID 115474, (2024). https://doi.org/10.1016/j.cam.2023.115474
24. Gupta, V., Ege, O., Saini, R.: Extended G_b-metric spaces and some fixed point results with an application to Fredholm Integral equation. J. Math. Anal. Appl. **10**(2), 219–227 (2022)
25. Kamran, T., Samreen, M., Ain, O.U.: A generalization of b-metric space and some fixed point theorems. Mathematics **5**(2), 1–7 (2017). https://doi.org/10.3390/math5020019
26. Banach, S.: Sur les opérations dans les ensembles abstraits et leur application aux equations integrales. Fundam. Math. **3**, 133–181 (1922)
27. Huang, L.G., Zhang, X.: Cone metric spaces and fixed point theorems of contractive mappings. J. Math. Anal. Appl. **33**, 1468–1476 (2007)
28. Jleli, M., Samet, B.: On a new generalization of metric spaces. J. Fix Point Theory A. **208**(128), 1–20 (2018). https://doi.org/10.1007/s11784-018-0606-6
29. Das, A., Kundu, A., Bag, T.: A new approach to generalize metric functions. Int. J. Nonlinear Anal. Appl. **14**(3), 279–298 (2023)
30. Das, A., Bag, T.: A generalization to parametric metric spaces. Int. J. Nonlinear Anal. Appl. **14**(1), 229–244 (2023)
31. Matthews, S.G.: The Topology of Partial Metric Spaces. Research Report RR. 222, University of Warwick (1992)
32. Mlaiki, N., Aydi, H., Souayah, N., Abdeljawad, T.: Controlled metric type spaces and the related contraction principle. Mathematics **6**(194), 1–7 (2018). https://doi.org/10.3390/math6100194
33. Ćirić, L.B.: On Some Maps with a non-unique fixed point. Publ. Inst. Math. (beograd)(N.S.) **17**(31), 52–58 (1974)

Some Common Coupled Fixed Point Theorems of Integral Type Conditions for Occasionally w-Compatible Mappings in S-Metric Spaces and an Application

G. S. Saluja and Reena Jain

Abstract This paper seeks to establish specific common coupled fixed point theorems that incorporate integral type conditions for a pair of occasionally w-compatible mappings within the framework of S-metric spaces. The validity of the results is demonstrated through an illustrative example. Furthermore, our findings serve to extend, enhance, and generalize several prominent results documented in the current literature. Finally, an application related to a Fredholm integral equation is presented.

Keywords Common coupled fixed point · Integral type condition · Occasionally w-compatible mapping · S-metric space

1 Introduction

In 1922, the Polish mathematician Stephen Banach [3] formulated a theorem that guarantees the existence and uniqueness of a fixed point under certain conditions. This significant result is commonly known as the Banach contraction principle or Banach's fixed point theorem. It serves as a valuable tool for solving a diverse array of applied problems in both engineering and mathematical sciences. Since its inception, many authors have sought to expand, generalize, and enhance Banach's fixed point theorem. Sedghi et al. contributed to this field by introducing the concept of S-metric space through modifications of G-metric and D^*-metric spaces, and they demonstrated several fixed point theorems. Their research included an examination of various properties of S-metric space.

In their research [4], which is also discussed in [7], Bhashkar and Lakshmikantham proposed the concept of coupled fixed points and established multiple theorems

G. S. Saluja
H.N. 3/1005, Geeta Nagar, Raipur, India

R. Jain (✉)
Mathematics Division, School of Advanced Sciences and Languages, VIT Bhopal University, Kothrikalan, Sehore, Madhya Pradesh, India
e-mail: reena.jain@vitbhopal.ac.in

H. Kalita et al. (eds.), *Nonlinear Analysis and Computational Techniques*,
Springer Proceedings in Mathematics & Statistics 501,
https://doi.org/10.1007/978-981-96-6038-4_11

concerning coupled fixed points in the context of partially ordered complete metric spaces. They were the first to validate the mixed monotone property and to introduce their classical coupled fixed point theorem for mappings that adhere to this property. Additionally, they examined the existence and uniqueness of solutions to a periodic boundary value problem linked to a first-order differential equation. This area of study has since been broadened and generalized by a number of researchers (see, for instance, [6, 12–14]).

In this research, we establish two principal coupled fixed point theorems characterized by integral type conditions for a pair of occasionally w-compatible (owc) mappings in the context of S-metric spaces. The results articulated in this paper broaden and enhance several prior studies documented in the literature.

2 Preliminaries

In this section, we revisit the definitions, lemmas, and auxiliary results necessary for demonstrating our primary findings.

Definition 1 (*see* [16]) Let $\Omega \neq \emptyset$ be a set and let $\mathcal{S}_{\mathcal{M}} : \Omega^3 \to [0, \infty)$ be a function satisfying the following conditions for all $\zeta, \eta, \theta, u \in \Omega$:

$(S1)$ $0 < \mathcal{S}_{\mathcal{M}}(\zeta, \eta, \theta)$ for all $\zeta, \eta, \theta \in \Omega$ with $\zeta \neq \eta \neq \theta$;
$(S2)$ $\mathcal{S}_{\mathcal{M}}(\zeta, \eta, \theta) = 0 \Leftrightarrow \zeta = \eta = \theta$;
$(S3)$ $\mathcal{S}_{\mathcal{M}}(\zeta, \eta, \theta) \leq \mathcal{S}_{\mathcal{M}}(\zeta, \zeta, u) + \mathcal{S}_{\mathcal{M}}(\eta, \eta, u) + \mathcal{S}_{\mathcal{M}}(\theta, \theta, u)$.

The function $\mathcal{S}_{\mathcal{M}}$ is referred to as an S-metric on the set Ω, and the combination $(\mathcal{S}_{\mathcal{M}}, \Omega)$ is designated as an S-metric space.

Example 1 (*see* [16]) Let $\Omega = \mathbb{R}^n$ and $\|\cdot\|$ a norm on Ω. Then

$(E1)$ $\mathcal{S}_{\mathcal{M}}(\zeta, \eta, \theta) = \|\eta + \theta - 2\zeta\| + \|\eta - \theta\|$ is an S-metric on Ω.
$(E2)$ $\mathcal{S}_{\mathcal{M}}(\zeta, \eta, \theta) = \|\zeta - \theta\| + \|\eta - \theta\|$ is an S-metric on Ω.

Example 2 (*see* [17]) Consider the set $\Omega = \mathbb{R}$, which represents the real line. The function defined as $\mathcal{S}_{\mathcal{M}}(\zeta, \eta, \theta) = |\zeta - \theta| + |\eta - \theta|$ for all elements $\zeta, \eta, \theta \in \Omega$ serves as an S-metric on Ω. This particular S-metric is referred to as the standard S-metric on Ω.

Example 3 (*see* [10]) Let Ω be a non-empty set and let ρ represent a standard metric on Ω. For all elements ζ, η, θ belonging to Ω, the function defined by $\mathcal{S}_{\mathcal{M}}(\zeta, \eta, \theta) = \rho(\zeta, \theta) + \rho(\eta, \theta)$ serves as an S-metric on Ω.

Example 4 (*see* [18]) Consider a non-empty set Ω and two standard metrics ρ_1 and ρ_2 defined on Ω. The expression $\mathcal{S}_{\mathcal{M}}(\zeta, \eta, \theta) = \rho_1(\zeta, \theta) + \rho_2(\eta, \theta)$, applicable for all elements ζ, η, θ in Ω, constitutes an S-metric on the set Ω.

Example 5 (*see* [16]) Define the set Ω to be $\mathbb{R}^2$ and let ρ denote a typical metric on this set. The function $\mathcal{S}_{\mathcal{M}}(\zeta, \eta, \theta)$ is given by the expression $\rho(\zeta, \eta) + \rho(\zeta, \theta) + \rho(\eta, \theta)$ for any points ζ, η, θ in Ω. Therefore, it can be concluded that $\mathcal{S}_{\mathcal{M}}$ serves as an S-metric on the space Ω.

Definition 2 Consider the set $\Omega = \mathbb{R}^2$ and let ρ represent a standard metric defined on Ω. Define the function $\mathcal{S}_{\mathcal{M}}(\zeta, \eta, \theta)$ as $\rho(\zeta, \eta) + \rho(\zeta, \theta) + \rho(\eta, \theta)$ for any points ζ, η, θ within Ω. It follows that $\mathcal{S}_{\mathcal{M}}$ constitutes an S-metric on the space Ω.

$$\mathcal{B}_{\mathcal{S}_{\mathcal{M}}}(\omega, c) = \{\omega_1 \in \Omega : \mathcal{B}_{\mathcal{S}_{\mathcal{M}}}(\omega_1, \omega_1, \omega) < c\},$$

$$\mathcal{B}_{\mathcal{S}_{\mathcal{M}}}[\omega, c] = \{\omega_1 \in \Omega : \mathcal{B}_{\mathcal{S}_{\mathcal{M}}}(\omega_1, \omega_1, \omega) \leq c\}.$$

Definition 3 (*see* [16, 17]) Consider $(\Omega, \mathcal{S}_{\mathcal{M}})$ as an S-metric space, where V is a subset of Ω.

(A_1) A subset V is classified as an open subset of Ω if, for each element $\omega \in V$, there exists a constant $c > 0$ such that the ball $\mathcal{B}_{\mathcal{S}_{\mathcal{M}}}(\omega, c)$ is entirely contained within V.

(A_2) A sequence $\{\omega_n\}$ within the set Ω is said to converge to an element $\omega \in \Omega$ if the expression $\mathcal{S}_{\mathcal{M}}(\omega_n, \omega_n, \omega)$ approaches zero as n approaches infinity. Specifically, for every $\mu > 0$, there exists a natural number n_0 such that for all $n \geq n_0$, the inequality $\mathcal{S}_{\mathcal{M}}(\omega_n, \omega_n, \omega) < \mu$ holds true. This convergence is denoted as $\lim_{n\to+\infty} \omega_n = \omega$ or alternatively, $\omega_n \to \omega$ as n tends to infinity.

(A_3) A sequence $\{\omega_n\}$ within the set Ω is defined as a Cauchy sequence if the expression $\mathcal{S}_{\mathcal{M}}(\omega_n, \omega_n, \omega_m)$ approaches 0 as both n and m tend to infinity. Specifically, for every $\mu > 0$, there exists a natural number n_0 such that for all indices n and m greater than or equal to n_0, the inequality $\mathcal{S}_{\mathcal{M}}(\omega_n, \omega_n, \omega_m) < \mu$ holds true.

(A_4) A metric space denoted as $(\Omega, \mathcal{S}_{\mathcal{M}})$ is referred to as complete if every Cauchy sequence within Ω converges to a limit that also resides in Ω.

(A_5) Define τ as the set of all subsets $V \subset \Omega$ that satisfy the condition that for each element $\omega \in V$, there exists a positive constant $c > 0$ such that the ball $\mathcal{B}_{\mathcal{S}_{\mathcal{M}}}(\omega, c)$ is entirely contained in V. Therefore, τ constitutes a topology on Ω, arising from the S-metric space.

(A_6) A non-empty subset V of Ω is considered S-closed if its closure is equal to the subset itself, which can be expressed as $\overline{V} = V$.

Definition 4 (*see* [16]) Let $(\Omega, \mathcal{S}_{\mathcal{M}})$ and $(\Omega', \mathcal{S}'_{\mathcal{M}})$ represent two S-metric spaces. A function $R\colon \Omega \to \Omega'$ is considered continuous at a point $\omega_0 \in \Omega$ if, for any sequence $\{\omega_n\}$ in Ω such that $\mathcal{S}_{\mathcal{M}}(\omega_n, \omega_n, \omega_0) \to 0$, it follows that $\mathcal{S}'_{\mathcal{M}}(R(\omega_n), R(\omega_n), R(\omega_0)) \to 0$ as n approaches infinity. The function R is classified as continuous on Ω if it maintains continuity at every point $\omega_0 \in \Omega$.

Definition 5 (*see* [2, 11])An element expressed as $(\zeta, \eta) \in \Omega^2$ is termed as

(i) a coupled fixed point of the mapping $\wp_1 : \Omega^2 \to \Omega$ is defined as a pair (ζ, η) such that $\wp_1(\zeta, \eta) = \zeta$ and $\wp_1(\eta, \zeta) = \eta$.;

(*ii*) the mappings $\wp_1 \colon \Omega^2 \to \Omega$ and $\wp_2 \colon \Omega \to \Omega$ exhibit a coupled fixed point if the conditions $\wp_1(\zeta, \eta) = \wp_2(\zeta)$ and $\wp_1(\eta, \zeta) = \wp_2(\eta)$ are satisfied.;

(*iii*) a common coupled fixed point for the mappings $\wp_1 \colon \Omega^2 \to \Omega$ and $\wp_2 \colon \Omega \to \Omega$ is defined by the equations $\zeta = \wp_1(\zeta, \eta) = \wp_2(\zeta)$ and $\eta = \wp_1(\eta, \zeta) = \wp_2(\eta)$.

Example 6 Let $\Omega = [0, +\infty)$ and $\wp_1 \colon \Omega^2 \to \Omega$ defined by $\wp_1(\zeta, \eta) = \frac{\zeta+\eta}{3}$ for all $\zeta, \eta \in \Omega$. Obviously (0, 0) is a unique coupled fixed point of $\wp_1$.

Example 7 Let $\Omega = [0, +\infty)$ and $\wp_1 \colon \Omega^2 \to \Omega$ be defined by $\wp_1(\zeta, \eta) = \frac{\zeta+\eta}{2}$ for all $\zeta, \eta \in \Omega$. Then $\wp_1(0, 0) = 0$ and $\wp_1(1, 1) = 1$.

Definition 6 (*see* [6]) Let $\Omega \neq \emptyset$ be a set. Then the mappings $\wp_1 \colon \Omega^2 \to \Omega$ and $\wp_2 \colon \Omega \to \Omega$ are commutative if $\wp_2(\wp_1(\zeta, \eta)) = \wp_1(\wp_2\zeta, \wp_2\eta)$.

Definition 7 (*see* [1]) The mappings $\wp_1 \colon \Omega^2 \to \Omega$ and $\wp_2 \colon \Omega \to \Omega$ are called weakly compatible if $\wp_2(\wp_1(\zeta, \eta)) = \wp_1(\wp_2\zeta, \wp_2\eta)$ and $\wp_2(\wp_1(\eta, \zeta)) = \wp_1(\wp_2\eta, \wp_2\zeta)$ for all $\zeta, \eta \in \Omega$, whenever $\wp_2(\zeta) = \wp_1(\zeta, \eta)$ and $\wp_2(\eta) = \wp_1(\eta, \zeta)$.

Example 8 Let $\Omega = [0, 3]$. Define $\wp_1 \colon \Omega^2 \to \Omega$ and $\wp_2 \colon \Omega \to \Omega$ by

$$\wp_1(\zeta, \eta) = \begin{cases} \zeta + \eta, \text{ if } \zeta, \eta \in [0, 1), \\ \quad 3, \quad \text{otherwise,} \end{cases}$$

for all $\zeta, \eta \in \Omega$ and

$$\wp_2(\zeta) = \begin{cases} \zeta, \text{ if } \zeta \in [0, 1), \\ 3, \text{ if } \zeta \in [1, 3], \end{cases}$$

for all $\zeta, \eta \in \Omega$. Then for any $\zeta, \eta \in [1, 3]$,

$$\wp_1(\wp_2\zeta, \wp_2\eta) = \wp_1(3, 3) = 3 = \wp_2(\wp_1(\zeta, \eta)) = \wp_2(3) = 3.$$

Similarly, we have

$$\wp_1(\wp_2\eta, \wp_2\zeta) = \wp_1(3, 3) = 3 = \wp_2(\wp_1(\eta, \zeta)) = \wp_2(3) = 3.$$

Thus,

$$\wp_1(\wp_2\zeta, \wp_2\eta) = \wp_2(\wp_1(\zeta, \eta)) \text{ and } \wp_1(\wp_2\eta, \wp_2\zeta) = \wp_2(\wp_1(\eta, \zeta)).$$

This shows that the mappings $\wp_1$ and $\wp_2$ are weakly compatible on [0, 3].

Example 9 Let $\Omega = \mathbb{R}$. Define $\wp_1 \colon \Omega^2 \to \Omega$ and $\wp_2 \colon \Omega \to \Omega$ by $\wp_1(\zeta, \eta) = \zeta + \eta$ and $\wp_2(\zeta) = \zeta^2$ for all $\zeta, \eta \in \Omega$. Then $\wp_1$ and $\wp_2$ are not weakly compatible maps on $\mathbb{R}$, since

$$\wp_1(\wp_2\zeta, \wp_2\eta) = \wp_1(\zeta^2, \eta^2) = \zeta^2 + \eta^2, \text{ but } \wp_2(\wp_1(\zeta, \eta)) = \wp_2(\zeta + \eta) = (\zeta + \eta)^2.$$

Therefore,

$$\wp_1(\wp_2\zeta, \wp_2\eta) \neq \wp_2(\wp_1(\zeta, \eta)).$$

Hence the mappings $\wp_1$ and $\wp_2$ are not weakly compatible on $\mathbb{R}$.

Now, we give the following concepts.

Definition 8 Let $(\Omega, \mathcal{S}_{\mathcal{M}})$ be an S-metric space and let the mappings $\wp_1 : \Omega^2 \to \Omega$ and $\wp_2 : \Omega \to \Omega$ are said to satisfy the $(E.A)$ property if there exist sequences $\{\omega_n\}$ and $\{\omega_n'\}$ in Ω such that

$$\lim_{n\to+\infty} \wp_1(\omega_n, \omega_n') = \lim_{n\to+\infty} \wp_2\omega_n = \omega_1,$$

and

$$\lim_{n\to+\infty} \wp_1(\omega_n', \omega_n) = \lim_{n\to+\infty} \wp_2\omega_n' = \omega_2,$$

for some $\omega_1, \omega_2 \in \Omega$.

Definition 9 Let $(\Omega, \mathcal{S}_{\mathcal{M}})$ be an S-metric space and let $\wp_1 : \Omega^2 \to \Omega$, $\wp_2 : \Omega \to \Omega$, $\mathcal{P} : \Omega \to \Omega$, $\mathcal{Q} : \Omega \to \Omega$ be four mappings. Then the pairs $(\wp_2, \mathcal{P})$ and $(\wp_1, \mathcal{Q})$ are said to satisfy the common $(E.A)$ property if there exist sequences $\{\sigma_n\}$, $\{\omega_n\}$, $\{\sigma_n'\}$, $\{\omega_n'\}$ in Ω such that

$$\lim_{n\to+\infty} \wp_1(\sigma_n, \omega_n) = \lim_{n\to+\infty} \mathcal{Q}\sigma_n = \lim_{n\to+\infty} \wp_2(\sigma_n', \omega_n') = \lim_{n\to+\infty} \mathcal{P}\sigma_n' = z_1,$$

and

$$\lim_{n\to+\infty} \wp_1(\omega_n, \sigma_n) = \lim_{n\to+\infty} \mathcal{Q}\omega_n = \lim_{n\to+\infty} \wp_2(\omega_n', \sigma_n') = \lim_{n\to+\infty} \mathcal{P}\omega_n' = z_2,$$

for some $z_1, z_2 \in \Omega$.

Definition 10 Let $(\Omega, \mathcal{S}_{\mathcal{M}})$ be an S-metric space. The mappings $\wp_1 : \Omega^2 \to \Omega$ and $\wp_2 : \Omega \to \Omega$ are said to satisfy the $(CLR_{\wp_2})$ property if there exist sequences $\{\sigma_n\}$ and $\{\omega_n\}$ in Ω such that

$$\lim_{n\to+\infty} \wp_1(\sigma_n, \omega_n) = \lim_{n\to+\infty} \wp_2\sigma_n = \wp_2 u_1,$$

and

$$\lim_{n\to+\infty} \wp_1(\omega_n, \sigma_n) = \lim_{n\to+\infty} \wp_2\omega_n = \wp_2 u_2,$$

for some $u_1, u_2 \in \Omega$.

Definition 11 Let $(\Omega, \mathcal{S}_{\mathcal{M}})$ be an S-metric space and let $\wp_1 : \Omega^2 \to \Omega$, $\wp_2 : \Omega \to \Omega$, $\mathcal{P} : \Omega \to \Omega$, $\mathcal{Q} : \Omega \to \Omega$ be four mappings. Then the pairs $(\wp_2, \mathcal{P})$ and $(\wp_1, \mathcal{Q})$

are said to satisfy the common $(CLR_{\mathcal{PQ}})$ property if there exist sequences $\{\sigma_n\}$, $\{\omega_n\}$, $\{\sigma'_n\}$, $\{\omega'_n\}$ in Ω such that

$$\lim_{n\to+\infty} \wp_1(\sigma_n, \omega_n) = \lim_{n\to+\infty} \mathcal{Q}\sigma_n = \lim_{n\to+\infty} \wp_2(\sigma'_n, \omega'_n) = \lim_{n\to+\infty} \mathcal{P}\sigma'_n = \upsilon_1,$$

and

$$\lim_{n\to+\infty} \wp_1(\omega_n, \sigma_n) = \lim_{n\to+\infty} \mathcal{Q}\omega_n = \lim_{n\to+\infty} \wp_2(\omega'_n, \sigma'_n) = \lim_{n\to+\infty} \mathcal{P}\omega'_n = \upsilon_2,$$

where $\upsilon_1, \upsilon_2 \in \mathcal{P}(\Omega) \cap \mathcal{Q}(\Omega)$.

Definition 12 Let $\wp_1 : \Omega^2 \to \Omega$ and $\wp_2 : \Omega \to \Omega$ be two self-mappings on Ω. Then the pair $(\wp_1, \wp_2)$ is called occasionally weakly compatible (owc) if $\wp_2(\wp_1(\zeta, \eta)) = \wp_1(\wp_2\zeta, \wp_2\eta)$ and $\wp_2(\wp_1(\eta, \zeta)) = \wp_1(\wp_2\eta, \wp_2\zeta)$ for some $\zeta, \eta \in \Omega$, whenever $\wp_2(\zeta) = \wp_1(\zeta, \eta)$ and $\wp_2(\eta) = \wp_1(\eta, \zeta)$.

Example 10 Let $\Omega = [0, 1]$. Define $\wp_1 : \Omega^2 \to \Omega$ and $\wp_2 : \Omega \to \Omega$ by

$$\wp_1(\zeta, \eta) = \begin{cases} \frac{1}{2}(\zeta + \eta), & \text{if } \zeta, \eta \in [0, \frac{1}{2}], \\ \frac{1}{4}, & \text{if } \zeta, \eta \in (\frac{1}{2}, 1], \end{cases}$$

for all $\zeta, \eta \in \Omega$ and

$$\wp_2(\zeta) = \begin{cases} \frac{1}{2}, & \text{if } \zeta \in [0, \frac{1}{2}], \\ 0, & \text{if } \zeta \in (\frac{1}{2}, 1], \end{cases}$$

for all $\zeta \in \Omega$. Then for $\zeta = \eta = \frac{1}{2}$, we have

$$\wp_1(\wp_2\zeta, \wp_2\eta) = \frac{1}{2}[\wp_2(\zeta) + \wp_2(\eta)] = \frac{1}{2}\Big[\frac{1}{2} + \frac{1}{2}\Big] = \frac{1}{2}$$

$$= \wp_2(\wp_1(\zeta, \eta)) = \wp_2\Big[\frac{1}{2}(\zeta + \eta)\Big] = \frac{1}{2}.$$

Similarly, we have

$$\wp_1(\wp_2\eta, \wp_2\zeta) = \frac{1}{2}\Big[\frac{1}{2} + \frac{1}{2}\Big] = \frac{1}{2} = \wp_2(\wp_1(\eta, \zeta)) = \wp_2\Big[\frac{1}{2}(\zeta + \eta)\Big] = \frac{1}{2}.$$

Thus,

$$\wp_1(\wp_2\zeta, \wp_2\eta) = \wp_2(\wp_1(\zeta, \eta)) \text{ and } \wp_1(\wp_2\eta, \wp_2\zeta) = \wp_2(\wp_1(\eta, \zeta)).$$

This shows that the mappings $\wp_1$ and $\wp_2$ are occasionally weakly compatible on [0, 1].

Lemma 1 (see [16], Lemma 2.5) *Let $(\Omega, \mathcal{S}_{\mathcal{M}})$ be an S-metric space. Then, we have $\mathcal{S}_{\mathcal{M}}(\zeta, \zeta, \eta) = \mathcal{S}_{\mathcal{M}}(\eta, \eta, \zeta)$ for all $\zeta, \eta \in \Omega$.*

One can refer [16, 17] for other related results.
In the following lemma we see the relationship between a metric and S-metric.

Lemma 2 (see [8]) *Let (Ω, ρ) be a metric space. Then the following properties are satisfied:*

(1) $\mathcal{S}_{\mathcal{M}_\rho}(\zeta, \eta, \theta) = \rho(\zeta, \theta) + \rho(\eta, \theta)$ *for all* $\zeta, \eta, \theta \in \Omega$ *is an S-metric on* Ω.
(2) $\omega_n \to \omega \in (\Omega, \rho)$ *iff* $\omega_n \to \omega \in (\Omega, \mathcal{S}_{\mathcal{M}_\rho})$.
(3) $\{\omega_n\}$ *is Cauchy in* (Ω, ρ) *iff* $\{\omega_n\}$ *is Cauchy in* $(\Omega, \mathcal{S}_{\mathcal{M}_\rho})$.
(4) (Ω, ρ) *is complete iff* $(\Omega, \mathcal{S}_{\mathcal{M}_\rho})$ *is complete.*

We denote the function $\mathcal{S}_{\mathcal{M}_\rho}$, referenced in Lemma 2 (1), as the S-metric associated with the metric ρ. An illustration of an S-metric that is not produced by any metric can be found in [8, 15].

In the year 2002, Branciari [5] examined the existence of fixed points for the mapping $\mathcal{R}$, established within a complete metric space (Ω, ρ) that meets a general contractive condition of integral type. The results are articulated as follows:

We define the collection Φ as follows: $\Phi := \{\varsigma : [0, +\infty) \to [0, +\infty)$ such that ς is Lebesgue-integrable, summable on all compact subsets of $[0, +\infty)$, and for any $\varepsilon > 0$, $\int_0^\varepsilon \varsigma(\hbar)d\hbar > 0\}$.

Theorem 1 (see [5]) *Let (Ω, ρ) be a complete metric space, $b \in [0, 1)$, and let $\mathcal{R} : \Omega \to \Omega$ be a mapping such that for each $\zeta, \eta \in \Omega$,*

$$\int_0^{\rho(\mathcal{R}\zeta, \mathcal{R}\eta)} \varsigma(\hbar)d\hbar \leq b \int_0^{\rho(\zeta,\eta)} \varsigma(\hbar)d\hbar,$$

where $\varsigma \in \Phi$, then $\mathcal{R}$ has a unique fixed point $\omega \in \Omega$, such that for each $u \in \Omega$, $\lim_{n\to\infty} \mathcal{R}^n(u) = \omega$.

3 Main Results

The first result is the following:

Theorem 2 *Let $(\Omega, \mathcal{S}_\mathcal{M})$ be a complete S-metric space. Let $\wp_1 : \Omega^2 \to \Omega$ and $\wp_2 : \Omega \to \Omega$ be two mappings satisfying the following conditions:*

(i) for all $\zeta, \eta, \zeta_1, \eta_1, \zeta_2, \eta_2 \in \Omega$:

$$\int_0^{[\mathcal{S}_\mathcal{M}(\wp_1(\zeta,\eta),\wp_1(\zeta_1,\eta_1),\wp_1(\zeta_2,\eta_2))]^2} \varsigma(\hbar)d\hbar$$
$$\leq L_1 \int_0^{[\mathcal{S}_\mathcal{M}(\wp_2\zeta,\wp_2\zeta_1,\wp_2\zeta_2)]^2} \varsigma(\hbar)d\hbar + L_2 \int_0^{[\mathcal{S}_\mathcal{M}(\wp_1(\zeta,\eta),\wp_1(\zeta,\eta),\wp_2\zeta)]^2} \varsigma(\hbar)d\hbar$$
$$+ L_3 \int_0^{[\mathcal{S}_\mathcal{M}(\wp_1(\zeta_1,\eta_1),\wp_1(\zeta_1,\eta_1),\wp_2\zeta_1)]^2} \varsigma(\hbar)d\hbar$$

$$+ L_4 \int_0^{[\mathcal{S}_{\mathcal{M}}(\wp_1(\zeta_2,\eta_2),\wp_1(\zeta_2,\eta_2),\wp_2\zeta_2)]^2} \varsigma(\hbar)d\hbar, \tag{1}$$

where L_1, L_2, L_3, L_4 are nonnegative constants such that $L_1 + L_2 + L_3 + L_4 < 1$, $\varsigma \in \Phi$;

(ii) the pair $(\wp_1, \wp_2)$ is occasionally w-compatible (owc);

(iii) the pair $(\wp_1, \wp_2)$ satisfies $(CLR_{\wp_2})$ property.

Then $\wp_1$ and $\wp_2$ have a coupled coincidence point in Ω. Moreover, there exists a unique point $\mu \in \Omega$ such that $\mu = \wp_1(\mu, \mu) = \wp_2(\mu)$.

Proof By hypothesis (iii), there exist sequences $\{\varpi_n\}$ and $\{\varpi_n'\}$ in Ω such that

$$\lim_{n\to+\infty} \wp_1(\varpi_n, \varpi_n') = \lim_{n\to+\infty} \wp_2(\varpi_n) = \wp_2(\vartheta),$$
$$\lim_{n\to+\infty} \wp_1(\varpi_n', \varpi_n) = \lim_{n\to+\infty} \wp_2(\varpi_n') = \wp_2(\varrho), \tag{2}$$

for some $\vartheta, \varrho \in \Omega$.

Part(1) Now, we show that $\wp_1$ and $\wp_2$ have a coupled coincidence point. Hence, from equation (1), we have

$$\int_0^{[\mathcal{S}_{\mathcal{M}}(\wp_1(\varpi_n,\varpi_n'),\wp_1(\varpi_n,\varpi_n'),\wp_1(\vartheta,\varrho))]^2} \varsigma(\hbar)d\hbar$$
$$\leq L_1 \int_0^{[(\wp_2\varpi_n,\wp_2\varpi_n,\wp_2\vartheta)]^2} \varsigma(\hbar)d\hbar$$
$$+L_2 \int_0^{[\mathcal{S}_{\mathcal{M}}(\wp_1(\varpi_n,\varpi_n'),\wp_1(\varpi_n,\varpi_n'),\wp_2\varpi_n)]^2} \varsigma(\hbar)d\hbar$$
$$+L_3 \int_0^{[\mathcal{S}_{\mathcal{M}}(\wp_1(\varpi_n,\varpi_n'),\wp_1(\varpi_n,\varpi_n'),\wp_2\varpi_n)]^2} \varsigma(\hbar)d\hbar$$
$$+L_4 \int_0^{[\mathcal{S}_{\mathcal{M}}(\wp_1(\vartheta,\varrho),\wp_1(\vartheta,\varrho),\wp_2\vartheta)]^2} \varsigma(\hbar)d\hbar. \tag{3}$$

Letting $n \to +\infty$ in (3), using (2) and condition $(S2)$,

$$\int_0^{[\mathcal{S}_{\mathcal{M}}(\wp_1(\vartheta,\varrho),\wp_1(\vartheta,\varrho),\wp_2\vartheta)]^2} \varsigma(\hbar)d\hbar \leq L_4 \int_0^{[\mathcal{S}_{\mathcal{M}}(\wp_1(\vartheta,\varrho),\wp_1(\vartheta,\varrho),\wp_2\vartheta)]^2} \varsigma(\hbar)d\hbar,$$

a contradiction, since by assumption $L_4 < 1$. Hence

$$\int_0^{[\mathcal{S}_{\mathcal{M}}(\wp_1(\vartheta,\varrho),\wp_1(\vartheta,\varrho),\wp_2\vartheta)]^2} \varsigma(\hbar)d\hbar = 0.$$

Therefore, by the property of integral ς, we have $\mathcal{S}_{\mathcal{M}}(\wp_1(\vartheta, \varrho), \wp_1(\vartheta, \varrho), \wp_2\vartheta) = 0$ and so $\wp_1(\vartheta, \varrho) = \wp_2\vartheta$. Similarly, we get $\wp_1(\varrho, \vartheta) = \wp_2\varrho$.

Let

$$\wp_1(\vartheta, \varrho) = \wp_2\vartheta = \mu \quad \text{and} \quad \wp_1(\varrho, \vartheta) = \wp_2\varrho = \nu. \tag{4}$$

By hypothesis (ii), we have

$$\wp_2\mu = \wp_2(\wp_1(\vartheta, \varrho)) = \wp_1(\wp_2\vartheta, \wp_2\varrho) = \wp_1(\mu, \nu), \tag{5}$$

and

$$\wp_2\nu = \wp_2(\wp_1(\varrho, \vartheta)) = \wp_1(\wp_2\varrho, \wp_2\vartheta) = \wp_1(\nu, \mu). \tag{6}$$

Hence $\wp_1$ and $\wp_2$ have a coupled coincidence point.

Part(2) To show that $\wp_2(\mu) = \nu$ and $\wp_2(\nu) = \mu$. From equation (1), we have

$$\int_0^{[\mathcal{S}_{\mathcal{M}}(\wp_2\varpi_n,\wp_2\varpi_n,\wp_2\nu]^2} \varsigma(\hbar)d\hbar = \int_0^{[\mathcal{S}_{\mathcal{M}}(\wp_1(\varpi_n,\varpi_n'),\wp_1(\varpi_n,\varpi_n'),\wp_1(\nu,\mu))]^2} \varsigma(\hbar)d\hbar$$

$$\begin{aligned}
&\leq L_1 \int_0^{[\mathcal{S}_{\mathcal{M}}(\wp_2\varpi_n,\wp_2\varpi_n,\wp_2\nu)]^2} \varsigma(\hbar)d\hbar \\
&\quad + L_2 \int_0^{[\mathcal{S}_{\mathcal{M}}(\wp_1(\varpi_n,\varpi_n'),\wp_1(\varpi_n,\varpi_n'),\wp_2\varpi_n)]^2} \varsigma(\hbar)d\hbar \\
&\quad + L_3 \int_0^{[\mathcal{S}_{\mathcal{M}}(\wp_1(\varpi_n,\varpi_n'),\wp_1(\varpi_n,\varpi_n'),\wp_2\varpi_n)]^2} \varsigma(\hbar)d\hbar \\
&\quad + L_4 \int_0^{[\mathcal{S}_{\mathcal{M}}(\wp_1(\nu,\mu),\wp_1(\nu,\mu),\wp_2\nu)]^2} \varsigma(\hbar)d\hbar.
\end{aligned} \tag{7}$$

Passing to the limit as $n \to +\infty$ in (7), using (2), (4) and condition $(S2)$, we get

$$\int_0^{[\mathcal{S}_{\mathcal{M}}(\mu,\mu,\wp_2\nu)]^2} \varsigma(\hbar)d\hbar \leq L_1 \int_0^{[\mathcal{S}_{\mathcal{M}}(\mu,\mu,\wp_2\nu)]^2} \varsigma(\hbar)d\hbar,$$

a contradiction, since by assumption $L_1 < 1$. Hence, we conclude that

$$\int_0^{[\mathcal{S}_{\mathcal{M}}(\mu,\mu,\wp_2\nu)]^2} \varsigma(\hbar)d\hbar = 0.$$

Thus, by the property of integral ς, we have $\mathcal{S}_{\mathcal{M}}(\mu, \mu, \wp_2\nu) = 0$ and so $\wp_2\nu = \mu$. Likewise, we can show that $\wp_2\mu = \nu$.

Part(3) Now, we show that $\mu = \nu$. From Eqs. (1), (5) and (6), we have

$$\int_0^{[\mathcal{S}_{\mathcal{M}}(\mu,\mu,\nu]^2} \varsigma(\hbar)d\hbar = \int_0^{[\mathcal{S}_{\mathcal{M}}(\wp_1(\vartheta,\varrho),\wp_1(\vartheta,\varrho),\wp_1(\varrho,\vartheta))]^2} \varsigma(\hbar)d\hbar$$

$$\begin{aligned}
&\leq L_1 \int_0^{[\mathcal{S}_{\mathcal{M}}(\wp_2\vartheta,\wp_2\vartheta,\wp_2\varrho)]^2} \varsigma(\hbar)d\hbar + L_2 \int_0^{[\mathcal{S}_{\mathcal{M}}(\wp_1(\vartheta,\varrho),\wp_1(\vartheta,\varrho),\wp_2\vartheta)]^2} \varsigma(\hbar)d\hbar \\
&+L_3 \int_0^{[\mathcal{S}_{\mathcal{M}}(\wp_1(\vartheta,\varrho),\wp_1(\vartheta,\varrho),\wp_2\vartheta)]^2} \varsigma(\hbar)d\hbar + L_4 \int_0^{[\mathcal{S}_{\mathcal{M}}(\wp_1(\varrho,\vartheta),\wp_1(\varrho,\vartheta),\wp_2\varrho)]^2} \varsigma(\hbar)d\hbar \\
&= L_1 \int_0^{[\mathcal{S}_{\mathcal{M}}(\mu,\mu,\nu)]^2} \varsigma(\hbar)d\hbar + L_2 \int_0^{[\mathcal{S}_{\mathcal{M}}(\mu,\mu,\mu)]^2} \varsigma(\hbar)d\hbar \\
&+L_3 \int_0^{[\mathcal{S}_{\mathcal{M}}(\mu,\mu,\mu)]^2} \varsigma(\hbar)d\hbar + L_4 \int_0^{[\mathcal{S}_{\mathcal{M}}(\nu,\nu,\nu)]^2} \varsigma(\hbar)d\hbar. \qquad (8)
\end{aligned}$$

Using condition $(S2)$ in Eq. (8), we obtain

$$\int_0^{[\mathcal{S}_{\mathcal{M}}(\mu,\mu,\nu)]^2} \varsigma(\hbar)d\hbar \leq L_1 \int_0^{[\mathcal{S}_{\mathcal{M}}(\mu,\mu,\nu)]^2} \varsigma(\hbar)d\hbar,$$

a contradiction, since by hypothesis $L_1 < 1$. Hence

$$\int_0^{[\mathcal{S}_{\mathcal{M}}(\mu,\mu,\nu)]^2} \varsigma(\hbar)d\hbar = 0,$$

which implies that $\mathcal{S}_{\mathcal{M}}(\mu, \mu, \nu) = 0$ by the property of integral ς and so $\mu = \nu$. From equation (5) and **Part(2)**, we can conclude that $\mu = \wp_1(\mu, \mu) = \wp_2(\mu)$.

Theorem 3 *Let* $(\Omega, \mathcal{S}_{\mathcal{M}})$ *be a complete S-metric space. Let* $\wp_1 \colon \Omega^2 \to \Omega$ *and* $\wp_2 \colon \Omega \to \Omega$ *be two mappings satisfying (ii)–(iii) of Theorem 2 with following condition:*

(iv) for all $\zeta, \eta, \zeta_1, \eta_1, \zeta_2, \eta_2 \in \Omega$*:*

$$\int_0^{[\mathcal{S}_{\mathcal{M}}(\wp_1(\zeta,\eta),\wp_1(\zeta_1,\eta_1),\wp_1(\zeta_2,\eta_2))]^2} \varsigma(\hbar)d\hbar \leq \gamma \int_0^{\mathcal{D}_{S_M}(\zeta,\eta,\zeta_1,\eta_1,\zeta_2,\eta_2)} \varsigma(\hbar)d\hbar, \quad (9)$$

where

$$\begin{aligned}
\mathcal{D}_{S_M}(\zeta, \eta, \zeta_1, \eta_1, \zeta_2, \eta_2) = \max \Big\{ &[\mathcal{S}_{\mathcal{M}}(\wp_2\zeta, \wp_2\zeta_1, \wp_2\zeta_2)]^2, \\
&[\mathcal{S}_{\mathcal{M}}(\wp_1(\zeta, \eta), \wp_1(\zeta, \eta), \wp_2\zeta)]^2, \\
&[\mathcal{S}_{\mathcal{M}}(\wp_1(\zeta_1, \eta_1), \wp_1(\zeta_1, \eta_1), \wp_2\zeta_1)]^2, \\
&[\mathcal{S}_{\mathcal{M}}(\wp_1(\zeta_2, \eta_2), \wp_1(\zeta_2, \eta_2), \wp_2\zeta_2)]^2,
\end{aligned}$$

$$\frac{[\mathcal{S}_{\mathcal{M}}(\wp_1(\zeta,\eta),\wp_1(\zeta,\eta),\wp_2\zeta)]^2}{1+[\mathcal{S}_{\mathcal{M}}(\wp_1(\zeta_1,\eta_1),\wp_1(\zeta_1,\eta_1),\wp_2\zeta_2)]^2}\Big\},$$

$\gamma \in [0, 1)$ is a constant and $\varsigma \in \Phi$.

Then $\wp_1$ and $\wp_2$ have a coupled coincidence point in Ω. Moreover, there exists a unique point $\mu \in \Omega$ such that $\mu = \wp_1(\mu, \mu) = \wp_2(\mu)$.

Proof By condition (iii), there exist sequences $\{\varpi_n\}$ and $\{\varpi_n'\}$ in Ω such that

$$\begin{aligned}\lim_{n\to+\infty}\wp_1(\varpi_n,\varpi_n') &= \lim_{n\to+\infty}\wp_2(\varpi_n)=\wp_2(\vartheta),\\ \lim_{n\to+\infty}\wp_1(\varpi_n',\varpi_n) &= \lim_{n\to+\infty}\wp_2(\varpi_n')=\wp_2(\varrho),\end{aligned} \tag{10}$$

for some $\vartheta, \varrho \in \Omega$.

Now from (9), we have

$$\int_0^{[\mathcal{S}_{\mathcal{M}}(\wp_1(\varpi_n,\varpi_n'),\wp_1(\varpi_n,\varpi_n'),\wp_1(\vartheta,\varrho))]^2}\varsigma(\hbar)d\hbar \le \gamma\int_0^{\mathcal{D}_{S_M}(\varpi_n,\varpi_n',\varpi_n,\varpi_n',\vartheta,\varrho)}\varsigma(\hbar)d\hbar, \tag{11}$$

where

$$\begin{aligned}\mathcal{D}_{S_M}(\varpi_n,\varpi_n',\varpi_n,\varpi_n',\vartheta,\varrho) = \max\Big\{&[\mathcal{S}_{\mathcal{M}}(\wp_2\varpi_n,\wp_2\varpi_n,\wp_2\vartheta)]^2,\\ &[\mathcal{S}_{\mathcal{M}}(\wp_1(\varpi_n,\varpi_n'),\wp_1(\varpi_n,\varpi_n'),\wp_2\varpi_n)]^2,\\ &[\mathcal{S}_{\mathcal{M}}(\wp_1(\varpi_n,\varpi_n'),\wp_1(\varpi_n,\varpi_n'),\wp_2\varpi_n)]^2,\\ &[\mathcal{S}_{\mathcal{M}}(\wp_1(\vartheta,\varrho),\wp_1(\vartheta,\varrho),\wp_2\vartheta)]^2,\\ &\frac{[\mathcal{S}_{\mathcal{M}}(\wp_1(\varpi_n,\varpi_n'),\wp_1(\omega_n,\varpi_n'),\wp_2\varpi_n)]^2}{1+[\mathcal{S}_{\mathcal{M}}(\wp_1(\varpi_n,\varpi_n'),\wp_1(\varpi_n,\varpi_n'),\wp_2\vartheta)]^2}\Big\}.\end{aligned}$$

Passing to the limit as $n \to +\infty$ in the above inequality, using (10), condition $(S2)$ and Lemma 1, we obtain

$$\begin{aligned}\mathcal{D}_{S_M}(\varpi_n,\varpi_n'),\varpi_n,\varpi_n',\vartheta,\varrho) &= \max\Big\{0,0,0,[\mathcal{S}_{\mathcal{M}}(\wp_1(\vartheta,\varrho),\wp_1(\vartheta,\varrho),\wp_2\vartheta)]^2,0\Big\}\\ &= [\mathcal{S}_{\mathcal{M}}(\wp_1(\vartheta,\varrho),\wp_1(\vartheta,\varrho),\wp_2\vartheta)]^2.\end{aligned} \tag{12}$$

Passing to the limit as $n \to +\infty$ in (11) and using (12), we obtain

$$\int_0^{[\mathcal{S}_{\mathcal{M}}(\wp_1(\vartheta,\varrho),\wp_1(\vartheta,\varrho),\wp_2\vartheta)]^2}\varsigma(\hbar)d\hbar \le \gamma\int_0^{[\mathcal{S}_{\mathcal{M}}(\wp_1(\vartheta,\varrho),\wp_1(\vartheta,\varrho),\wp_2\vartheta)]^2}\varsigma(\hbar)d\hbar, \tag{13}$$

a contradiction, since by hypothesis $0 \le \gamma < 1$. Hence, we can conclude that

$$\int_0^{[\mathcal{S}_{\mathcal{M}}(\wp_1(\vartheta,\varrho),\wp_1(\vartheta,\varrho),\wp_2\vartheta)]^2}\varsigma(\hbar)d\hbar = 0.$$

Therefore, by the property of integral ς, we have $\mathcal{S}_{\mathcal{M}}(\wp_1(\vartheta,\varrho),\wp_1(\vartheta,\varrho),\wp_2\vartheta)=0$ and so $\wp_1(\vartheta,\varrho)=\wp_2\vartheta$. By similar fashion, one can prove that $\wp_1(\varrho,\vartheta)=\wp_2\varrho$.

Let

$$\wp_1(\vartheta,\varrho)=\wp_2\vartheta=\mu \quad \text{and} \quad \wp_1(\varrho,\vartheta)=\wp_2\varrho=\nu. \tag{14}$$

By assumption (ii),

$$\wp_2\mu=\wp_2(\wp_1(\vartheta,\varrho))=\wp_1(\wp_2\vartheta,\wp_2\varrho)=\wp_1(\mu,\nu), \tag{15}$$

and

$$\wp_2\nu=\wp_2(\wp_1(\varrho,\vartheta))=\wp_1(\wp_2\varrho,\wp_2\vartheta)=\wp_1(\nu,\mu). \tag{16}$$

Hence $\wp_1$ and $\wp_2$ have a coupled coincidence point.

Now, we have to show that $\wp_2(\mu)=\nu$ and $\wp_2(\nu)=\mu$. From (9), we have

$$\begin{aligned}\int_0^{[\mathcal{S}_{\mathcal{M}}(\wp_2\varpi_n,\wp_2\varpi_n,\wp_2\nu)]^2}\varsigma(\hbar)d\hbar &= \int_0^{[\mathcal{S}_{\mathcal{M}}(\wp_1(\varpi_n,\varpi_n'),\wp_1(\varpi_n,\varpi_n'),\wp_1(\nu,\mu))]^2}\varsigma(\hbar)d\hbar\\ &\le \gamma\int_0^{\mathcal{D}_{S_M}(\varpi_n,\varpi_n',\varpi_n,\varpi_n',\nu,\mu)}\varsigma(\hbar)d\hbar,\end{aligned} \tag{17}$$

where

$$\begin{aligned}\mathcal{D}_{S_M}(\varpi_n,\varpi_n',\varpi_n,\varpi_n',\nu,\mu) = \max\Big\{&[\mathcal{S}_{\mathcal{M}}(\wp_2\varpi_n,\wp_2\varpi_n,\wp_2\nu)]^2,\\ &[\mathcal{S}_{\mathcal{M}}(\wp_1(\varpi_n,\varpi_n'),\wp_1(\varpi_n,\varpi_n'),\wp_2\varpi_n)]^2,\\ &[\mathcal{S}_{\mathcal{M}}(\wp_1(\varpi_n,\varpi_n'),\wp_1(\varpi_n,\varpi_n'),\wp_2\varpi_n)]^2,\\ &[\mathcal{S}_{\mathcal{M}}(\wp_1(\nu,\mu),\wp_1(\nu,\mu),\wp_2\nu)]^2,\\ &\frac{[\mathcal{S}_{\mathcal{M}}(\wp_1(\varpi_n,\varpi_n'),\wp_1(\varpi_n,\varpi_n'),\wp_2\varpi_n)]^2}{1+[\mathcal{S}_{\mathcal{M}}(\wp_1(\varpi_n,\varpi_n'),\wp_1(\varpi_n,\varpi_n'),\wp_2\nu)]^2}\Big\}.\end{aligned}$$

Passing to the limit as $n\to+\infty$ in the above and using (2), (15), (16) and condition $(S2)$, we get

$$\begin{aligned}\mathcal{D}_{S_M}(\varpi_n,\varpi_n',\varpi_n,\varpi_n',\nu,\mu) &= \max\Big\{[\mathcal{S}_{\mathcal{M}}(\mu,\mu,\wp_2\nu)]^2,0,0,0,0\Big\}\\ &= [\mathcal{S}_{\mathcal{M}}(\mu,\mu,\wp_2\nu)]^2.\end{aligned} \tag{18}$$

Passing to the limit as $n\to+\infty$ in (17) and using (18), we obtain

$$\int_0^{[\mathcal{S}_{\mathcal{M}}(\mu,\mu,\wp_2\nu)]^2}\varsigma(\hbar)d\hbar \le \gamma\int_0^{[\mathcal{S}_{\mathcal{M}}(\mu,\mu,\wp_2\nu)]^2}\varsigma(\hbar)d\hbar,$$

a contradiction, since $0 \leq \gamma < 1$. Hence

$$\int_0^{[\mathcal{S}_{\mathcal{M}}(\mu,\mu,\wp_2\nu)]^2} \varsigma(\hbar)d\hbar = 0.$$

Hence, by the property of integral ς, we have $\mathcal{S}_{\mathcal{M}}(\mu, \mu, \wp_2\nu) = 0$ and so $\wp_2\nu = \mu$. By similar way, we can show that $\wp_2\mu = \nu$.

Finally, we show that $\mu = \nu$. The proof follows from Theorem 2, so we omit it.

If we take $L_1 = L_4 = 0$, $L_2 = k$, $L_3 = l$ and $\wp_2 = I$ (the identity map on Ω) in Theorem 2, then we have the following result.

Corollary 1 *Let $(\Omega, \mathcal{S}_{\mathcal{M}})$ be a complete S-metric space. Let $\wp_1 \colon \Omega^2 \to \Omega$ be a mapping satisfying the following inequality:*

$$\int_0^{[\mathcal{S}_{\mathcal{M}}(\wp_1(\zeta,\eta),\wp_1(\zeta_1,\eta_1),\wp_1(\zeta_2,\eta_2))]^2} \varsigma(\hbar)d\hbar \leq k \int_0^{[\mathcal{S}_{\mathcal{M}}(\wp_1(\zeta,\eta),\wp_1(\zeta,\eta),\zeta)]^2} \varsigma(\hbar)d\hbar$$
$$+l \int_0^{[\mathcal{S}_{\mathcal{M}}(\wp_1(\zeta_1,\eta_1),\wp_1(\zeta_1,\eta_1),\zeta_1)]^2} \varsigma(\hbar)d\hbar,$$

for all $\zeta, \eta, \zeta_1, \eta_1, \zeta_2, \eta_2 \in \Omega$, where k, l are nonnegative constants such that $k + l < 1$ and $\varsigma \in \Phi$. Then $\wp_1$ has a unique coupled point in Ω.

Remark 1 Corollary 1 extends and generalizes Theorem 2.4 of Aydi [2] from partial metric spaces to the setting of S-metric spaces and integral type contraction.

Remark 2 If we take $\varsigma(\hbar) = 1$ for all $t \in [0, +\infty)$ in Corollary 1, then it extends and generalizes Theorem 2.4 of Aydi [2] from complete partial metric spaces to complete S-metric spaces.

If we take $L_1 = k$, $L_2 = L_3 = L_4 = 0$ and $\wp_2 = I$ (the identity map on Ω) in Theorem 2, then we have the following result.

Corollary 2 *Let $(\Omega, \mathcal{S}_{\mathcal{M}})$ be a complete S-metric space. Let $\wp_1 \colon \Omega^2 \to \Omega$ be a mapping satisfying the following inequality:*

$$\int_0^{[\mathcal{S}_{\mathcal{M}}(\wp_1(\zeta,\eta),\wp_1(\zeta_1,\eta_1),\wp_1(\zeta_2,\eta_2))]^2} \varsigma(\hbar)d\hbar \leq k \int_0^{[\mathcal{S}_{\mathcal{M}}(\zeta,\zeta_1,\zeta_2)]^2} \varsigma(\hbar)d\hbar, \qquad (19)$$

for all $\zeta, \eta, \zeta_1, \eta_1, \zeta_2, \eta_2 \in \Omega$, where $k \in [0, 1)$ is a constant and $\varsigma \in \Phi$. Then $\wp_1$ has a unique coupled point in Ω.

If we define $\wp_1(\zeta, \zeta) = \mathcal{T}\zeta$ in Corollary 2, then we have the following result.

Corollary 3 *Let $(\Omega, \mathcal{S}_{\mathcal{M}})$ be a complete S-metric space. Let $\mathcal{T} \colon \Omega \to \Omega$ be a mapping satisfying the following inequality:*

$$\int_0^{[\mathcal{S}_{\mathcal{M}}(\mathcal{T}\zeta,\mathcal{T}\zeta_1,\mathcal{T}\zeta_2)]^2} \varsigma(\hbar)d\hbar \leq k \int_0^{[\mathcal{S}_{\mathcal{M}}(\zeta,\zeta_1,\zeta_2)]^2} \varsigma(\hbar)d\hbar, \tag{20}$$

for all $\zeta, \eta, \zeta_1, \eta_1, \zeta_2, \eta_2 \in \Omega$, *where* $k \in [0, 1)$ *is a constant and* $\varsigma \in \Phi$. *Then* $\mathcal{T}$ *has a unique fixed point in* Ω.

Proof Taking $\zeta = \eta$, $\zeta_1 = \eta_1$ and $\zeta_2 = \eta_2$ in Corollary 2, then inequality (19) coincides with the inequality (20). Thus, we have the conclusion of Corollary 3 from Corollary 2.

Remark 3 Corollary 3 extends and generalizes Theorem 1 of Branciari [5] from complete metric spaces to complete S-metric spaces.

Remark 4 If we take $\varsigma(\hbar) = 1$ for all $t \in [0, +\infty)$ in Corollary 3, then it generalizes Theorem 3.1 of *Sedghi et al.* [16] for square inequality.

Now, we give an example to validate the result.

Example 11 Let $\Omega = [0, 1]$ and the function $\mathcal{S}_{\mathcal{M}} \colon \Omega^3 \to [0, \infty)$ be defined as $\mathcal{S}_{\mathcal{M}}(\zeta, \eta, \theta) = |\eta - \theta| + |\eta + \theta - 2\zeta|$ for all $\zeta, \eta, \theta \in \Omega$. Then the function $\mathcal{S}_{\mathcal{M}}$ is an S-metric on Ω and $(\Omega, \mathcal{S}_{\mathcal{M}})$ is an S-metric space. Define mappings $\wp_1 \colon \Omega^2 \to \Omega$ and $\wp_2 \colon \Omega \to \Omega$ by

$$\wp_1(\zeta, \eta) = \begin{cases} \frac{1}{25}(\zeta + \eta), & \text{if } \zeta, \eta \in [0, \frac{1}{2}], \\ \frac{1}{4}, & \text{if } \zeta, \eta \in (\frac{1}{2}, 1], \end{cases}$$

for all $\zeta, \eta \in \Omega$ and

$$\wp_2(\zeta) = \begin{cases} \frac{\zeta}{5}, & \text{if } \zeta \in [0, \frac{1}{2}], \\ 0, & \text{if } \zeta \in (\frac{1}{2}, 1], \end{cases}$$

for all $\zeta \in \Omega$. Let $\varsigma(\hbar) = 1$ for all $t \in [0, +\infty)$. Then we show that the following:

(1) for $\zeta = \eta = \zeta_1 = \eta_1 = \zeta_2 = \eta_2 = \frac{1}{2}$, then we have

$$[\mathcal{S}_{\mathcal{M}}(\wp_1(\zeta, \eta), \wp_1(\zeta_1, \eta_1), \wp_1(\zeta_2, \eta_2))]^2$$

$$\begin{aligned}
&= [|\wp_1(\zeta_1, \eta_1) + \wp_1(\zeta_2, \eta_2) - 2\wp_1(\zeta, \eta)| + |\wp_1(\zeta_1, \eta_1) - \wp_1(\zeta_2, \eta_2)|]^2 \\
&= \Big[\Big|\frac{\zeta_1 + \eta_1}{25} + \frac{\zeta_2 + \eta_2}{25} - \frac{2(\zeta + \eta)}{25}\Big| + \Big|\frac{\zeta_1 + \eta_1}{25} - \frac{\zeta_2 + \eta_2}{25}\Big|\Big]^2 \\
&= \Big[\frac{1}{25}|\zeta_1 + \zeta_2 - 2\zeta| + \frac{1}{25}|\eta_1 + \eta_2 - 2\eta| + \frac{1}{25}|\zeta_1 - \zeta_2| + \frac{1}{25}|\eta_1 - \eta_2|\Big]^2 \\
&= \Big[\frac{1}{25}\big(|\zeta_1 + \zeta_2 - 2\zeta| + |\zeta_1 - \zeta_2|\big) + \frac{1}{25}\big(|\eta_1 + \eta_2 - 2\eta| + |\eta_1 - \eta_2|\big)\Big]^2 \\
&= \Big[\frac{1}{5}\Big(\Big|\frac{\zeta_1}{5} + \frac{\zeta_2}{5} - \frac{2\zeta}{5}\Big| + \Big|\frac{\zeta_1}{5} - \frac{\zeta_2}{5}\Big|\Big) + \frac{1}{5}\Big(\Big|\frac{\eta_1}{5} + \frac{\eta_2}{5} - \frac{2\eta}{5}\Big| + \Big|\frac{\eta_1}{5} - \frac{\eta_2}{5}\Big|\Big)\Big]^2 \\
&= \frac{1}{25}\big[\mathcal{S}_{\mathcal{M}}(\wp_2\zeta, \wp_2\zeta_1, \wp_2\zeta_2) + \mathcal{S}_{\mathcal{M}}(\wp_2\eta, \wp_2\eta_1, \wp_2\eta_2)\big]^2
\end{aligned}$$

$$
\begin{aligned}
&\le \frac{1}{12}\Big([(\mathcal{S}_{\mathcal{M}}(\wp_2\zeta, \wp_2\zeta_1, \wp_2\zeta_2)]^2 + [\mathcal{S}_{\mathcal{M}}(\wp_2\eta, \wp_2\eta_1, \wp_2\eta_2)]^2\Big) \\
&\le \frac{1}{5}\,[(\mathcal{S}_{\mathcal{M}}(\wp_2\zeta, \wp_2\zeta_1, \wp_2\zeta_2)]^2 \\
&\le \frac{1}{5}\Big([\mathcal{S}_{\mathcal{M}}(\wp_2\zeta, \wp_2\zeta_1, \wp_2\zeta_2)]^2 + [\mathcal{S}_{\mathcal{M}}(\wp_1(\zeta, \eta), \wp_1(\zeta, \eta), \wp_2\zeta)]^2 \\
&\qquad + [\mathcal{S}_{\mathcal{M}}(\wp_1(\zeta_1, \eta_1), \wp_1(\zeta_1, \eta_1), \wp_2\zeta_1)]^2 + [\mathcal{S}_{\mathcal{M}}(\wp_1(\zeta_2, \eta_2), \wp_1(\zeta_2, \eta_2), \wp_2\zeta_2)]^2\Big),
\end{aligned}
$$

holds for all $\zeta, \eta, \zeta_1, \eta_1, \zeta_2, \eta_2 \in \Omega$, where $L_1 = L_2 = L_3 = L_4 = \frac{1}{5}$ with $L_1 + L_2 + L_3 + L_4 = \frac{4}{5} < 1$.

(2) To show that the pair $(\wp_1, \wp_2)$ satisfies the occasionally w-compatible (owc) property.

Case (i): If $\zeta = \eta = 0$, then we have

$$\wp_2(\wp_1(0, 0)) = \wp_2(0) = 0 \quad \text{and} \quad \wp_1(\wp_2(0), \wp_2(0)) = \wp_1(0, 0) = 0.$$

Thus, $\wp_2(\wp_1(0, 0)) = \wp_1(\wp_2(0), \wp_2(0)) = 0$.

Case (ii): If $\zeta = \eta = \frac{1}{2}$, then we have

$$\wp_2(\wp_1(1/2, 1/2)) = \wp_2(1/25) = 1/125 \text{ and } \wp_1(\wp_2(1/2), \wp_2(1/2)) = \wp_1(1/10, 1/10) = 1/125.$$

Thus, $\wp_2(\wp_1(1/2, 1/2)) = \wp_1(\wp_2(1/2), \wp_2(1/2)) = 1/125$.

Case (iii): If $\zeta = \eta = 1$, then we have

$$\wp_2(\wp_1(1, 1)) = \wp_2(1/4) = 1/20 \quad \text{and} \quad \wp_1(\wp_2(1), \wp_2(1)) = \wp_1(0, 0) = 0.$$

Thus, $\wp_2(\wp_1(1, 1)) \neq \wp_1(\wp_2(1), \wp_2(1))$. This shows that $\wp_2(\wp_1(\zeta, \eta)) = \wp_1(\wp_2\zeta, \wp_2\eta)$ for some $\zeta, \eta \in \Omega$.

Hence, we see that the pair $(\wp_1, \wp_2)$ satisfies the occasionally w-compatible (owc) property.

(3) Now, we show that the pair $(\wp_1, \wp_2)$ satisfies $(CLR_{\wp_2})$ property.
Consider two sequences $\{\varpi_n\} = \{\frac{1}{n^2}\}$ and $\{\varpi_n'\} = \{\frac{1}{n^3}\}$ for all $n \ge 1$. Then

$$
\begin{aligned}
\mathcal{S}_{\mathcal{M}}(\wp_1(\varpi_n, \varpi_n'), \wp_1(\varpi_n, \varpi_n'), 0) &= \mathcal{S}_{\mathcal{M}}\Big(\frac{\varpi_n}{25} + \frac{\varpi_n'}{25}, \frac{\varpi_n}{25} + \frac{\varpi_n'}{25}, 0\Big) \\
&= \Big|\frac{\varpi_n}{25} + \frac{\varpi_n'}{25} + 0 - \frac{2\varpi_n}{25} - \frac{2\varpi_n'}{25}\Big| + \Big|\frac{\varpi_n}{25} + \frac{\varpi_n'}{25} - 0\Big| \\
&= 2\Big|\frac{\varpi_n}{25} + \frac{\varpi_n'}{25}\Big| \to 0 \text{ as } n \to +\infty,
\end{aligned}
$$

and

$$\mathcal{S}_{\mathcal{M}}(\wp_2(\varpi_n), \wp_2(\varpi_n), 0) = \mathcal{S}_{\mathcal{M}}\Big(\frac{\varpi_n}{5}, \frac{\varpi_n}{5}, 0\Big)$$

$$= \left|\frac{\varpi_n}{5} + 0 - \frac{2\varpi_n}{5}\right| + \left|\frac{\varpi_n}{5} - 0\right| = 2\left|\frac{\varpi_n}{5}\right|$$
$$\to 0 \text{ as } n \to +\infty.$$

This shows that

$$\lim_{n\to+\infty} \mathcal{S}_{\mathcal{M}}(\wp_1(\varpi_n, \varpi_n'), \wp_1(\varpi_n, \varpi_n'), 0) = \lim_{n\to+\infty} \mathcal{S}_{\mathcal{M}}(\wp_2(\varpi_n), \wp_2(\varpi_n), 0) = 0 = \wp_2(0),$$

where $0 \in \Omega$.

Thus the pair $(\wp_1, \wp_2)$ satisfies $(CLR_{\wp_2})$ property.

Hence the mappings $\wp_1$ and $\wp_2$ satisfy all the conditions of Theorem 2 and consequently $\wp_1$ and $\wp_2$ have a unique common coupled fixed point, namely $\wp_1(0, 0) = 0$ and $\wp_2(0) = 0$.

4 An Application to the Fredholm Integral Equation

Utilizing Corollary 2, we establish a result concerning the existence and uniqueness for a specific category of the subsequent system of Fredholm integral equations.

$$\zeta(\hbar) = \int_E A(t, s, \zeta(s), \eta(s))ds + q(\hbar), \quad t, s \in E,$$
$$\eta(\hbar) = \int_E A(t, s, \eta(s), \zeta(s))ds + q(\hbar), \quad t, s \in E, \tag{21}$$

where E is measurable, $A\colon E \times E \times \mathbb{R} \times \mathbb{R} \to \mathbb{R}$ and $q \in \mathcal{L}^\infty(E)$. Let $\Omega = \mathcal{L}^\infty(E)$. Now, we define $\wp_1 : \Omega^2 \to \Omega$ by

$$\wp_1(\zeta, \eta)(\hbar) = \int_E A(t, s, \zeta(s), \eta(s))ds + q(\hbar), \quad t, s \in E.$$

Obviously, $(\zeta(\hbar), \eta(\hbar))$ is a solution of the system of Fredholm integral equations (21) if and only if $(\zeta(\hbar), \eta(\hbar))$ is a coupled fixed point of $\wp_1$. Now, we define the function $\mathcal{S}_{\mathcal{M}}\colon \Omega^3 \to [0, +\infty)$ by

$$\mathcal{S}_{\mathcal{M}}(\zeta, \eta, \theta) = \sup_{s\in E} |\eta(s) - \theta(s)| + \sup_{s\in E} |\eta(s) + \theta(s) - 2\zeta(s)|, \tag{22}$$

for all $\zeta, \eta, \theta \in \Omega$. Then the function $\mathcal{S}_{\mathcal{M}}$ is an S-metric on Ω. Now, we show that this S-metric cannot be generated by any metric d. We assume that this S-metric is generated by any metric d, that is, there exists a metric d such that

$$\mathcal{S}_{\mathcal{M}}(\zeta, \eta, \theta) = d(\zeta, \theta) + d(\eta, \theta), \tag{23}$$

for all $\zeta, \eta, \theta \in \Omega$. Then we get

$$\mathcal{S}_{\mathcal{M}}(\zeta, \zeta, \theta) = 2d(\zeta, \theta) = 2 \sup_{s \in E} |\zeta(s) - \theta(s)|,$$

and

$$d(\zeta, \theta) = \sup_{s \in E} |\zeta(s) - \theta(s)|. \tag{24}$$

Likewise, we obtain

$$\mathcal{S}_{\mathcal{M}}(\eta, \eta, \theta) = 2d(\eta, \theta) = 2 \sup_{s \in E} |\eta(s) - \theta(s)|,$$

and

$$d(\eta, \theta) = \sup_{s \in E} |\eta(s) - \theta(s)|. \tag{25}$$

From Eqs. (23), (24), and (25), we get

$$\sup_{s \in E} |\eta(s) - \theta(s)| + \sup_{s \in E} |\eta(s) + \theta(s) - 2\zeta(s)|$$

$$= \sup_{s \in E} |\zeta(s) - \theta(s)| + \sup_{s \in E} |\eta(s) - \theta(s)|,$$

which is a contradiction. Hence this S-metric is not generated by any metric d. Thus, $(\Omega, \mathcal{S}_{\mathcal{M}})$ is a complete S-metric space.

Now, we state and prove our result as follows.

Theorem 4 *Suppose the following:*

(a) There exists a continuous function $K : E \times E \to \mathbb{R}$ such that

$$\big|A(t, s, \zeta(s), \eta(s)) - A(t, s, \zeta_1(s), \eta_1(s))\big|$$

$$\leq |K(t, s)|\big[|\zeta(s) - \zeta_1(s)| + |\eta(s) - \eta_1(s)|\big],$$

for all $\zeta, \zeta_1, \eta, \eta_1 \in \Omega$ and $t, s \in E$.

(b)

$$\int_E |K(t, s)| ds \leq \frac{1}{4}.$$

Then the integral equation (21) has a unique solution in Ω.

Proof Consider

$$\big[\mathcal{S}_{\mathcal{M}}(\wp_1(\zeta,\eta),\wp_1(\zeta,\eta),\wp_1(\zeta_1,\eta_1))\big]^2=\Big[2\left|\wp_1(\zeta,\eta)-\wp_1(\zeta_1,\eta_1)\right|\Big]^2$$

$$\begin{aligned}
&=\Big[2\Big|\int_E A(t,s,\zeta(s),\eta(s))ds+q(\hbar)\\
&\quad-\Big(\int_E A(t,s,\zeta_1(s),\eta_1(s))ds+q(\hbar)\Big)\Big|\Big]^2\\
&=\Big[2\Big|\int_E\big[A(t,s,\zeta(s),\eta(s))-A(t,s,\zeta_1(s),\eta_1(s))\big]ds\Big|\Big]^2\\
&\le 4\Big[\int_E|K(t,s)|\big[|\zeta(s)-\zeta_1(s)|+|\eta(s)-\eta_1(s)|\big]ds\Big]^2\\
&\le\frac{1}{4}\Big[|\zeta(s)-\zeta_1(s)|+|\eta(s)-\eta_1(s)|\Big]^2\\
&=\frac{1}{16}\Big[2(|\zeta(s)-\zeta_1(s)|)+2(|\eta(s)-\eta_1(s)|)\Big]^2\\
&=\frac{1}{16}\Big[\mathcal{S}_{\mathcal{M}}(\zeta,\zeta,\zeta_1)+\mathcal{S}_{\mathcal{M}}(\eta,\eta,\eta_1)\Big]^2\\
&\le\frac{1}{8}\Big(\big[\mathcal{S}_{\mathcal{M}}(\zeta,\zeta,\zeta_1)\big]^2+\big[\mathcal{S}_{\mathcal{M}}(\eta,\eta,\eta_1)\big]^2\Big)\\
&\le\frac{1}{4}\big[\mathcal{S}_{\mathcal{M}}(\zeta,\zeta,\zeta_1)\big]^2=k\,\big[\mathcal{S}_{\mathcal{M}}(\zeta,\zeta,\zeta_1)\big]^2,
\end{aligned}$$

which implies

$$\int_0^{\left[\mathcal{S}_{\mathcal{M}}(\wp_1(\zeta,\eta),\wp_1(\zeta,\eta),\wp_1(\zeta_1,\eta_1))\right]^2}\varsigma(\hbar)d\hbar\le k\int_0^{\left[\mathcal{S}_{\mathcal{M}}(\zeta,\zeta,\zeta_1)\right]^2}\varsigma(\hbar)d\hbar,$$

for all $\zeta,\zeta_1,\eta,\eta_1\in\Omega$, where $0\le k=\frac{1}{4}<1$. Hence, all the hypothesis of Corollary 2 are satisfied and consequently, the integral equation (21) has a unique solution.

5 Conclusion

By employing the framework of S-metric spaces, we presented various well-known linked fixed point theorems that necessitate integral type conditions for mappings that are sometimes w-compatible. Additionally, we discussed some implications of the results obtained. An application concerning the Fredholm integral equation was also included. Our results serve to extend and generalize numerous findings previously documented in the literature. For instances,

(1) Corollary 1 extends and generalizes Theorem 2.4 of *Aydi* [2] form partial metric spaces to S-metric spaces and integral type contraction.
(2) Corollary 3 broadens and generalizes Theorem 2.1 from the work of *Branciari* [5], extending its applicability from metric spaces to S-metric spaces.
(3) If we take $\varsigma(\hbar) = 1$ for all $t > 0$, then Corollary 3 also generalizes Theorem 3.1 of *Sedghi et al.* [16] for quadratic terms and many others.

Acknowledgements We appreciate the editor's generous support. The authors are grateful to the reviewer for his helpful suggestions.

References

1. Abbas, M., Ali Khan, M., Radenović, S.: Common coupled fixed point theorems in cone metric spaces for w-compatible mappings. Appl. Math. Comput. **217**(1), 195–202 (2010)
2. Aydi, H.: Some coupled fixed point results on partial metric spaces. Int. J. Math. Math. Sci. **2011**, Article ID 647091, 11. https://doi.org/10.1155/2011/647091
3. Banach, S.: Surles operation dans les ensembles abstraits et leur application aux equation integrals. Fund. Math. **3**, 133–181 (1922)
4. Bhaskar, T.G., Lakshmikantham, V.: Fixed point theorems in partially ordered metric spaces and applications. Nonlinear Anal.: TMA **65**(7), 1379–1393 (2006)
5. Branciari, A.: A fixed point theorem for mappings satisfying a general contractive condition of integral type. Int. J. Math. Math. Sci. **29**(9), 531–536 (2002)
6. Ciric, L., Lakshmikantham, V.: Coupled fixed point theorems for nonlinear contractions in partially ordered metric spaces. Nonlinear Anal.: TMA **70**(12), 4341–4349 (2009)
7. Guo, D., Lakshmikantham, V.: Coupled fixed point of nonlinear operator with application. Nonlinear Anal.: (TMA.) **11**, 623–632 (1987)
8. Hieu, N.T., Ly, N.T., Dung, N.V.: A generalization of Ćirić quasi-contractions for maps on S-metric spaces. Thai J. Math. **13**(2), 369–380 (2015)
9. Jungck, G., Rhoades, B.E.: Fixed points for set valued functions without continuity. Indian J. Pure Appl. Math. **29**, 227–238 (1998)
10. Kim, J.K., Sedghi, S., Gholidahneh, A., Rezaee, M.M.: Fixed point theorems in S-metric spaces. East Asian Math. J. **32**(5), 677–684 (2016)
11. Kim, J.K., Okeke, G.A., Lim, W.H.: Common coupled fixed point theorems for w-compatible mappings in partial metric spaces. Global J. Pure Appl. Math. **13**(2), 519–536 (2017)
12. Luong, N.V., Thuan, N.X.: Coupled fixed points theorems for mixed monotone mappings and an application to integral equations. Comput. Math. Appl. **62**, 4238–4248 (2011)
13. Nashine, H.K., Kim, J.K., Sharma, A.K., Saluja, G.S.: Some coupled fixed point without mixed monotone mappings. Nonlinear Funct. Anal. Appl. **21**(2), 235–247 (2016)
14. Olaleru, J.O., Okeke, G.A., Akewe, H.: Coupled fixed point theorems for generalized φ-mappings satisfying contractive condition of integral type on cone metric spaces. Int. J. Math. Model. Comput. **2**(2), 87–98 (2012)
15. *Özgür*, N.Y., Tas, N.: Some new contractive mappings on S-metric spaces and their relationships with the mapping (**S25**). Math. Sci. **11**(7), 7–16 (2017)
16. Sedghi, S., Shobe, N., Aliouche, A.: A generalization of fixed point theorems in S-metric spaces. Mat. Vesnik **64**(3), 258–266 (2012)
17. Sedghi, S., Dung, N.V.: Fixed point theorems on S-metric spaces. Mat. Vesnik **66**(1), 113–124 (2014)
18. Sedghi, S., Shobkolaei, N., Shahraki, M., Došenović, T.: Common fixed point of four maps in S-metric space. Math. Sci. **12**, 137–143 (2018)

Fixed Point Theorems that are Applicable to Compatible Mappings of Type (P) in Dynamic Programming

Shiv Kant Tiwari, Lakshmi Narayan Mishra, Vishnu Narayan Mishra, and Mohd Raiz

Abstract We construct a few fixed point theorems in this chapter for compatible mappings of type (P), which are extensions of compatible mapping. Additionally, we are considering applications such as dynamic programming's use of solutions for a class of functional equations.

Keywords Compatible mappings of type (P) · Iterative sequence · System of functional equation

1 Introduction

The very existence of a solution to many nonlinear problems has been guaranteed by the fixed point theorems, which contribute sufficient conditions. The solution to equations of the form Tx = x, where T is self-mapping, is given by fixed point theorems, on the other hand, there is no accuracy that a solution will be found if the mapping T is non-self. We have seen in [1–3, 6–8, 12, 16], the concepts of compatible mappings and compatible mappings of type (C), type (K), type (A), type (P), etc., are generalizations of commuting mappings and weekly commuting mappings and

These authors contributed equally to this work.

S. K. Tiwari (✉)
Department of Mathematics, Lukhdhirji Engineering College, Morbi, Gujarat, India
e-mail: shivkant.math@gmail.com

L. N. Mishra
Department of Mathematics, School of Advanced Sciences, Vellore Institute of Technology, Vellore, Tamil Nadu, India

V. N. . Mishra
Department of Mathematics, Faculty of Science, Indira Gandhi National Tribal University, Lalpur, Amarkantak, Madhya Pradesh, India

M. Raiz
Department of Applied Sciences, Global Institute of Technology and Management, 5KM Milestone, Kheda Khurampur, Farrukhnagar, Haily Mandi Road, Gurugram, Haryana-122506, India

H. Kalita et al. (eds.), *Nonlinear Analysis and Computational Techniques*,
Springer Proceedings in Mathematics & Statistics 501,
https://doi.org/10.1007/978-981-96-6038-4_12

they are equivalent under some conditions. There are a few researchers including Sintunavarat et al. [17], Sanjay et al. [13], Chang [15], Hadzic [9], Jungck [4, 5], Liu [18, 19], Liu and Kim [20], Pathak, Cho, Kang and Lee [1] and others, who proved some common fixed point theorems concerning the compatible mappings and their types and established the existence and uniqueness of solution and common solutions for some classes of functional equations and system of functional equations arising in dynamic programming. For example, Pathak, Cho, Kang and Lee [1] studied the existence of common fixed point for the compatible mappings of type (P)

$$e(Ex, Fy) \leq \varphi(max\{e(Vx, Sx), e(Vx, Ex), e(Sy, Fy), \frac{1}{2}[e(Vx, Fy) + e(Sy, Ex)]\})$$

for all $x, y \in X$ and $e(, :,) = d(, :,)$ throughout the paper. Inspired by the many results cited in references, we demonstrate two related fixed point theorems for contractive mappings of type (P) that are compatible with mappings. As a application, we give the existence and uniqueness of common solution for a system of functional of equations arising in dynamic programming, which improve and generalize the corresponding results in [1, 10, 11, 18, 23].

Now, we remember the usable definitions and propositions, where N is the collection of positive numbers.

Definition 1 ([1, 5]) Let $R, S : W \to W$ on a complete metric space (W, d) are called compatible, if $\lim_{n\to\infty} d(RSx_n, SRx_n) = 0$ and compatible of type (P), if $\lim_{n\to\infty} d(SSx_n, TTx_n) = 0$ and (hold for both) $\lim_{n\to\infty} Rx_n = \lim_{n\to\infty} Sx_n = z$ for some $z \in W$, where $\{x_n\}_{n\in N}$ is any sequence in X.

□

Example 1 ([12]) Let $W = [0, 1]$ be a complete metric space and two self-mappings $E, R : W \to W$ on W defined by $E(x) = x$, $R(x) = \frac{1}{4}$ for $x \in [0, \frac{1}{2}] - \{\frac{1}{4}\}$, $A(x) = \frac{1}{8}$, $R(x) = x$ for $x = \frac{1}{4}$ and $A(x) = \frac{1}{2} - \frac{x}{2}$, $S(x) = \frac{x}{2}$ for $x = (\frac{1}{2}, 1]$. Choose a sequence $\{x_n\}$ such that $x_n = \frac{1}{2} + \frac{1}{n}$, then as $n \to \infty$, $A(x_n)$ and $S(x_n)$ both converges to $\frac{1}{4}$. Also, $EE(x_n) \to \frac{1}{4}$, $ES(x_n) \to \frac{1}{4}$, $RR(x_n) \to \frac{1}{4}$ and $RE(x_n) \to \frac{1}{4}$, then mappings are compatible and compatible of type (P).

□

Proof ([1]) Consider $R, S : W \to W$ on a complete metric space (W, d), which are compatible mappings of type (P) and $Rx_n, Sx_n \to z$ as $n \to \infty$ for some $z \in W$, then holds the followings:
(1) If R is continuous at z then $\lim_{n\to\infty} SSx_n = Rz$.
(2) If S is continuous at z then $\lim_{n\to\infty} RRx_n = Sz$.
(3) If R and S are continuous at z then $RSz = SRz$ and $Sz = Sz$.

□

2 Common Fixed Point Theorems

Two popular fixed point theorems pertaining to compatible mappings of type (P) in a full metric space (X, d) are presented in this section. Now, define a non-decreasing, upper semi-continuous function $\Phi = \{\varphi; \varphi : [0, +\infty) \to [0, +\infty)$, such that $\varphi(t) < t$ for all $t > 0\}$.

Theorem 1 *Consider four self-mappings $E, F, S, U : X \to X$ on a complete metric space (X, d) with two mappings S and U are continuous and satisfying the followings*
 (a) there exists a sequence $\{y_m\}_{m\in N}$ such that

$$Ty_{2m-1} = Ey_{2m-2} \;\; and \;\; Vy_{2m} = Fx_{2m-1,\ \forall m\in N;} \tag{1}$$

 (b) two pairs $\{E, S\}$ and $\{F, U\}$ are compatible of type (P);
 (c) $e(Ey, Fz) - min\{e(Vy, Fz), e(Uz, Ey), max\{e(Vy, Ey), e(Uz, Fz)\}\}$

$$\begin{aligned}\le&\varphi(max\{e(Vy, Uz), e(Vy, Ey), e(Uz, Fz),\\ &\frac{1}{2}[e(Vy, Fz) + e(Uz, Ey)], \frac{1}{2}[e(Ey, Fz) + e(Vy, Uz)],\\ &\frac{1}{2}[e(Ey, Fz) + e(Vy, Ey)], \frac{1}{2}[e(Ey, Fz) + e(Uz, Fz)]\})\end{aligned} \tag{2}$$

for all $y, z \in Y$, where $\varphi \in \Phi$, then there exist a sequence $\{z_m\}_{m\in N}$ generated by

$$z_{2m-1} = Ty_{2m-1} = Ey_{2m-2} \;\; and \;\; z_{2m} = Vy_{2m} = Fy_{2m-1,\ \forall m\in N} \tag{3}$$

converges to a unique common fixed point of E, F, S and T in Y.
Proof Define $e_m = e(z_m, z_{m+1})$ for all $m \in N$. From (2.2), we get

$$\begin{aligned}e_{2m+1} &= e(Ey_{2m}, Fy_{2m+1}) - min\{e(Vy_{2m}, Fy_{2m+1}), e(Uy_{2m+1}, Ey_{2m}),\\ &\qquad max\{e(Vy_{2m}, Ey_{2m}), e(Uy_{2m+1}, Fy_{2m+1})\}\}\\ &\le \varphi\ (max\{e(Vy_{2m}, Uy_{2m+1}), e(Vy_{2m}, Ey_{2m}), e(Uy_{2m+1}, Fy_{2m+1}),\\ &\qquad \tfrac{1}{2}[e(Vy_{2m}, Fy_{2m+1}) + e(Uy_{2m+1}, Ey_{2m})],\\ &\qquad \tfrac{1}{2}[e(Ey_{2m}, Fy_{2m+1}) + e(Vy_{2m}, Uy_{2m+1})],\\ &\qquad \tfrac{1}{2}[e(Ey_{2m}, Fy_{2m+1}) + e(Vy_{2m}, Ey_{2m})],\\ &\qquad \tfrac{1}{2}[e(Ey_{2m}, Fy_{2m+1}) + e(Uy_{2m+1}, Fy_{2m+1})]\})\\ &= \varphi\quad (max\{e_{2m}, e_{2m}, e_{2m+1}, \tfrac{1}{2}[e(z_{2m}, z_{2m+2}) + 0], \tfrac{1}{2}[e_{2m+1} + e_{2m}]\\ &\qquad \tfrac{1}{2}[e_{2m+1} + e_{2m}], \tfrac{1}{2}[e_{2m+1} + e_{2m+1}]\})\\ &\le \varphi \qquad (max\{e_{2m}, e_{2m+1}\}).\end{aligned}$$

it is claimed for all $m \in N$, $e_{2n+1} \le e_{2m}$. In fact, if $e_{2m+1} > e_{2m}$ for some $m \in N$, then (3) ensures that $e_{2n+1} \le \varphi(e_{2n+1}) < e_{2n+1}$, which is impossible. Similarly, we can show for all $m \in N$, $e_{2m+2} \leq e_{2m+1}$. Therefore, $e_{m+1} \leq e_m$ for all $m \in N$. Since the sequence $\{e_m\}_{n \in N}$ is decreasing, therefore it converges to 0 as $n \to \infty$.

In order to prove that the sequence $\{z_m\}_{n \in N}$ is a Cauchy sequence. By contrary, we assume $\{z_{2m}\}_{m \in N}$ is not a Cauchy sequence. Thus, it follows that for some $\varepsilon_0 > 0$ and there exists even integers $2n(j)$ and $2m(j)$ for every even number $2j$, such that $2n(j) > 2m(j) > j$ and $e(z_{2n(j), y_{2m(j)}}) > \varepsilon_0$. Let $2n(j)$ be the smallest even integer which satisfies the conditions $2n(j) > 2n(j) > j$, $e(y_{2n(j)}, y_{2m(j)}) > \varepsilon_0$ and $e(z_{2n(j-1)}, z_{2m(j)}) \leq \varepsilon_0$. It follows that

$$\varepsilon_0 < e(z_{2m(j)}, z_{2n(j)}) \leq e(z_{2m(j)}, z_{2n(j)-2}) + e_{2n(j)-2} + e_{2n(j)-1}, \quad \forall n \in N,$$

which implies that

$$\lim_{m \to \infty} e(z_{2m(j)}, z_{n(j)}) = \varepsilon_0. \tag{4}$$

From (4) and the following inequalities:

$$|e(y_{2m(j)}, z_{2n(j)-1}) - e(z_{2m(j)}, z_{2n(j)})| \leq e_{2n(j)-1}, \quad \forall n \in N,$$

$$|e(z_{2m(j)+1}, z_{2n(j)}) - e(y_{2m(j)}, z_{2n(j)})| \leq e_{2m(j)}, \quad \forall m \in N$$

and

$$|e(z_{2m(j)+1}, z_{2n(j)-1}) - e(z_{2m(j)}, z_{2n(j)})| \leq e_{2n(j)-1} + e_{2m(j)}, \quad \forall m \in N,$$

it follows that

$$\begin{aligned} \lim_{j \to \infty} e(z_{2m(j)}, z_{2n(j)-1}) &= \lim_{j \to \infty} e(z_{2m(j)+1}, z_{2n(j)}) \\ &= \lim_{j \to \infty} e(z_{2m(j)+1}, z_{2n(j)-1}) = \varepsilon_0. \end{aligned} \tag{5}$$

By (3), (4) and (5), one derives that

$$\begin{aligned} e(z_{2m(j)}, z_{2n(j)}) \leq & e_{2m(j)} + e(Ey_{2m(j)}, Fy_{2n(j)-1}) \\ \leq & e_{2m(j)} + \varphi(max\{e(Sy_{2m(j)}, Ty_{2n(j)-1}), \\ & e(Vy_{2m(j)}, Ey_{2m(j)}), e(Uy_{2n(j)-1}, Fy_{2n(j)-1}), \\ & \frac{1}{2}[e(Vy_{2m(j)}, Fy_{2n(j)-1}) + e(Uy_{2n(j)-1}, Ey_{2m(j)})], \\ & \frac{1}{2}[e(Ey_{2m(j)}, Fy_{2n(j)-1}) + e(Sy_{2m(j)}, Uy_{2n(j)-1})], \\ & \frac{1}{2}[e(Ey_{2m(j)}, Fy_{2n(j)-1}) + e(Vy_{2m(j)}, Ey_{2m(j)})], \end{aligned}$$

$$
\begin{aligned}
&\frac{1}{2}[e(Ey_{2m(j)}, Fy_{2n(j)-1}) + e(Ty_{2n(j)-1}, Fy_{2n(j)-1})]\}) \\
&+min\{e(Sy_{2m(j)}, Fy_{2n(j)-1}), e(Ty_{2n(j)-1}, Ey_{2m(j)}), \\
&\quad max\{e(Sy_{2m(j)}, Ey_{2m(j)}), e(Ty_{2n(j)-1}, Fy_{2n(j)-1})\}\} \\
&\le e_{2m(j)} + \varphi(max\{e(z_{2m(j)}, z_{2n(j)-1}), e_{2m(j)}, e_{2n(j)-1}, \\
&\quad \frac{1}{2}[e(z_{2m(j)}, z_{2n(j)}) + e(z_{2n(j)-1}, z_{2m(j)+1})], \\
&\quad \frac{1}{2}[e(z_{2m(j)+1}, z_{2n(j)}) + e(z_{2m(j)}, z_{2n(j)-1})], \\
&\quad \frac{1}{2}[e(z_{2m(j)+1}, z_{2n(j)}) + e_{2m(j)}], \\
&\quad \frac{1}{2}[e(z_{2m(j)+1}, z_{2n(j)}) + e_{2n(j)-1}]\}) \\
&+min\{e(z_{2m(j)}, z_{2n(j)}), e(z_{2n(j)-1}, z_{2m(j)+1}), \\
&\quad max\{e_{2m(j)}, e_{2n(j)-1}\}\}, \quad \forall m \in N.
\end{aligned}
$$

Let $k \to \infty$ in the inequalities. It is easy to gain that $\varepsilon_0 \le \varphi(\varepsilon_0) < \varepsilon_0$, a contradiction. Therefore $\{z_m\}_{n\in N}$ is a Cauchy sequence. By completeness of X, sequence $\{z_m\}_{m\in N}$ converges to a point $z \in X$ as $m \to \infty$. Thus the subsequences $\{Ex_{2m-2}\}_{m\in N}$, $\{Fy_{2m-1}\}_{m\in N}$, $\{Vy_{2m}\}_{m\in N}$ and $\{Uy_{2m-1}\}_{m\in N}$ of $\{z_m\}_{m\in N}$ also converges to the point z as $n \to \infty$.

We are given that the pair $\{E, V\}$ and $\{F, T\}$ are compatible of type (P), and S and T are continuous. From the equation (3) and proposition (1), we get

$$
\begin{aligned}
&Uz_{2n} \to Uz, Fz_{2m} = FFy_{2m} \to Tz \\
&Vz_{2n-1} \to Vz, Ez_{2m-1} = EEy_{2m_2} \to Sz \text{ as } m \to \infty.
\end{aligned} \tag{6}
$$

By (3) and (4), it is derived that

$$
\begin{aligned}
e(Ez_{2m+1}, Fz_{2m}) =&e(Ez_{2m-1}, Fz_{2m}) - min\{e(Vz_{2m-1}, Fz_{2m}), e(Uz_{2m}, Ez_{2m-1}), \\
&max\{e(Vz_{2m-1}, Ez_{2m-1}), e(Uz_{2m}, Fz_{2m})\}\} \\
\le& \varphi(max\{e(Vz_{2m-1}, Uz_{2m}), e(Vz_{2m-1}, Ez_{2m-1}), e(Uz_{2m}, Fz_{2m}), \\
&\frac{1}{2}[e(Vz_{2m-1}, Fz_{2m}) + e(Uz_{2m}, Ez_{2m-1})], \\
&\frac{1}{2}[e(Ez_{2m-1}, Fz_{2m}) + e(Vz_{2m-1}, Uz_{2m})], \\
&\frac{1}{2}[e(Ez_{2m-1}, Fz_{2m}) + e(Vz_{2m-1}, Ez_{2m-1})], \\
&\frac{1}{2}[e(Ez_{2m-1}, Fz_{2m}) + e(Uz_{2m}, Fz_{2m})]\}).
\end{aligned} \tag{7}
$$

From (3), (6), (7) and upper semi-continuity of φ, we obtain

$$\begin{aligned} e(Vz, Uz) &\leq \varphi(max\{e(Vz, Uz), 0, 0, \frac{1}{2}[e(Vz, Uz) + e(Uz, Vz)], \\ &\qquad \frac{1}{2}[e(Vz, Uz) + e(Vz, Uz)], \frac{1}{2}[e(Vz, Uz) + 0], \frac{1}{2}[e(Vz, Uz) + 0]\}) \\ &\leq \varphi(e(Vz, Uz)), \end{aligned}$$

which implies that $Sz = Tz$. Similarly, we can be show $Sz = Fz$ and $Tz = Ez$. Thus

$$Ez = Fz = Sz = Uz. \tag{8}$$

From (2) and (3), it follows that

$$\begin{aligned} &e(Ey_{2n}, Fz) - min\{e(Vy_{2m}, Fz), e(Uz, Ey_{2m}), \\ &\qquad max\{e(Vy_{2m}, Ey_{2m}), e(Uz, Fz)\}\} \\ &\quad \leq \varphi(max\{e(Vy_{2m}, Uz), e(Vy_{2m}, Ey_{2m}), e(Uz, Fz), \\ &\qquad \frac{1}{2}[e(Vy_{2m}, Fz) + e(Uz, Ey_{2m})], \frac{1}{2}[e(Ey_{2m}, Fz) + e(Vy_{2m}, Uz)], \\ &\qquad \frac{1}{2}[e(Ey_{2m}, Fz) + e(Vy_{2m}, Ey_{2m})], \frac{1}{2}[e(Ey_{2m}, Fz) + e(Uz, Fz)]\}). \end{aligned}$$

Let $n \to \infty$ in above inequality, we have

$$\begin{aligned} &e(z, Fz) - min\{e(z, Fz), e(Uz, z), max\{e(z, z), e(Uz, Fz)\}\} \\ &\quad \leq \varphi(max\{e(z, Uz), e(z, z), e(Uz, Fz), \\ &\qquad \frac{1}{2}[e(z, Fz) + e(Uz, z)], \frac{1}{2}[e(z, Fz) + e(z, Uz)], \\ &\qquad \frac{1}{2}[e(z, Fz) + e(z, z)], \frac{1}{2}[e(z, Fz) + e(Uz, Fz)]\}) \\ &\quad = \varphi(e(z, Fz)), \end{aligned}$$

which implies that $z = Fz$. Thus $z = Ez = Fz = Sz = Uz$. Assume that E, F, V, and U have another common fixed point $w \in X$, which differ from z. It follows from (2) that

$$\begin{aligned} &e(Ez, Fw) - min\{e(Vz, Fw), e(Uw, Ez), max\{e(Vz, Ez), e(Uw, Fw)\}\} \\ &\quad \leq \varphi(max\{e(Vz, Uw), e(Vz, Ez), e(Uw, Fw), \\ &\qquad \frac{1}{2}[e(Vz, Fw) + e(Uw, Ez)], \frac{1}{2}[e(Ez, Fw) + e(Vz, Uw)], \\ &\qquad \frac{1}{2}[e(Ez, Fw) + e(Vz, Ez)], \frac{1}{2}[e(Ez, Fw) + e(Uw, Fw)]\}) \\ &\quad = \varphi(e(z, w)), \end{aligned}$$

which implies that $z = w$. Thus E, F, V, and U have a unique common fixed point $z \in X$.

□

Theorem 2 *Consider four self-mappings $E, F, V, U : X \to X$, S and T are continuous on a complete metric space (X, d) with two mappings, satisfying (b) and (c) of theorem 2.1 and additionally (d) $E(X) \subseteq U(X)$ and $F(X) \subseteq V(X)$. Then four mappings E, F, V & U have a unique common fixed point in X.*

Proof Consider any point $x_0 \in X$. By condition (d), mappings satisfying $E(X) \subseteq U(X)$ and $F(X) \subseteq V(X)$. For each $n \in N$, we can select a path. $\{y_n\}_{n \in N}$ in X such that $Vy_{2n} = Fy_{2n-1}$ and $Ux_{2n-1} = Ey_{2n-2}$. As a result, Theorem 2.1 leads to Theorem 2.2. □

Remark 1 Theorem 2.2 improves and generalizes Theorem 2.1 of Liu [18] and Theorem 3.1 of Pathak, Cho, Kang, and Lee [1].

□

3 An Application in Dynamic Programming

Here, we consider $(X, \|.\|)$ and $(Y, \|.\|^{'})$ be the real Banach spaces and $V \subseteq X$ and $D \subseteq Y$ be the state space and the decision space respectively. The set of all bounded real-value functions on V is denoted by $F(V)$ and a metric defined as $d(f, g) = sup\{|f(x_1) - g(x_1)| : x \in S\}$. Then $(E(V), D)$ is a complete metric space.

It is generally known that many functional equations arising in dynamic programming have both practical and theoretical relevance in their existence and uniqueness solutions. In the past 20 years or so, many authors, including Bhakta and Mitra [10], Chang [15], Kang, Guan Liu and Shim [14], Liu [18, 19], Liu and Ume [21], Liu and Kang [23], Liu, Ume and Kang [22], Liu and Kim [20] and Pathak, Cho, Kangh and Lee [1] and others, by using various fixed point, common fixed point and coincidence point theorems, studied and investigated the existence or uniqueness of solutions, common solutions or coincidence solutions for several classes of functional equations and system of functional equations arising in dynamic programming. For example, in 1984, Bhakta and Mitra [10] established an existence and uniqueness of solution for the following functional equation

$$f(x) = \sup_{y \in D}\{r(x_1, x_2) + f(c(x_1, x_2))\}, \quad \forall x_1 \in S.$$

In 1995, Pathak, Cho, Kang, and Lee [1] investigated an existence and uniqueness of common solution for the following system of functional equations

$$f_i(x) = \sup_{y \in D} H_i(x_1, x_2, f_i(U(x_1, x_2))),$$

$$g_i(x) = \sup_{y \in D} T_i(x_1, x_2, g_i(U(x_1, x_2))), \quad \forall x \in V, \; i \in \{1, 2, 3, 4\}.$$

In 1999, Liu [18] obtained the existence and uniqueness of common solution for the following system of functional equations

$$f_i(x) = \sup_{y \in D}\{u(x_1, x_2) + H_i(x_1, x_2, f_i(U(x_1, x_2)))\}, \quad \forall x \in V, \; i \in \{1, 2, 3, 4\}.$$

In 2001, Liu [19] gained the existence and uniqueness of nonnegative solution for the following functional equations

$$f(x_1) = \inf_{y \in D} max\{p(x_1, x_2) + f(a(x_1, x_2))\}, \quad \forall x \in S.$$

In 2003, Liu and Ume [21] presented some sufficient conditions which ensure the existence and uniqueness of solution for the following functional equation

$$f(x) = opt_{y \in D}\{u[p(x_1, x_2) + f(a(x_1, x_2))] + v.opt_{\in D}\{q(x_1, x_2), f(b(x_1, x_2))\}\}, \quad \forall x \in S,$$

where u and v are nonnegative constants with $u + v = 1$. In 2006, Liu and Kang [23] studied the following functional equation

$$\begin{aligned} f(x_1) = opt_{y \in D}\{&u(x_1, x_2)max\{p(x_1, x_2), f(a(x_1, x_2))\} \\ &+v(x_1, x_2)min\{q(x_1, x_2), f(b(x_1, x_2))\} \\ &+w(x_1, x_2)[r(x_1, x_2) + f(c(x_1, x_2))]\} + k(x), \quad \forall x \in S, \end{aligned}$$

where *opt* denotes sup or inf. The above works motive us to investigate the following system of functional equations arising in dynamic programming:

$$\begin{aligned} f_i(x_1) = opt_{y \in D}\{&u(x_1, x_2)max\{p(x_1, x_2), H_i(x_1, x_2, f_i(a(x_1, x_2)))\} \\ &+v(x_1, x_2)min\{q(x_1, x_2), H_i(x_1, x_2, f_i(b(x_1, x_2)))\} \\ &+w(x_1, x_2)[r(x_1, x_2) + H_i(x_1, x_2, f_i(c(x_1, x_2)))]\} + k(x), \\ &\forall x \in S, i \in \{1, 2, 3, 4\}, \end{aligned} \tag{9}$$

where *opt* denotes sup or inf, x_1 and X_2 denote the state and decision vectors, respectively, $a, b, c : V \times D \to V$ denote the transformations of the process, $f_i(x_1)$ denote the optimal return function with the initial state x_1 and $H_i : V \times D \times R \to R$ for $i \in \{1, 2, 3, 4\}$, $u, v, w, o, p, r : V \times D \to R$ and $J \in F(V)$. □

Theorem 2.1 is used in this section to show that the common solution to the system of functional equations (3.1) exists and is distinct from all other solutions. [13]. Let a, b, c, d be in R. Then

$$|opt\{a,b\} - opt\{c,d\}| \leq max\{|a-c|, |b-d|\}.$$

□

Theorem 3 *Assume the following facts exist:*

(a) $J \in F(V)$, and H_i, for $i \in \{1,2,3\}$ and O, P, r are bounded and u, v, w are nonnegative and $u(x_1, X_2) + v(x_1, X_2) + w(x_1, X_2) \leq 1$ for all $(x_1, X_2) \in V \times D$;

(b) $|H_1(x_1, X_2, g(t)) - H_2(x_1, X_2, h(t))|$

$$\begin{aligned}
&\leq \varphi(max\{e(E_3g, E_4h), e(E_3g, E_1g), e(E_4h, E_2h),\\
&\quad \frac{1}{2}[e(E_3g, E_2h) + e(E_4h, E_1g)], \frac{1}{2}[e(E_1g, E_2h) + e(E_3g, E_4h)],\\
&\quad \frac{1}{2}[e(E_1g, E_2h) + e(E_3g, E_1g)], \frac{1}{2}[e(E_1g, E_2h) + e(E_4h, E_2h)]\})\\
&\quad +min\{e(E_3g, E_2h), e(E_4h, E_1g), max\{[e(E_3g, E_1g), e(A_4h, E_2h)\}\}
\end{aligned}$$

for all $(x_1, X_2) \in V \times D$, $\varphi \in \phi$, $g, h \in F(V)$ & $t \in V$ and the mappings E_i are defined as follows:

$$\begin{aligned}
E_ig(x_1) = opt_{y\in D}\{&u(x_1,x_2)max\{p(x_1,x_2), H_i(x_1,x_2,g(a(x_1,x_2)))\}\\
&+v(x_1,x_2)min\{q(x_1,x_2), H_i(x_1,x_2,g(b(x_1,x_2)))\}\\
&+w(x_1,x_2)[r(x_1,x_2) + H_i(x_1,x_2,g(c(x_1,x_2)))]\} + k(x_1)
\end{aligned}$$

for all $(g, x_1) \in F(V) \times V$, $i \in \{1,2,3,4\}$;

(c) $A_1(F(V)) \subseteq E_4(F(V))$ and $E_2(F(V)) \subseteq E_3(F(V))$;

(d) For any sequence $\{h_n\}_{n\in N} \subseteq F(V)$ and $h \in F(V)$,

$$\lim_{n\to\infty} \sup_{x\in S} |h_n(x_1) - h(x_1)| = 0 \Rightarrow \lim_{n\to\infty} \sup_{x_1\in V} |E_ih_n(x) - E_ih(w)| = 0, \quad i \in \{3,4\};$$

(e) For any sequence $\{h_n\}_{n\in N} \subseteq F(V)$, if there exists $h \in F(V)$ such that

$$\lim_{n\to\infty} \sup_{x\in S} |E_ih_n(x_1) - h(x_1)| = \lim_{n\to\infty} \sup_{x_1\in V} |E_{i+2}h_n(x_1) - h(x_1)| = 0, \quad i \in \{1,2\},$$

then

$$\lim_{n\to\infty} \sup_{x\in S} |E_{i+2}E_{i+2}h_n(x) - E_iE_ih_n(x_1)| = 0, \quad i \in \{1,2\}.$$

Then, for the set of functional equations (9), *$E(S)$ has the one and only common solution.*

Proof From $(a) - (e)$, it can be deduced that E_1, E_2, E_3 and E_4 are self-mappings of $F(V)$ and E_3, E_4 are continuous and the mapping pairs E_i, $E_i + 2$ are compatible of type (P) for $i \in \{1,2\}$. Let's assume that $opt = sup$. For any $g, h \in F(V), x \in V$ and $\varepsilon > 0$, there exist $s, t \in D$, such that

$$
\begin{aligned}
E_1 g(x_1) <& u(x_1, q_2) max\{o(x_1, q_2), H_1(x_1, q_2, g(a(x_1, q_2)))\} \\
&+ t(x_1, q_2) min\{q(x_1, q_2), H_1(x_1, q_2, g(b(x_1, q_2)))\} \\
&+ s(x_1, q_2)[r(x_1, q_2) + H_1(x_1, q_2, g(c(x_1, q_2)))] + k(x_1) + \varepsilon
\end{aligned} \tag{10}
$$

and

$$
\begin{aligned}
E_2 h(x_1) <& u(x_1, q) max\{p(x_1, q), H_2(x_1, q, h(a(x_1, q)))\} \\
&+ v(x_1, q) min\{q(x_1, q), H_2(x_1, q, h(b(x_1, q)))\} \\
&+ w(x_1, q)[r(x_1, q) + H_2(x_1, q, h(c(x_1, q)))] + k(x_1) + \varepsilon.
\end{aligned} \tag{11}
$$

Obviously,

$$
\begin{aligned}
A_1 g(x_1) >& u(x_1, q) max\{p(x_1, q), H_1(x_1, q, g(a(x_1, q)))\} \\
&+ v(x_1, q) min\{p(x_1, q), H_1(x_1, q, g(b(x_1, q)))\} \\
&+ w(x_1, q)[r(x_1, q) + H_1(x_1, q, g(c(x_1, q)))] + k(x_1)
\end{aligned} \tag{12}
$$

and

$$
\begin{aligned}
E_2 h(x_1) <& u(x_1, q) max\{p(x_1, q_2), H_2(x_1, q, h(a(x_1, q_2)))\} \\
&+ v(x_1, q_2) min\{q(x_1, q), H_2(x_1, q_2, h(b(x_1, q_2)))\} \\
&+ w(x_1, q_2)[r(x_1, q_2) + H_2(x_1, q_2, h(c(x_1, q_2)))] + k(x_1).
\end{aligned} \tag{13}
$$

In light of (12), (13), (b) and Lemma 3.1, it is deduced that

$$
\begin{aligned}
& E_1 g(x_1) - E_2 h(x_1) \\
&\quad < u(x_1, q)[max\{p(x_1, q), H_1(x_1, q, g(a(x_1, q)))\} \\
&\qquad\qquad - max\{o(x_1, q), H_2(x_1, q, h(a(x_1, q)))\}] \\
&\quad + v(x_1, q)[min\{q(x_1, q), H_1(x_1, q, g(b(x_1, q)))\} \\
&\qquad\qquad - min\{p(x_1, q), H_2(x_1, q, h(b(x_1, q)))\}] \\
&\quad + w(x_1, q)[H_1(x_1, q, g(c(x_1, q))) - H_2(x_1, q, h(c(x_1, q)))] + \varepsilon
\end{aligned}
$$

$$
\begin{aligned}
&\le u(x_1, q) max\{|p(x_1, q) - p(x_1, q)|, |H_1(x_1, q, g(a(x_1, q))) - H_2(x_1, q, h(a(x_1, q)))|\} \\
&+ v(x_1, q) max\{|q(x_1, q) - p(x_1, q)|, |H_1(x_1, q, g(b(x_1, q))) - H_2(x_1, q, h(b(x_1, q)))|\} \\
&+ w(x_1, q)|H_1(x_1, q, g(c(x_1, q))) - H_2(x_1, q, h(c(x_1, q)))| + \varepsilon
\end{aligned} \tag{14}
$$

$$
\begin{aligned}
\le (u(x_1, q) + v(x_1, q) + x_1(x_1, q)) max\{&|H_1(x_1, q, g(a(x_1, q))) - H_2(x_1, q, h(a(x_1, q)))|, \\
&|H_1(x_1, q, g(b(x_1, q))) - H_2(x_1, q, h(b(x_1, q)))|, \\
&|H_1(x_1, q, g(c(x_1, q))) - H_2(x_1, q, h(c(x_1, q)))|\} + \varepsilon
\end{aligned}
$$

$$\begin{aligned}
&\le \varphi(max\{d(E_3g, E_4h), d(E_3g, E_1g), d(E_4h, E_2h),\\
&\quad \frac{1}{2}[d(E_3g, E_2h) + d(E_4h, E_1g)], \frac{1}{2}[d(E_1g, E_2h) + d(E_3g, E_4h)],\\
&\quad \frac{1}{2}[d(E_1g, E_2h) + d(E_3g, E_1g)], \frac{1}{2}[d(E_1g, E_2h) + d(E_4h, E_2h)]\})\\
&\quad +min\{d(E_3g, E_2h), d(E_4h, E_1g), max\{[d(E_3g, E_1g), d(E_4h, E_2h)\}\} + \varepsilon.
\end{aligned}$$

Similarly, (12), (13), (b) and Lemma 3.1 yield that

$$\begin{aligned}
&E_1g(x_1) - E_2h(x_1)\\
&\quad > u(x_1, q)[max\{o(x_1, q), H_1(x_1, q, g(a(x_1, q)))\}\\
&\qquad - max\{o(x_1, q), H_2(x_1, q, h(a(x_1, q)))\}]\\
&\quad +v(x_1, q)[min\{q(x_1, q), H_1(x_1, q, g(b(x_1, q)))\}\\
&\qquad - min\{p(x_1, q), H_2(x_1, q, h(b(x_1, q)))\}]\\
&\quad +w(x_1, q)[H_1(x_1, q, g(c(x_1, q))) - H_2(x_1, q, h(c(x_1, q)))] - \varepsilon\\
&\ge -p(x_1, q)max\{|o(x_1, q) - p(x_1, q)|, |H_1(x_1, q, g(a(x_1, q))) - H_2(x_1, q, h(a(x_1, q)))|\}\\
&\quad -v(x, s)max\{|p(x_1, q) - o(x_1, q)|, |H_1(x_1, q, g(b(x_1, q))) - H_2(x_1, q, h(b(x_1, q)))|\} \qquad (15)\\
&\quad -w(x_1, q)|H_1(x_1, q, g(c(x_1, q))) - H_2(x_1, q, h(c(x_1, q)))| - \varepsilon\\
&\ge -(u(x_1, q)+v(x_1, q) + x_1(x_1, q))max\{|H_1(x_1, q, g(a(x_1, q))) - H_2(x_1, q, h(a(x_1, q)))|,\\
&\qquad |H_1(x_1, q, g(b(x_1, q))) - H_2(x_1, q, h(b(x_1, q)))|,\\
&\qquad |H_1(x_1, q, g(c(x_1, q))) - H_2(x_1, q, h(c(x_1, q)))|\} - \varepsilon\\
&\ge -\varphi(max\{e(E_3g, E_4h), e(E_3g, E_1g), e(E_4h, E_2h),\\
&\quad \frac{1}{2}[e(E_3g, E_2h) + e(E_4h, E_1g)], \frac{1}{2}[d(E_1g, E_2h) + e(E_3g, E_4h)],\\
&\quad \frac{1}{2}[e(E_1g, E_2h) + e(E_3g, E_1g)], \frac{1}{2}[e(E_1g, E_2h) + e(E_4h, E_2h)]\})\\
&\quad -min\{e(E_3g, E_2h), e(E_4h, E_1g), max\{[e(E_3g, E_1g), e(E_4h, E_2h)\}\} - \varepsilon.
\end{aligned}$$

It follows from (14) and (15) that

$$
\begin{aligned}
&e(E_1g,E_2h)\\
&\quad= \sup_{x\in V} |E_1g(x_1) - E_2h(x_1)|\\
&\quad\leq \varphi(max\{e(E_3g, E_4h), e(E_3g, E_1g), e(E_4h, E_2h),\\
&\qquad \frac{1}{2}[e(E_3g, E_2h) + e(E_4h, E_1g)], \frac{1}{2}[e(E_1g, E_2h) + e(E_3g, E_4h)],\\
&\qquad \frac{1}{2}[e(E_1g, E_2h) + e(E_3g, E_1g)], \frac{1}{2}[e(E_1g, E_2h) + e(E_4h, E_2h)]\})\\
&\quad + min\{e(E_3g, E_2h), e(E_4h, E_1g), max\{e(E_3g, E_1g), e(E_4h, E_2h)\}\} + \varepsilon.
\end{aligned} \tag{16}
$$

Thus (3.8) implies that

$$
\begin{aligned}
&e(E_1g,E_2h) - min\{e(E_3g, E_2h), e(E_4h, E_1g),\\
&\qquad max\{e(E_3g, E_1g), e(E_4h, E_2h)\}\}\\
&\quad\leq \varphi(max\{e(E_3g, E_4h), e(E_3g, E_1g), e(E_4h, E_2h),\\
&\qquad \frac{1}{2}[e(E_3g, E_2h) + e(E_4h, E_1g)], \frac{1}{2}[e(E_1g, E_2h) + e(E_3g, E_4h)],\\
&\qquad \frac{1}{2}[e(E_1g, E_2h) + e(E_3g, E_1g)], \frac{1}{2}[e(E_1g, E_2h) + e(E_4h, E_2h)]\}).
\end{aligned} \tag{17}
$$

Suppose that $opt = inf$. Similarly, (17) holds also. By Theorem 2.2, E_1, E_2, E_3 & E_4 have a unique common fixed point z. Therefore the functional equations (3.1) have a unique common solution $z \in B(S)$. □

Remark 2 Theorem 3.1 extends and improves Theorem 2.4 in [10], Theorem 2.1 in [23] and Theorem 5.1 in [1].

□

Example 2 ([11]) Let $X = [2, 20]$ and d be a usual metric. Let E, F, V, U be the four self-mappings on X defined by

$$
A(x) = \begin{cases} 2 : if x \in [2, 4], \\ \frac{27}{10} : if x \in [4, 20], \end{cases} \qquad B(x) = \begin{cases} 2 \quad : if x = 2, \\ \frac{12}{5} : \text{if } x \in (0, 20], \end{cases}
$$

$$
S(x) = \begin{cases} 2 : if x \in [2, 4), \\ \frac{12}{5} : if x = 4, \\ x - \frac{1}{2} : if x \in (4, 20], \end{cases} \qquad T(x) = \begin{cases} x \quad : if x = 2, \frac{27}{10}, \\ 6 : if x \in (2, 20] - \{\frac{27}{10}\}. \end{cases}
$$

Let $\varphi : [0, \infty) \to [0, \infty)$ be a function defined by $\varphi(t) = \frac{t}{16}$ for $t \geq 0$. Then, one can easily verify that all the conditions of Theorems 2.1 and 2.2 are satisfied and 2 is the unique common fixed point of A, B, S, T.

□

References

1. Pathak, H.K., Cho, Y.J., Kang, S.M., Lee, B.S.: Fixed point theorems for compatible mappings of type (P) and applications to dynamic programming. Matematiche(Catania) **50**, 15–33 (1995)
2. Shukla, D.P., Tiwari, S.K., Shukla, S.K.: Fixed point theorems for a pair of compatible mappings in integral type equation. Int. J. Math. Sci. Eng. Appl. **7**(VI), 413–419 (2013)
3. Shukla, D.P., Tiwari, S.K., Shukla, S.K.: Unique common fixed point theorems for compatible mappings in complete metric spaces. Gen. Math. Notes **18**(1), 13–23 (2013)
4. Jungck, G.: Commuting mappings and fixed points. Am. Math. Monthly **83**, 261–263 (1976)
5. Jungck, G.: Compatible mappings and common fixed points (2). Int. J. Math. Math. Sci. **11**, 285–288 (1988)
6. Meitei, M.K., Rohen, Y., Verma, R.S.: Some common fixed point theorems for two pairs of weak compatible mappings of type (A) in G_b-metric Space. Am. J. Appl. Math. Stat. **6**(4), 135–140 (2018)
7. Zhou, M., Jain, M.K., Khan, Md.S., Secelean, N.A.: Semi-compatible mappings and common fixed point theorems of an implicit relation via inverse C-class functions. AIMS Math. **6**(3), 2636–2652 (2020)
8. Ege, O., Jain, D., Kumar, S., Park, C., Shin, D.Y.: Commuting and compatible mappings in digital metric spaces. J. Fixed Point Theory Appl. 1–16 (2020); **50**, 15–33 (1995). https://doi.org/10.1007/s11784-019-0744-5
9. Had, O.: Common fixed point theorems for a family of mappings in complete metric spaces. Math. Japon. **29**, 127–134 (1984)
10. Bhakta, P.C., Mitra, S.: Some existence theorems for functional equations arising in dynamic programming. J. Math. Anal. Appl. **98**, 348–362 (1984)
11. Murthy, P.P., Kumar, S. Kumar, R., Sahu, P., Mitrovic, Z.D., George, R.: New results for weakly compatible (WC) and R-weakly commuting (RWC) mappings with an application in dynamic programming, fractal and fractional **6,733**, 1–13 (2022). https://doi.org/10.3390/fractalfract6120733
12. Rao, R.U., Reddy, B.V.: Compatible mappings of type (K) and common fixed point of a fuzzy metric space. Adv. Theor. Appl. Math. **11**(4), 443–449 (2016)
13. Kumar, S., Vashistha, S., Anamika, Singh, B.: Compatible mapping and its variants for generalized $\psi - \phi$ weak contration condition. J. Math. Comput. Sci. **10**(6), 2686–2712 (2020)
14. Kang, S.M., Guan, H., Liu, Z., Shim, S.H.: Common coincidence point theorems in topological spaces with application in dynamic programming. Int. J. Pure Appl. Math. **15**(4), 535–546 (2004)
15. Chang, S.S.: Some existence theorems of common and coincidence solutions for a class of functional equations arising in dynamic programming. Appl. Math. Mech. **12**, 31–37 (1991)
16. Pragadeeswarar, V., Gopi, R., la Sen, M.D., Radenovic, S.: Proximally compatible mappings and common best proximity points. Symmetry **12**(353), 1–13 (2020). https://doi.org/10.3390/sym12030353
17. Sintunavarat, W., Kumam, P.: Common fixed point theorems for a pair of weakly compatible mappings in fuzzy metric spaces. J. Appl. Math. **2011**, Article ID 637958, 14 (2011). https://doi.org/10.1155/2011/637958
18. Liu, Z.: Compatible mappings and fixed points. Acta. Sci. Math. **65**, 371–383 (1999)
19. Liu, Z.: Existence theorems of solutions for certain classes of functional equations arising in dynamic programming. J. Math. Anal. Appl. **262**, 529–553 (2001)
20. Liu, Z., Kim, J.K.: A common fixed point theorem with applications in dynamic programming. Nonlinear Funct. Anal. Appl. **11**(1), 11–19 (2006)
21. Liu, Z., Ume, J.S.: On properties of solutions for a class of functional equations arising in dynamic programming. J. Optim. Theory Appl. **117**(3), 533–551 (2003)
22. Liu, Z., Ume, J.S., Kang, S.M.: Some existence theorems for functional equations arising in dynamic programming. J. Korean Math. Soc. **43**(1), 11–28 (2006)
23. Liu, Z., Kang, S.M.: Properties of solutions for certain functional equations arising in dynamic programming. J. Global Optim. **34**, 273–392 (2006)

Impulsive Fractal Sturm–Liouville Equations

Bilender P. Allahverdiev, Hüseyin Tuna, and Alireza Khalili Golmankhaneh

Abstract In this study, impulsive fractal Sturm–Liouville problems are considered. Firstly, the symmetric operator corresponding to the problem is obtained. Later, an existence theorem is proved. Finally, an eigenfunction expansion is obtained.

Keywords Fractal differential equations · Discontinuous equations · Green's functions · Eigenfunction expansions

1 Introduction

It is well known that fractal calculus represents a broad extension of the fundamental principles of ordinary calculus. Fractal sets are irregular shapes whose Hausdorff dimension exceeds their topological dimension [9]. In 2009, the notion of F^{α}-calculus on fractal subsets is defined by Parvate and Gangal [24]. In [25], the relationship between classical calculus and F^{α}-calculus is analysed. Generally, fractal calculus is often confused with fractional calculus. These two calculi differ mostly in the following ways: A F^{α}-derivative is a local operator with many properties analogous to those of the classical derivative. However, a fractional derivative is

B. P. Allahverdiev
Department of Mathematics, Khazar University, AZ1096, Baku, Azerbaijan

Research Center of Econophysics, UNEC-Azerbaijan State University of Economics, Baku, Azerbaijan

H. Tuna (✉)
Department of Mathematics, Mehmet Akif Ersoy University, Burdur, Turkey
e-mail: hustuna@gmail.com

Research Center of Econophysics, UNEC-Azerbaijan State University of Economics, Baku, Azerbaijan

A. Khalili Golmankhaneh
Department of Physics, Ur. C., Islamic Azad University, West Azerbaijan, Urmia, Iran
e-mail: alireza.khalili@iau.ac.ir

Department of Mathematics, Faculty of Sciences, Van Yuzuncu Yil University, Van, Turkey

H. Kalita et al. (eds.), *Nonlinear Analysis and Computational Techniques*,
Springer Proceedings in Mathematics & Statistics 501,
https://doi.org/10.1007/978-981-96-6038-4_13

not local. In light of these merits, a significant number of scholars are currently pursuing research on this topic (see [8–11, 25]). Furthermore, a number of earlier articles have discussed the topic of local fractional derivatives (see [15–18]). In [10], the authors examined a fractal stochastic differential equation. The existence and uniqueness theorems for both linear and nonlinear fractal differential equations are given in [11]. In a recent publication, Çetinkaya and Golmankhaneh [8] presented a study on the regular fractal Sturm–Liouville problems. In [1], Allahverdiev and Tuna studied fractal Sturm–Liouville problems. They obtained the existence and uniqueness of solutions to equations of this type. On the other hand, the Fourier method is a technique utilized in mathematics to find solutions to partial differential equations, which frequently arise in mathematical physics. Using this method, the partial differential equation is reduced to the Sturm–Liouville problem. Hence, these problems under various conditions are intensively studied (see [5, 6, 13, 20, 23]). One of these conditions is impulsive boundary conditions ([2–7, 13, 14]). These problems occur in radio science (see [22]), geophysics (see [19]), and the theory of heat and mass transfer (see [21]). Therefore, it is necessary to analyse impulsive Sturm–Liouville problems under fractal framework. For this purpose, we investigate the impulsive fractal Sturm–Liouville problems defined as

$$- D_F^{2\alpha} u(\zeta) + q(\zeta) u(\zeta) = \lambda u(\zeta), \zeta \in [0, 1], \ \zeta \in [0, \varsigma) \cup (\varsigma, 1], \tag{1}$$

$$\begin{aligned} u(0) + k_1 D_F^{\alpha} u(0) &= 0, &(2)\\ u(\varsigma-) - k_2 u(\varsigma+) &= 0, &(3)\\ D_F^{\alpha} u(\varsigma-) - k_3 D_F^{\alpha} u(\varsigma+) &= 0, &(4)\\ u(1) + k_4 D_F^{\alpha} u(1) &= 0, &(5) \end{aligned}$$

where k_1, k_2, k_3, k_4 are real numbers, $k_2 k_3 = \vartheta > 0$ and μ is a complex parameter. To the authors' knowledge, there are no studies on this subject in the literature.

2 Preliminaries

The objective of this section is to present the fundamental concepts of fractal calculus (for further details, please refer to sources [8, 9, 24, 25]). In the remainder of this paper, we will assume that F is a fractal subset of the real numbers.

Definition 1 ([24])Given a partition

$$P_J = \{c = \zeta_0 < \zeta_1 < \zeta_2 < \dots < \zeta_n = d\}$$

of $J = [c, d] \subset \mathbb{R}$, $\sigma^{\alpha}[F, P_J]$ is defined by

$$\sigma^{\alpha}[F, P_J] = \sum_{i=0}^{n-1} \Gamma(\alpha+1)(\zeta_{i+1}-\zeta_i)^{\alpha}\,\theta(F, [\zeta_i, \zeta_{i+1}]),$$

where $0 < \alpha \leq 1$ and

$$\theta(F, J) = \begin{cases} 1, & \text{if } F \cap J \neq \emptyset, \\ 0, & \text{otherwise.} \end{cases}$$

If $c = d$, $\sigma^{\alpha}[F, P_J] = 0$.

Definition 2 ([24])The coarse-grained mass $\gamma_{\delta}^{\alpha}(F, c, d)$ is defined by

$$\gamma_{\delta}^{\alpha}(F, c, d) = \inf_{\{P_J : |P_J| \leq \delta\}} \sigma^{\alpha}[F, P_J],$$

where $\delta > 0$, $c \leq d$ and

$$|P_J| = \max_{0 \leq i \leq n-1} (\zeta_{i+1} - \zeta_i)$$

for a partition P_J.

At this point, an analogue of the Hausdorff dimension is defined.

Definition 3 ([24]) The γ-dimension of $F \cap J$ is defined as

$$\dim_{\gamma}(F \cap J) = \inf\{\alpha : \gamma^{\alpha}(F, c, d) = 0\} = \sup\{\alpha : \gamma^{\alpha}(F, c, d) = \infty\},$$

where

$$\gamma^{\alpha}(F, c, d) = \lim_{\delta \to 0} \gamma_{\delta}^{\alpha}(F, c, d).$$

Definition 4 ([24]) Let us consider a function $\Pi : \mathbb{R} \to \mathbb{R}$ and a point ζ belonging to the set F, where F is a subset of $\mathbb{R}$. Then

$$A = F - \lim_{\psi \to \zeta} \Pi(\psi) \Leftrightarrow$$

given $\varepsilon > 0$, we can find $\delta > 0$ such that, if $|\psi - \zeta| < \delta$, then

$$|\Pi(\psi) - A| < \varepsilon.$$

Definition 5 ([24]) A function Π is the F-continuous at $\zeta \in F$ if

$$F - \lim_{\psi \to \zeta} \Pi(\psi) = \Pi(\zeta)$$

holds.

Definition 6 ([24]) Let $\zeta \in (\eta, \beta) \subset \mathbb{R}$. ζ is referred to as the point of change of Π if a function Π is not constant over (η, β). Let

$$Sch(\Pi) = \{\zeta : \zeta \text{ is the point of change of } \Pi\}.$$

If $Sch\left(S_F^\alpha\right)$ is closed and each of its points is a limit point, then it is referred to as an α-perfect set.

Definition 7 ([24]) Let F be an α-perfect set and

$$S_F^\alpha(\zeta) = \begin{cases} \gamma^\alpha(F, c_0, \zeta), & \text{if } \zeta \geq c_0, \\ -\gamma^\alpha(F, \zeta, c_0) & \text{otherwise,} \end{cases}$$

where $c_0 \in \mathbb{R}$ is a fixed constant. Then $D_F^\alpha \Pi$ is defined by

$$D_F^\alpha \Pi(\zeta) = \begin{cases} F - \lim \frac{\Pi(x)-\Pi(\zeta)}{S_F^\alpha(x)-S_F^\alpha(\zeta)}, & \text{if } \zeta \in F, \\ 0, & \text{otherwise,} \end{cases}$$

if the limit exists.

Theorem 1 ([24]) *Let $\Pi, \Sigma : \mathbb{R} \to \mathbb{R}$ be F^α-differentiable functions. Then we get*

$$i)\ D_F^\alpha(\Xi\Sigma)(\zeta) = \Sigma(\zeta) D_F^\alpha \Pi(\zeta) + \Pi(\zeta) D_F^\alpha \Sigma(\zeta),$$

$$ii)\ D_F^\alpha(a\Xi + b\Sigma)(\zeta) = a D_F^\alpha \Pi(\zeta) + b D_F^\alpha \Sigma(\zeta),$$

where $c, d \in \mathbb{R}$.

Theorem 2 ([24]) *The definite fractal integral of Π is defined by*

$$\Sigma(\zeta) = \int_0^\zeta \Pi(x)\, d_F^\alpha x,$$

where $\zeta \in J$, Π is a bounded function on $F \cap J$, and

$$D_F^\alpha \int_0^\zeta \Pi(t)\, d_F^\alpha t = \Pi(\zeta)\, \chi_F(\zeta).$$

Theorem 3 ([24]) *Suppose $\Omega : \mathbb{R} \to \mathbb{R}$ is a F^α-differentiable function and $Sch(\Pi) \subset F$, where F is an α-perfect. Then we have*

$$\int_c^d \Sigma(\zeta)\, d_F^\alpha \zeta = \Omega(d) - \Omega(c),$$

where $\Sigma : \mathbb{R} \to \mathbb{R}$ is F-continuous and

$$\Sigma(\zeta)\chi_F(\zeta) = D_F^\alpha \Omega(\zeta).$$

Theorem 4 ([24]) *Let* $\Pi, \Sigma \in C[c, d]$, *where* $Sch(\Pi) \subset F$. *Further* $D_F^{\alpha}\Pi$ *exists and is* F*-continuous on* J. *Then we have*

$$\int_c^d \Pi(\zeta)\,\Sigma(\zeta)\,d_F^{\alpha}\zeta + \int_c^d D_F^{\alpha}\Pi(\zeta)\int_c^{\zeta}\Sigma(\psi)\,d_F^{\alpha}y d_F^{\alpha}\zeta$$

$$= \left[\Pi(\zeta)\int_c^{\zeta}\Sigma(\psi)\,d_F^{\alpha}\psi\right]_c^d. \tag{6}$$

The F^{α}-Wronskian W^{α} of Π and Σ is given as

$$W^{\alpha}(\zeta) = \Pi(\zeta)\,D_F^{\alpha}\Sigma(\zeta) - \Sigma(\zeta)\,D_F^{\alpha}\Pi(\zeta),\ \zeta \in [0, 1]\ ([8]).$$

Let us introduce the following space:

$H = L_2^{\alpha}(0, \varsigma) \dot{+} L_2^{\alpha}(\varsigma, 1)$ is a Hilbert space endowed with the following inner product

$$\langle \Phi, \Psi \rangle_H := \int_0^{\varsigma} \Phi^{(1)}\overline{\Psi^{(1)}}d_F^{\alpha}\zeta + \vartheta\int_{\varsigma}^1 \Phi^{(2)}\overline{\Psi^{(2)}}d_F^{\alpha}\zeta,$$

where

$$\Phi(\zeta) = \begin{cases} \Phi^{(1\pi)}(\zeta), & \zeta \in [0, \varsigma) \\ \Phi^{(2)}(\zeta), & \zeta \in (\varsigma, 1], \end{cases} \quad \Psi(\zeta) = \begin{cases} \Psi^{(1)}(\zeta), & \zeta \in [0, \varsigma) \\ \Psi^{(2)}(\zeta), & \zeta \in (\varsigma, 1]. \end{cases}$$

3 Main Results

We consider the system

$$\Gamma u := -D_F^{2\alpha}u(\zeta) + q(\zeta)\,u(\zeta) = \lambda u(\zeta),\ \zeta \in [0, 1],\ \zeta \in [0, \varsigma) \cup (\varsigma, 1], \tag{7}$$

$$u(0) + k_1 D_F^{\alpha}u(0) = 0, \tag{8}$$

$$u(\varsigma-) - k_2 u(\varsigma+) = 0, \tag{9}$$

$$D_F^{\alpha}u(\varsigma-) - k_3 D_F^{\alpha}u(\varsigma+) = 0, \tag{10}$$

$$u(1) + k_4 D_F^{\alpha}u(1) = 0, \tag{11}$$

where k_1, k_2, k_3, k_4 are real numbers, $k_2k_3 = \vartheta > 0$ and μ is a complex parameter.

Our basic assumption throughout the paper is the following:

(K1) $q(.)$ is a real valued-function and has finite limits, $q(\varsigma\pm)$, and $q \in H$.

Consider the following sets

$$D_{\max} = \left\{ u \in H : \begin{array}{c} \text{one-sided limits } u(\varsigma\pm) \text{ and} \\ D_F^{\alpha} u(\varsigma\pm) \text{ exist and finite,} \\ u(\varsigma-) - k_2 u(\varsigma+) = 0, \\ D_F^{\alpha} u(\varsigma-) - k_3 D_F^{\alpha} u(\varsigma+) = 0, \\ \text{and } \Gamma u \in H. \end{array} \right\},$$

$$D_{\min} = \left\{ u \in D_{\max} : u(0) = D_F^{\alpha} u(0) = u(1) = D_F^{\alpha} u(1) = 0 \right\}.$$

The maximal operator $\mathcal{L}_{\max}$ on $D_{\max}$ is therefore defined as follows:

$$\mathcal{L}_{\max} u = \Gamma u.$$

If we restrict $\mathcal{L}_{\max}$ to $D_{\min}$, then we get the *minimal operator* $\mathcal{L}_{\min}$.

Let $u, v \in D_{\max}$. Then Green's formula is given as

$$\int_0^1 \left[(\Gamma u)(\zeta)\overline{v(\zeta)} - u(\zeta)\overline{(\Gamma v)(\zeta)} \right] d_F^{\alpha} \zeta$$
$$= [u, v](1) - [u, v](\varsigma+) + [u, v](\varsigma-) - [u, v](0), \tag{12}$$

where

$$[u, v] := u(\overline{D_F^{\alpha} v}) - (D_F^{\alpha} u)\overline{v}.$$

Let

$$\mathcal{D} = \left\{ u \in D_{\max} : \begin{array}{c} u(0) + k_1 D_F^{\alpha} u(0) = 0, \\ u(\varsigma-) - k_2 u(\varsigma+) = 0, \\ D_F^{\alpha} u(\varsigma-) - k_3 D_F^{\alpha} u(\varsigma+) = 0, \\ u(1) + k_4 D_F^{\alpha} u(1) = 0. \end{array} \right\}.$$

Let us define the operator $\mathcal{L}$ with

$$\mathcal{L} u = \Gamma u,$$

where $u \in \mathcal{D}$

Theorem 5 *Operator $\mathcal{L}$ is symmetric.*

Corollary 1 *All eigenvalues of Problem (7)–(11) are real.*

Corollary 2 *If ξ_1 and ξ_2 are two different eigenvalues of Problem (7)–(11), then the corresponding eigenfunctions u_1 and u_2 are orthogonal.*

Theorem 6 *For any $\mu \in \mathbb{C}$, Eq. (7) has a solution $\varphi(\zeta, \mu)$ satisfying Conditions (8)–(10).*

Proof. It follows from [1] that the following fractal boundary-value problem

$$-D_F^{2\alpha} u(\zeta) + q(\zeta) u(\zeta) = \lambda u(\zeta), \ \zeta \in [0, \varsigma),$$

$$u(0) = -k_1,$$

$$D_F^{\alpha} u(0) = 1,$$

has a unique solution $\varphi_1(\zeta, \mu)$.

Consider the following fractal problem

$$-D_F^{2\alpha} u(\zeta) + q(\zeta) u(\zeta) = \lambda u(\zeta), \ \zeta \in (\varsigma, 1], \tag{13}$$

$$u(\varsigma+) = \frac{1}{k_2} \varphi_1(\varsigma-, \mu), \tag{14}$$

$$D_F^{\alpha} u(\varsigma+) = \frac{1}{k_3} D_F^{\alpha} \varphi_1(\varsigma-, \mu). \tag{15}$$

Let

$$U_n(\zeta, \mu) = U_0(\zeta, \mu)$$

$$+ \int_{\varsigma}^{\zeta} \left(\frac{\sin\left(\sqrt{\mu} S_F^{\alpha}(\zeta)\right)}{\sqrt{\mu}} \cos(\sqrt{\mu} S_F^{\alpha}(\gamma)) - \cos(\sqrt{\mu} S_F^{\alpha}(\zeta)) \frac{\sin\left(\sqrt{\mu} S_F^{\alpha}(\gamma)\right)}{\sqrt{\mu}} \right) \times$$

$$\times q(\gamma) U_{n-1}(\gamma, \mu) \, d_F^{\alpha} \gamma. \tag{16}$$

$$U_0(\zeta, \mu) = \frac{1}{k_2} \varphi_1(\varsigma-, \mu) + \frac{1}{k_3} (\zeta - \varsigma) D_F^{\alpha} \varphi_1(\varsigma-, \mu), \ \zeta \in (\varsigma, 1],$$

and the functions $\frac{\sin(\sqrt{\mu} S_F^{\alpha}(\zeta))}{\sqrt{\mu}}$, $\cos(\sqrt{\mu} S_F^{\alpha}(\zeta))$ are the fundamental solutions of

$$-D_F^{2\alpha} u(\zeta) = \lambda u(\zeta). \tag{17}$$

By induction, the basic estimate

$$|U_{n+1}(\zeta, \mu) - U_n(\zeta, \mu)| \leq \widetilde{\sigma(\mu)} \frac{\left(A \sigma(\mu) S_F^{\alpha}(\zeta)\right)^n}{n!} \tag{18}$$

is proven. For $n = 1$,

$$|U_1(\zeta,\mu) - U_0(\zeta,\mu)|$$

$$\leq \left| \int_{\varsigma}^{\zeta} \begin{pmatrix} \frac{\sin(\sqrt{\mu} S_F^{\alpha}(\zeta))}{\sqrt{\mu}} \cos(\sqrt{\mu} S_F^{\alpha}(\gamma)) \\ -\cos(\sqrt{\mu} S_F^{\alpha}(\zeta)) \frac{\sin(\sqrt{\mu} S_F^{\alpha}(\gamma))}{\sqrt{\mu}} \end{pmatrix} q(\gamma) U_0(\gamma,\mu)\, d_F^{\alpha}\gamma \right|$$

$$\leq \sigma(\mu)\, A \widetilde{\sigma(\mu)} \left| \int_0^{\zeta} \chi_F(\gamma)\, d_F^{\alpha}\gamma \right| = \sigma(\mu)\, A \widetilde{\sigma(\mu)} S_F^{\alpha}(\zeta),$$

where $\mu \in \mathbb{C}$; $\sigma(\mu)$, $\widetilde{\sigma(\mu)}$ and A are positive numbers such that

$$\left| \begin{pmatrix} \frac{\sin(\sqrt{\mu} S_F^{\alpha}(\zeta))}{\sqrt{\mu}} \cos(\sqrt{\mu} S_F^{\alpha}(\gamma)) \\ -\cos(\sqrt{\mu} S_F^{\alpha}(\zeta)) \frac{\sin(\sqrt{\mu} S_F^{\alpha}(\gamma))}{\sqrt{\mu}} \end{pmatrix} \right| \leq \sigma(\mu),$$

$$\max_{\zeta \in (\varsigma, 1]} |q(\zeta)| = A, \; |U_0(\zeta,\mu)| \leq \widetilde{\sigma(\mu)}, \;\; \zeta \in (\varsigma, 1].$$

Based on the induction hypothesis

$$|U_n(\zeta,\mu) - U_{n-1}(\zeta,\mu)|$$

$$\leq \left| \begin{matrix} \int_{\varsigma}^{\zeta} \begin{pmatrix} \frac{\sin(\sqrt{\mu} S_F^{\alpha}(\zeta))}{\sqrt{\mu}} \cos(\sqrt{\mu} S_F^{\alpha}(\gamma)) \\ -\cos(\sqrt{\mu} S_F^{\alpha}(\zeta)) \frac{\sin(\sqrt{\mu} S_F^{\alpha}(\gamma))}{\sqrt{\mu}} \end{pmatrix} \times \\ \times q(\gamma) \left[U_{n-1}(\gamma,\mu) - U_{n-2}(\gamma,\mu) \right] d_F^{\alpha}\gamma \end{matrix} \right|$$

$$\leq \widetilde{\sigma(\mu)} \frac{\left(A\sigma(\mu) \left[S_F^{\alpha}(\zeta) \right] \right)^{n-1}}{(n-1)!}, \;\; n = 1, 2, \ldots.$$

The rest of the argument follows a simple pattern.

$$U_1(\zeta,\mu) + \sum_{n=1}^{\infty} \{ U_{n+1}(\zeta,\mu) - U_n(\zeta,\mu) \} \tag{19}$$

is uniformly convergent with respect to variable ζ on $(\varsigma, 1]$, due to the series

$$\sum_{n=1}^{\infty} \widetilde{\sigma(\mu)} \frac{\left(A\sigma(\mu) \left[S_F^{\alpha}(\zeta) \right] \right)^{n}}{n!}$$

is convergent.

Let

$$\varphi_2(\zeta,\mu) = U_1(\zeta,\mu) + \sum_{n=1}^{\infty}\left\{U_{n+1}(\zeta,\mu) - U_n(\zeta,\mu)\right\}.$$

Then we obtain

$$\lim_{n\to\infty} U_n(\zeta,\mu) = \varphi_2(\zeta,\mu).$$

It follows from (16) that

$$D_F^{\alpha}U_{n+1}(\zeta,\mu) - D_F^{\alpha}U_n(\zeta,\mu)$$

$$= \int_{\varsigma}^{\zeta}\begin{pmatrix} D_F^{\alpha}\frac{\sin(\sqrt{\mu}S_F^{\alpha}(\zeta))}{\sqrt{\mu}}\cos(\sqrt{\mu}S_F^{\alpha}(\gamma)) \\ -D_F^{\alpha}\cos(\sqrt{\mu}S_F^{\alpha}(\zeta))\frac{\sin(\sqrt{\mu}S_F^{\alpha}(\gamma))}{\sqrt{\mu}}\end{pmatrix}\times$$

$$\times q(\gamma)\begin{bmatrix} U_n(\gamma,\mu) \\ -U_{n-1}(\gamma,\mu)\end{bmatrix}d_F^{\alpha}\gamma,$$

and

$$-D_F^{2\alpha}U_{n+1}(\zeta,\mu) + -D_F^{2\alpha}U_n(\zeta,\mu)$$

$$= \int_{\varsigma}^{\zeta}\begin{pmatrix} -D_F^{2\alpha}\frac{\sin(\sqrt{\mu}S_F^{\alpha}(\zeta))}{\sqrt{\mu}}\cos(\sqrt{\mu}S_F^{\alpha}(\gamma)) \\ +D_F^{2\alpha}\cos(\sqrt{\mu}S_F^{\alpha}(\zeta))\frac{\sin(\sqrt{\mu}S_F^{\alpha}(\gamma))}{\sqrt{\mu}}\end{pmatrix}\times$$

$$\times q(\gamma)\begin{bmatrix} U_n(\gamma,\mu) \\ -U_{n-1}(\gamma,\mu)\end{bmatrix}d_F^{\alpha}\gamma$$

$$-q(\zeta)\left[U_n(\zeta,\mu) - U_{n-1}(\zeta,\mu)\right].$$

By (18), we deduce that, for every $\mu \in \mathbb{C}$,

$$\sum_{n=1}^{\infty}\left(D_F^{\alpha}U_{n+1}(\zeta,\mu) - D_F^{\alpha}U_n(\zeta,\mu)\right)$$

and

$$\sum_{n=1}^{\infty}\left(-D_F^{2\alpha}U_{n+1}(\zeta,\mu) + D_F^{2\alpha}U_n(\zeta,\mu)\right)$$

are uniformly convergent series on $(\varsigma, 1]$ with respect to variable ζ. By (17), we see that

$$-D_F^{2\alpha}\varphi_2(\zeta,\mu)$$

$$= \sum_{n=1}^{\infty} \left(-D_F^{2\alpha} U_{n+1}(\zeta, \mu) + D_F^{2\alpha} U_n(\zeta, \mu) \right)$$

$$= (\mu - q(\zeta)) \sum_{n=1}^{\infty} \left(U_n(\zeta, \mu) - U_{n-1}(\zeta, \mu) \right) = (\mu - q(\zeta)) \varphi_2(\zeta, \mu).$$

It is evident that φ_2 satisfies (14)–(15). Thus, we deduce that

$$\varphi(\zeta, \mu) = \begin{cases} \varphi_1(\zeta, \mu), & \zeta \in [0, \varsigma) \\ \varphi_2(\zeta, \mu), & \zeta \in (\varsigma, 1] \end{cases} \tag{20}$$

satisfies Problem (7)–(10).

Similarly, we can derive the subsequent theorem.

Theorem 7 *For any $\mu \in \mathbb{C}$, Eq. (7) has a solution*

$$\chi(\zeta, \mu) = \begin{cases} \chi_1(\zeta, \mu), & \zeta \in [0, \varsigma) \\ \chi_2(\zeta, \mu), & \zeta \in (\varsigma, 1] \end{cases} \tag{21}$$

satisfying Conditions (9)–(11).

Now, we can define the following functions

$$\omega_1(\mu) = W^\alpha(\varphi_1, \chi_1)(\zeta), \omega_2(\mu) = W^\alpha(\varphi_2, \chi_2)(\zeta),$$

due to these Wronskians are independent of ζ for $\zeta \in [0, \varsigma)$ and $\zeta \in (\varsigma, 1]$, respectively. By (9)–(10), we get

$$\omega_1(\mu) = \vartheta \omega_2(\mu).$$

Hence, the *characteristic function* of Problem (7)–(11) is given by the formula

$$\omega(\mu) := \omega_1(\mu) = \vartheta \omega_2(\mu).$$

Lemma 1 *Let*

$$\Delta(\mu) := \begin{vmatrix} \Upsilon_1\varphi_1 & \Upsilon_1\chi_1 & \Upsilon_1\varphi_2 & \Upsilon_1\chi_2 \\ \Upsilon_2\varphi_1 & \Upsilon_2\chi_1 & \Upsilon_2\varphi_2 & \Upsilon_2\chi_2 \\ \Upsilon_3\varphi_1 & \Upsilon_3\chi_1 & \Upsilon_3\varphi_2 & \Upsilon_3\chi_2 \\ \Upsilon_4\varphi_1 & \Upsilon_4\chi_1 & \Upsilon_4\varphi_2 & \Upsilon_4\chi_2 \end{vmatrix},$$

where

$$\Upsilon_1 u := u(0) + k_1 D_F^\alpha u(0),$$
$$\Upsilon_2 u := u(1) + k_4 D_F^\alpha u(1),$$
$$\Upsilon_3 u := u(\varsigma-) - k_2 u(\varsigma+),$$

$$\Upsilon_4 u := D_F^\alpha u(\varsigma-) - k_3 D_F^\alpha u(\varsigma+).$$

Then, for every $\mu \in \mathbb{C}$, *we obtain*

$$\Delta(\mu) = -\frac{1}{\vartheta}\omega^3(\mu).$$

Proof. It follows from (20) and (21) that

$$\Delta(\mu)$$

$$= \begin{vmatrix} 0 & \omega_1(\mu) & 0 & 0 \\ 0 & 0 & -\omega_2(\mu) & 0 \\ \varphi_1(\varsigma-,\mu) & \chi_1(\varsigma-,\mu) & -k_2\varphi_2(\varsigma+,\mu) & -k_2\chi_2(\varsigma+,\mu) \\ D_F^\alpha\varphi_1(\varsigma-,\mu) & D_F^\alpha\chi_1(\varsigma-,\mu) & -k_3 D_F^\alpha\varphi_2(\varsigma+,\mu) & -k_3 D_F^\alpha\chi_2(\varsigma+,\mu) \end{vmatrix}$$

$$= \omega_1(\mu)\begin{vmatrix} 0 & -\omega_2(\mu) & 0 \\ \varphi_1(\varsigma-,\mu) & -k_2\varphi_2(\varsigma+,\mu) & -k_2\chi_2(\varsigma+,\mu) \\ D_F^\alpha\varphi_1(\varsigma-,\mu) & -k_3 D_F^\alpha\varphi_2(\varsigma+,\mu) & -k_3 D_F^\alpha\chi_2(\varsigma+,\mu) \end{vmatrix}$$

$$= \omega_1(\mu)\,\omega_2(\mu)\begin{vmatrix} \varphi_1(\varsigma-,\mu) & -k_2\chi_2(\varsigma+,\mu) \\ D_F^\alpha\varphi_1(\varsigma-,\mu) & -k_3 D_F^\alpha\chi_2(\varsigma+,\mu) \end{vmatrix}$$

$$= -\omega_1(\mu)\,\omega_2(\mu)\begin{vmatrix} \varphi_1(\varsigma-,\mu) & \chi_1(\varsigma-,\mu) \\ D_F^\alpha\varphi_1(\varsigma-,\mu) & D_F^\alpha\chi_1(\varsigma-,\mu) \end{vmatrix}$$

$$= -\omega_1^2(\mu)\,\omega_2(\mu) = -\frac{1}{k_2k_3}\omega^3(\mu)$$

Theorem 8 *The eigenvalues of (7)–(11) same as the zeros of* $\omega(\mu)$.

Proof. Let $\mu^{(0)}$ be a zero of $\omega(\mu)$. Then $\omega_2\left(\mu^{(0)}\right) = W^\alpha(\varphi_2, \chi_2) = 0$, i.e. $\varphi_2 = \xi\chi_2$ for some $\xi \neq 0$. Thus φ_2 satisfies (11). *Therefore the function*

$$\varphi\left(\zeta,\mu^{(0)}\right) = \begin{cases} \varphi_1\left(\zeta,\mu^{(0)}\right), & \zeta \in [0,\varsigma) \\ \varphi_2\left(\zeta,\mu^{(0)}\right), & \zeta \in (\varsigma,1] \end{cases}$$

satisfies (7)–(11), *i.e.* $\mu^{(0)}$ is an eigenvalue.

Let $\mu^{(0)}$ be an eigenvalue and $\eta\left(\zeta,\mu^{(0)}\right)$ be any corresponding eigenfunction. We want to show that $\omega\left(\mu^{(0)}\right) = 0$. Assume that $\omega\left(\mu^{(0)}\right) \neq 0$. Then we see that $\omega_1\left(\mu^{(0)}\right) \neq 0$ and $\omega_2\left(\mu^{(0)}\right) \neq 0$. Thus there exist constants $\xi_i,\ i = 1, 2, 3, 4,$ at least one of which is not zero, such that

$$\eta\left(\zeta,\mu^{(0)}\right) = \begin{cases} \xi_1\varphi_1\left(\zeta,\mu^{(0)}\right) + \xi_2\chi_1\left(\zeta,\mu^{(0)}\right), & \zeta \in [0,\varsigma) \\ \xi_3\varphi_2\left(\zeta,\mu^{(0)}\right) + \xi_4\chi_2\left(\zeta,\mu^{(0)}\right), & \zeta \in (\varsigma,1]. \end{cases}$$

Therefore,

$$\Upsilon_i \eta\left(\zeta, \mu^{(0)}\right) = 0, \ i = 1, 2, 3, 4,$$

since $\eta\left(\zeta, \mu^{(0)}\right)$ is the eigenfunction. Hence, we find

$$\det\left(\Upsilon_i \eta\left(\zeta, \mu^{(0)}\right)\right) = \Delta(\mu) = 0,$$

due to at least one of the constants $\zeta_i, \ i = 1, 2, 3, 4$ is not zero. On the other hand, by Lemma 1, we conclude that $\Delta(\mu) \neq 0$, a contradiction.

Let us consider the following problem

$$\begin{aligned} &- D_F^{2\alpha} u(\zeta) + \{-\mu + q(\zeta)\} u(\zeta) \\ &= \Phi(\zeta), \ \zeta \in [0, \varsigma) \cup (\varsigma, 1], \ \mu \in \mathbb{C}, \ \Phi \in H, \end{aligned} \tag{22}$$

which satisfies (8)–(11).

Using the methods of variation of the constants, the general solution of (22) can be obtained by

$$\eta(\zeta, \mu) = \begin{cases} \xi_1(\zeta, \mu)\varphi_1(\zeta, \mu) + \xi_2(\zeta, \mu)\chi_1(\zeta, \mu), \zeta \in [0, \varsigma) \\ \xi_3(\zeta, \mu)\varphi_2(\zeta, \mu) + \xi_4(\zeta, \mu)\chi_2(\zeta, \mu), \zeta \in (\varsigma, 1], \end{cases}$$

where

$$D_F^{\alpha}\xi_1(\zeta, \mu) = \frac{1}{\omega(\mu)}\Phi(\zeta)\chi_1(\zeta, \mu), \ \zeta \in [0, \varsigma), \tag{23}$$

$$D_F^{\alpha}\xi_2(\zeta, \mu) = -\frac{1}{\omega(\mu)}\Phi(\zeta)\varphi_1(\zeta, \mu), \ \zeta \in [0, \varsigma), \tag{24}$$

$$D_F^{\alpha}\xi_3(\zeta, \mu) = \frac{1}{\omega(\mu)}\Phi(\zeta)\chi_2(\zeta, \mu), \ \zeta \in (\varsigma, 1], \tag{25}$$

$$D_F^{\alpha}\xi_4(\zeta, \mu) = -\frac{1}{\omega(\mu)}\Phi(\zeta)\varphi_2(\zeta, \mu), \ \zeta \in (\varsigma, 1]. \tag{26}$$

By (23)–(26), we find

$$\xi_1(\zeta, \mu) = \frac{1}{\omega(\mu)}\int_{\zeta}^{\varsigma} \Phi(\gamma)\chi_1(\gamma, \mu)\, d_F^{\alpha}\gamma + \xi_1, \ \zeta \in [0, \varsigma),$$

$$\xi_2(\zeta, \mu) = \frac{1}{\omega(\mu)}\int_{0}^{\zeta} \Phi(\gamma)\varphi_1(\gamma, \mu)\, d_F^{\alpha}\gamma + \xi_2, \ \zeta \in [0, \varsigma),$$

$$\xi_3(\zeta, \mu) = \frac{1}{\omega(\mu)}\int_{\zeta}^{1} \Phi(\gamma)\chi_2(\gamma, \mu)\, d_F^{\alpha}\gamma + \xi_3, \ \zeta \in (\varsigma, 1],$$

$$\xi_4(\zeta,\mu)=\frac{1}{\omega(\mu)}\int_{\varsigma}^{\zeta}\Phi(\gamma)\chi_2(\gamma,\mu)\,d_F^{\alpha}\gamma+\xi_4,\ \zeta\in(\varsigma,1],$$

where $\zeta_i\ \ (i=1,2,3,4)$ is an arbitrary constant. Hence we see that

$$\eta(\zeta,\mu)=\begin{cases}\xi_1\varphi_1(\zeta,\mu)+\xi_2\chi_1(\zeta,\mu)\\ +\frac{1}{\omega(\mu)}\chi_1(\zeta,\mu)\int_0^{\zeta}\Phi(\gamma)\varphi_1(\gamma,\mu)\,d_F^{\alpha}\gamma\\ +\frac{1}{\omega(\mu)}\varphi_1(\zeta,\mu)\int_{\zeta}^{\varsigma}\Phi(\gamma)\chi_1(\gamma,\mu)\,d_F^{\alpha}\gamma,\ \zeta\in[0,\varsigma)\\ \xi_3\varphi_2(\zeta,\mu)+\xi_4\chi_2(\zeta,\mu)\\ +\frac{1}{\omega(\mu)}\varphi_2(\zeta,\mu)\int_{\zeta}^{1}\Phi(\gamma)\chi_2(\gamma,\mu)\,d_F^{\alpha}\gamma\\ +\frac{1}{\omega(\mu)}\chi_2(\zeta,\mu)\int_{\varsigma}^{\zeta}\Phi(\gamma)\varphi_2(\gamma,\mu)\,d_F^{\alpha}\gamma,\ \zeta\in(\varsigma,1],\end{cases}\tag{27}$$

where $\zeta_i\ \ (i=1,2,3,4)$ is an arbitrary constant. By (27), we deduce that

$$D_F^{\alpha}\eta(\zeta,\mu)=\begin{cases}\xi_1 D_F^{\alpha}\varphi_1(\zeta,\mu)+\xi_2 D_F^{\alpha}\chi_1(\zeta,\mu)\\ +\frac{1}{\omega(\mu)}D_F^{\alpha}\chi_1(\zeta,\mu)\int_0^{\zeta}\Phi(\gamma)\varphi_1(\gamma,\mu)\,d_F^{\alpha}\gamma\\ +\frac{1}{\omega(\mu)}D_F^{\alpha}\varphi_1(\zeta,\mu)\int_{\zeta}^{\varsigma}\Phi(\gamma)\chi_1(\gamma,\mu)\,d_F^{\alpha}\gamma,\ \zeta\in[0,\varsigma)\\ \xi_3 D_F^{\alpha}\varphi_2(\zeta,\mu)+\xi_4 D_F^{\alpha}\chi_2(\zeta,\mu)\\ +\frac{1}{\omega(\mu)}D_F^{\alpha}\varphi_2(\zeta,\mu)\int_{\zeta}^{1}\Phi(\gamma)\chi_2(\gamma,\mu)\,d_F^{\alpha}\gamma\\ +\frac{1}{\omega(\mu)}D_F^{\alpha}\chi_2(\zeta,\mu)\int_{\varsigma}^{\zeta}\Phi(\gamma)\varphi_2(\gamma,\mu)\,d_F^{\alpha}\gamma,\ \zeta\in(\varsigma,1].\end{cases}$$

Hence

$$\Upsilon_1\eta=\eta(0)+k_1D_F^{\alpha}\eta(0)=\xi_1\left[\varphi_1(0,\mu)+k_1D_F^{\alpha}\varphi_1(0,\mu)\right]$$
$$+\xi_2\left[\chi_1(0,\mu)+k_1\chi_1D_F^{\alpha}(0,\mu)\right]$$
$$+\frac{1}{\omega(\mu)}\left[\varphi_1(0,\mu)+k_1D_F^{\alpha}\varphi_1(0,\mu)\right]\int_0^{\varsigma}\Phi(\gamma)\chi_1(\gamma,\mu)\,d_F^{\alpha}\gamma,$$

Since

$$\varphi_1(0,\mu)+k_1D_F^{\alpha}\varphi_1(0,\mu)=0$$

and

$$\chi_1(0,\mu)+k_1\chi_1 D_F^{\alpha}(0,\mu)=\omega(\mu)\neq 0$$

we infer that $\xi_2 = 0$. Likewise, we obtain

$$\Upsilon_2\eta = \eta(1)+k_4 D_F^{\alpha}\eta(1)$$

$$=\xi_3\left[\varphi_2(1,\mu)+k_4 D_F^{\alpha}\varphi_2(1,\mu)\right]$$

$$+\xi_4\left[\chi_2(1,\mu)+k_4 D_F^{\alpha}\chi_2(1,\mu)\right]$$

$$+\frac{1}{\omega(\mu)}\left[\chi_2(1,\mu)+k_4 D_F^{\alpha}\chi_2(1,\mu)\right]\int_{\varsigma}^{1}\Phi(\gamma)\varphi_2(\gamma,\mu)\,d_F^{\alpha}\gamma$$

Since

$$\chi_2(1,\mu)+k_4 D_F^{\alpha}\chi_2(1,\mu)=0$$

and

$$\varphi_2(1,\mu)+k_4 D_F^{\alpha}\varphi_2(1,\mu)=\omega(\mu)\neq 0,$$

we see that $\xi_3 = 0$. Similarly, we obtain

$$\Upsilon_3\eta = \eta(\varsigma-)-k_2\eta(\varsigma+)$$

$$=\xi_1\varphi_1(\varsigma-,\mu)-k_2\xi_4\chi_2(\varsigma+,\mu)$$

$$+\frac{1}{\omega(\mu)}\chi_1(\varsigma-,\mu)\int_0^{\varsigma}\Phi(\gamma)\varphi_1(\gamma,\mu)\,d_F^{\alpha}\gamma$$

$$-k_2\frac{1}{\omega(\mu)}\varphi_2(\varsigma+,\mu)\int_{\varsigma}^{1}\Phi(\gamma)\chi_2(\gamma,\mu)\,d_F^{\alpha}\gamma$$

and

$$\Upsilon_4\eta = D_F^{\alpha}\eta(\varsigma-)-k_3 D_F^{\alpha}\eta(\varsigma+)=\xi_1 D_F^{\alpha}\varphi_1(\varsigma-,\mu)$$

$$+\frac{1}{\omega(\mu)}D_F^{\alpha}\chi_1(\varsigma-,\mu)\int_0^{\varsigma}\Phi(\gamma)\varphi_1(\gamma,\mu)\,d_F^{\alpha}\gamma-k_3\xi_4 D_F^{\alpha}\chi_2(\varsigma+,\mu)$$

$$-k_3\frac{1}{\omega(\mu)}D_F^{\alpha}\varphi_2(\varsigma+,\mu)\int_{\varsigma}^{1}\Phi(\gamma)\chi_2(\gamma,\mu)\,d_F^{\alpha}\gamma.$$

It follows from (9) and (10) that

$$\begin{cases} \xi_1\varphi_1(\varsigma-,\mu) - k_2\xi_4\chi_2(\varsigma+,\mu) \\ = k_2\frac{1}{\omega(\mu)}\varphi_2(\varsigma+,\mu)\int_\varsigma^1 \Phi(\gamma)\chi_2(\gamma,\mu)\,d_F^\alpha\gamma \\ -\frac{1}{\omega(\mu)}\chi_1(\varsigma-,\mu)\int_0^\varsigma \Phi(\gamma)\varphi_1(\gamma,\mu)\,d_F^\alpha\gamma \\ \xi_1 D_F^\alpha\varphi_1(\varsigma-,\mu) - k_3\xi_4 D_F^\alpha\chi_2(\varsigma+,\mu) \\ = k_3\frac{1}{\omega(\mu)}D_F^\alpha\varphi_2(\varsigma+,\mu)\int_\varsigma^1 \Phi(\gamma)\chi_2(\gamma,\mu)\,d_F^\alpha\gamma \\ -\frac{q1}{\omega(\mu)}D_F^\alpha\chi_1(\varsigma-,\mu)\int_0^\varsigma \Phi(\gamma)\varphi_1(\gamma,\mu)\,d_F^\alpha\gamma. \end{cases} \tag{28}$$

By (28), we get

$$\xi_1 = \frac{1}{\omega(\mu)}\int_\varsigma^1 \Phi(\gamma)\chi_2(\gamma,\mu)\,d_F^\alpha\gamma$$

and

$$\xi_4 = \frac{1}{\omega(\mu)}\int_0^\varsigma \Phi(\gamma)\varphi_1(\gamma,\mu)\,d_F^\alpha\gamma.$$

Finally, we obtain

$$\eta(\zeta,\mu) = \frac{1}{\omega(\mu)}\chi(\zeta,\mu)\int_0^\zeta \Phi(\gamma)\varphi(\gamma,\mu)\,d_F^\alpha\gamma$$

$$+\frac{1}{\omega(\mu)}\varphi(\zeta,\mu)\int_\zeta^1 \Phi(\gamma)\chi(\gamma,\mu)\,d_F^\alpha\gamma,$$

i.e.

$$\eta(\zeta,\mu) = \int_0^1 G(\zeta,\gamma,\mu)\,\Phi(\gamma)d_F^\alpha\gamma,$$

where

$$G(\zeta,\gamma,\mu) = \begin{cases} \frac{1}{\omega(\mu)}\chi(\zeta,\mu)\varphi(\gamma,\mu), 0 \le \gamma \le \zeta \le 1, \zeta \ne \varsigma,\ \gamma \ne \varsigma, \\ \frac{1}{\omega(\mu)}\chi(\gamma,\mu)\varphi(\zeta,\mu), 0 \le \zeta \le \gamma \le 1, \zeta \ne \varsigma,\ \gamma \ne \varsigma. \end{cases} \tag{29}$$

Theorem 9 *Suppose that $\mu = 0$ is not an eigenvalue of (7)–(11). Then we have*

$$\int_0^\varsigma\int_0^\varsigma |G(\zeta,\gamma)|^2 d_F^\alpha\zeta\, d_F^\alpha\gamma < +\infty,$$

and

$$\int_1^\varsigma\int_1^\varsigma |G(\zeta,\gamma)|^2 d_F^\alpha\zeta\, d_F^\alpha\gamma < +\infty,$$

where $G(\zeta,\gamma)$ $(\mu=0)$ *defined as (29).*

Proof. From (29), we have

$$\int_0^{\varsigma} d_F^{\alpha}\zeta \int_0^{\varsigma} |G(\zeta,\gamma)|^2 d_F^{\alpha}\gamma < +\infty,$$

$$\int_{\varsigma}^{1} d_F^{\alpha}\zeta \int_{\varsigma}^{1} |G(\zeta,\gamma)|^2 d_F^{\alpha}\gamma < +\infty,$$

due to $\chi\,(.,\mu)\,,\varphi\,(.,\mu)\in H$. Consequently, we obtain

$$\int_0^{\varsigma}\int_0^{\varsigma} |G(\zeta,\gamma)|^2 d_F^{\alpha}\zeta\, d_F^{\alpha}\gamma < +\infty,$$

and

$$\int_1^{\varsigma}\int_1^{\varsigma} |G(\zeta,\gamma)|^2 d_F^{\alpha}\zeta\, d_F^{\alpha}\gamma < +\infty. \tag{30}$$

Theorem 10 ([23]) *Let*

$$A\,\{t_i\}=\{x_i\}\,,\, i\in\mathbb{N}:=\{1,2,3,\ldots\},$$

where

$$x_i=\sum_{k=1}^{\infty}\eta_{ik}t_k,\; i,k\in\mathbb{N}. \tag{31}$$

If

$$\sum_{i,k=1}^{\infty}|\eta_{ik}|^2<+\infty, \tag{32}$$

then A *is compact in* l^2.

Theorem 11 *Let* $\mathcal{T}\colon H\to H$ *be the integral operator defined as*

$$(\mathcal{T}\Phi)(\zeta)=\begin{cases}\int_0^{\varsigma} G\,(\zeta,\gamma)\,\Phi^{(1)}(\gamma)d_F^{\alpha}\gamma,\; \zeta\in[0,\varsigma)\\[2ex] \int_{\varsigma}^{1} G\,(\zeta,\gamma)\,\Phi^{(2)}(\gamma)d_F^{\alpha}\gamma,\; \zeta\in(\varsigma,1],\end{cases}$$

where

$$\Phi(\zeta)=\begin{cases}\Phi^{(1)}(\zeta),\;\zeta\in[0,\varsigma)\\ \Phi^{(2)}(\zeta),\;\zeta\in(\varsigma,1].\end{cases}$$

Then $\mathcal{T}$ *is a self-adjoint and compact operator in* H.

Proof. Let

$$\phi_i = \phi_i\,(\zeta) = \begin{cases} \phi_i^{(1)}(\zeta),\ \zeta \in [0, \varsigma) \\ \phi_i^{(2)}(\zeta),\ \zeta \in (\varsigma, 1] \end{cases} \quad (i \in \mathbb{N})$$

be a complete, orthonormal basis of H. Let $i, k \in \mathbb{N}$. Set

$$t_i = \langle \Phi, \phi_i \rangle_H = \int_0^{\varsigma} \Phi^{(1)}\,(\zeta)\,\overline{\phi_i^{(1)}\,(\zeta)} d_F^{\alpha}\zeta$$

$$+\,\vartheta \int_{\varsigma}^{1} \Phi^{(2)}\,(\zeta)\,\overline{\phi_i^{(2)}\,(\zeta)} d_F^{\alpha}\zeta,$$

$$x_i = \langle \Psi, \phi_i \rangle_H = \int_0^{\varsigma} \Psi^{(1)}\,(\zeta)\,\overline{\phi_i^{(1)}\,(\zeta)} d_F^{\alpha}\zeta$$

$$+\,\vartheta \int_{\varsigma}^{1} \Psi^{(2)}\,(\zeta)\,\overline{\phi_i^{(2)}\,(\zeta)} d_F^{\alpha}\zeta,$$

$$\eta_{ik} = \int_0^{\varsigma}\int_0^{\varsigma} G\,(\zeta, \gamma)\,\phi_i^{(1)}(\zeta)\overline{\phi_k^{(1)}\,(\gamma)} d_F^{\alpha}\zeta\, d_F^{\alpha}\gamma$$

$$+\,\vartheta \int_{\varsigma}^{1}\int_{\varsigma}^{1} G\,(\zeta, \gamma)\,\phi_i^{(2)}(\zeta)\overline{\phi_k^{(2)}\,(\gamma)} d_F^{\alpha}\zeta\, d_F^{\alpha}\gamma,$$

After that, H is isometrically mapped onto l^2. Through this mapping, (30) becomes (32), and $\mathcal{T}$ becomes the operator A described by (31) in l^2. It is evident from Theorems 9 and 10 that A and $\mathcal{T}$ are compact.

Let $h, \Psi \in H$. Then we obtain

$$\langle \mathcal{T}h, \Psi \rangle_H = \int_0^{\varsigma} (\mathcal{T}h^{(1)})(\zeta)\overline{\Psi^{(1)}(\zeta)} d_F^{\alpha}\zeta$$

$$+\,\vartheta \int_{\varsigma}^{1} (\mathcal{T}h^{(2)})(\zeta)\overline{\Psi^{(2)}(\zeta)} d_F^{\alpha}\zeta$$

$$= \int_0^{\varsigma}\int_0^{\varsigma} G\,(\zeta, \gamma)\,h^{(1)}(\gamma) d_F^{\alpha}\gamma\,\overline{\Psi^{(1)}(\zeta)} d_F^{\alpha}\zeta$$

$$+\,\vartheta \int_{\varsigma}^{1}\int_{\varsigma}^{1} G\,(\zeta, \gamma)\,h^{(2)}(\gamma) d_F^{\alpha}\gamma\,\overline{\Psi^{(2)}(\zeta)} d_F^{\alpha}\zeta$$

$$= \int_0^{\varsigma} h^{(1)}(\gamma) \left(\int_0^{\varsigma} G\,(\gamma, \zeta)\,\overline{\Psi^{(1)}(\zeta)} d_F^{\alpha}\zeta \right) d_F^{\alpha}\gamma$$

$$+\vartheta\int_{\varsigma}^{1}h^{(2)}(\gamma)\left(\int_{\varsigma}^{1}G(\gamma,\zeta)\,\overline{\Psi^{(2)}(\zeta)}d_F^{\alpha}\zeta\right)d_F^{\alpha}\gamma=\langle h,\mathcal{T}\Psi\rangle_H.$$

due to $G(\zeta,\gamma)$ is a symmetric function.

Without loss of generality, we can assume that $\mu=0$ is not an eigenvalue of $\mathcal{L}$. Then, $\ker\mathcal{L}=\{0\}$ and $\mathcal{T}=\mathcal{L}^{-1}$.

Theorem 12 *The operator $\mathcal{L}$ has an infinite countable set $\{\mu_n\}$ $(n\in\mathbb{N})$ of real eigenvalues which can be ordered as*

$$|\mu_1|<|\mu_2|<\cdots<|\mu_n|<\cdots,\ |\mu_n|\to\infty \text{ as } n\to\infty.$$

The set of all normalized eigenfunctions of $\mathcal{L}$ forms an orthonormal basis for the space H and for $z\in H$, $\mathcal{T}z=h$, $\mathcal{L}h=z$, $\mathcal{L}\chi_n=\mu_n\chi_n$ $(n\in\mathbb{N})$ the eigenfunction expansion formula

$$\mathcal{L}h=\sum_{n=1}^{\infty}\mu_n\langle h,\chi_n\rangle_H\chi_n$$

is valid.

Proof. It follows from the Hilbert–Schmidt theorem and Theorem 11 that $\mathcal{T}$ has an infinite sequence of non-zero real eigenvalues $\{\xi_n\}_{n=1}^{\infty}$ with

$$\lim_{n\to\infty}\xi_n=0.$$

Thus, we get

$$|\mu_n|=\frac{1}{|\xi_n|}\to\infty,\mathrm{n}\to\infty.$$

Let $\{\chi_n\}_{n=1}^{\infty}$ denote an orthonormal set of eigenfunctions corresponding to $\{\xi_n\}_{n=1}^{\infty}$ .Then, for $z\in H$, we see that

$$\mathcal{T}z=h,\ \mathcal{L}h=z,\ \mathcal{L}\chi_n=\mu_n\chi_n\ (n\in\mathbb{N})$$

and

$$\mathcal{L}h=z=\sum_{n=1}^{\infty}\langle z,\chi_n\rangle_H\chi_n=\sum_{n=1}^{\infty}\langle\mathcal{L}h,\chi_n\rangle_H\chi_n$$

$$=\sum_{n=1}^{\infty}\langle h,\mathcal{L}\chi_n\rangle_H\chi_n=\sum_{n=1}^{\infty}\mu_n\langle h,\chi_n\rangle_H\chi_n.$$

Statements and Declarations. This work does not have any conflict of interest.

References

1. Allahverdiev, B.P., Tuna, H.: Existence theorem for a fractal Sturm-Liouville problem. Vladikavkaz Math. J. **26**(1), 27–35 (2024)
2. Amirov, R., Ergun, A., Durak, S.: Half-inverse problems for the quadratic pencil of the Sturm-Liouville equations with impulse. Numer. Meth. Partial Differ. Equ. **37**, 915–924 (2020)
3. Amirov, R.K., Ozkan, A.S.: Discontinuous Sturm-Liouville problems with eigenvalue dependent boundary condition. Math. Phys. Anal. Geom. **17**, 483–491 (2014)
4. Aydemir, K., Olğar, H., Mukhtarov, OSh.: The principal eigenvalue and the principal eigenfunction of a boundary-value-transmission problem. Turkish J. Math. Comput. Sci. **11**(2), 97–100 (2019)
5. Aydemir, K., Mukhtarov, OSh.: A new type Sturm-Liouville problem with an abstract linear operator contained in the equation. Quaest. Math. **45**(12), 1931–1948 (2022)
6. Aydemir, K., Mukhtarov, O.S.: Qualitative analysis of eigenvalues and eigenfunctions of one boundary value-transmission problem, Bound. Value Probl. **2016**, Paper No. 82, 16 (2016)
7. Cebesoy, S., Bairamov, E., Aygar, Y.: Scattering problems of impulsive Schrödinger equations with matrix coefficients. Ric. Mat. **72**(1), 399–415 (2023)
8. Çetinkaya, F.A., Golmankaneh, A.K.: General characteristics of a fractal Sturm-Liouville problem. Turkish J. Math. **45**(4), 1835–1846 (2021)
9. Golmankhaneh, A.K.: Fractal Calculus and its Applications. World Scientific (2022)
10. Golmankhaneh, A.K., Tunç, C.: Stochastic differential equations on fractal sets. Stochastics **92**(8), 1244–1260 (2020)
11. Golmankhaneh, A.K., Tunc, C.: On the Lipschitz condition in the fractal calculus. Chaos, Solitons Fractals **95**, 140–147 (2017)
12. Golmankhaneh, A.K., Cattani, C., Kalita, H., Furuichi, S., Jorgensen, P.E.T.: Fractal green function theory (2024). https://doi.org/10.13140/RG.2.2.33233.54887
13. Karahan, D., Mamedov, K.R.: On a q-boundary value problem with discontinuity conditions. Vestn. Yuzhno-Ural. Gos. Un-ta. Ser. Matem. Mekh. Fiz. **13**(4), 5–12 (2021)
14. Karahan, D., Mamedov, Kh. R.: On a q-analogue of the Sturm–Liouville operator with discontinuity conditions. Vestn. Samar. Gos. Tekh. Univ., Ser. Fiz.-Mat. Nauk. **26**(3), 407–418 (2022)
15. Kolwankar, K.M., Gangal, A.D.: Fractional differentiability of nowhere differentiable functions and dimensions. Chaos **6**(4), 505–513 (1996)
16. Kolwankar, K.M., Gangal, A.D.: Holder exponents of irregular signals and local fractional derivatives. Pramana - J. Phys. **48**, 49–68 (1997)
17. Kolwankar, K.M., Gangal, A.D.: Local fractional Fokker-Planck equation. Phys. Rev. Lett. **80**(2), 214–217 (1998)
18. Kolwankar, K.M., Gangal, A.D.: Local fractional derivatives and fractal functions of several variables. Physics 9801010 (1998)
19. Lapwood, F.R., Usami, T.: Free Oscillations of the Earth. Cambridge University Press, Cambridge (1981)
20. Levitan, B.M., Sargsjan, I.S.: Sturm–Liouville and Dirac operators. Mathematics and its Applications (Soviet Series). Kluwer Academic Publishers Group, Dordrecht (1991) (translated from the Russian)
21. Likov, A.V., Mikhailov, Y.A.: The Theory of Heat and Mass Transfer. Translated from Russian by I. Shechtman, Israel Program for Scientific Translations, Jerusalem (1965)
22. Litvinenko, O.N., Soshnikov, V.I.: The Theory of Heteregenous Lines and their Applications in Radio Engineering. Radio, Moscow (1964). (in Russian)
23. Naimark, M.A.: Linear Differential Operators. 2nd edn. 1968, Nauka, Moscow, English transl. of 1st edn. 1, 2. New York (1969)
24. Parvate, A., Gangal, A.D.: Calculus on fractal subsets of real line I: formulation. Fractals **17**(1), 53–81 (2009)
25. Parvate, A., Gangal, A.D.: Calculus on fractal subsets of real line II: conjugacy with ordinary calculus. Fractals **19**(3), 271–290 (2011)

MIX
Papier aus verantwortungsvollen Quellen
Paper from responsible sources
FSC® C105338

If you have any concerns about our products,
you can contact us on
ProductSafety@springernature.com

In case Publisher is established outside the EU,
the EU authorized representative is:
Springer Nature Customer Service Center GmbH
Europaplatz 3, 69115 Heidelberg, Germany

Printed by Libri Plureos GmbH
in Hamburg, Germany